ISACA®

CISA®

Review Questions, Answers & Explanations Manual

12th Edition

Updated for 2019 Job Practice

About ISACA
Nearing its 50th year, ISACA® (isaca.org) is a global association helping individuals and enterprises achieve the positive potential of technology. Today's world is powered by technology, and ISACA equips professionals with the knowledge, credentials, education and community to advance their careers and transform their organizations. Among those credentials, ISACA advances and validates business-critical skills and knowledge through the globally respected Certified Information Systems Auditor® (CISA®), Certified in Risk and Information Systems Control™ (CRISC™), Certified Information Security Manager® (CISM®) and Certified in the Governance of Enterprise IT® (CGEIT®) credentials.

ISACA leverages the expertise of its half-million engaged professionals in information and cybersecurity, governance, assurance, risk and innovation, as well as its enterprise performance subsidiary, CMMI® Institute, to help advance innovation through technology. ISACA has a presence in more than 188 countries, including more than 215 chapters worldwide and offices in both the United States and China.

Disclaimer
ISACA has designed and created *CISA® Review Questions, Answers & Explanations Manual 12th Edition* primarily as an educational resource to assist individuals preparing to take the CISA certification exam. It was produced independently from the CISA exam and the CISA Certification Committee, which has had no responsibility for its content. Copies of past exams are not released to the public and were not made available to ISACA for preparation of this publication. ISACA makes no representations or warranties whatsoever with regard to these or other ISACA publications assuring candidates' passage of the CISA exam.

ISACA
1700 E. Golf Road, Suite 400
Schaumburg, IL 60173, USA
Phone: +1.847.660.5505
Fax: +1.847.253.1755
Contact us: https://support.isaca.org
Website: www.isaca.org

Participate in the ISACA Online Forums: https://engage.isaca.org/onlineforums

Twitter: http://twitter.com/ISACANews
LinkedIn: http://linkd.in/ISACAOfficial
Facebook: www.facebook.com/ISACAHQ
Instagram: www.instagram.com/isacanews/

ISBN 978-1-60420-768-2
CISA® Review Questions, Answers & Explanations Manual 12th Edition
Printed in the United States of America

PREFACE

ISACA is pleased to offer the 1,000 review questions in the *CISA® Review Questions, Answers & Explanations Manual 12th Edition*. The purpose of this manual is to provide the Certified Information Systems Auditor (CISA) candidate with sample questions and testing topics to help prepare and study for the CISA exam.

This manual includes 1,000 multiple-choice study questions, answers and explanations, which are organized according to the newly revised CISA job practice domains. These questions, answers and explanations are intended to introduce CISA candidates to the types of questions that may appear on the CISA exam. They are not actual questions from the exam. This manual also contains a 150-question sample exam, which has the same proportion of questions related to each CISA job practice domain as the actual exam.

The candidate also may want to obtain a copy of the *CISA® Review Manual 27th Edition*, which provides the foundational knowledge of a CISA. The CISA® Review Questions, Answers & Explanations Database–12-Month Subscription contains the same questions found in this manual in a web-based application. Finally, the candidate may also want to use the CISA® Online Review course or CISA virtual or live instructor-led training for exam preparation.

A job practice study is conducted periodically to ensure that the CISA certification is current and relevant. Further details regarding the new job practice is in the "New–CISA Job Practice" section in this manual.

ISACA created this publication as an educational resource to assist individuals who are preparing to take the CISA exam. It was produced independently from the CISA Certification Working Group, which has no responsibility for its content. Copies of past exams are not released to the public and are not made available to candidates. ISACA makes no representations or warranties whatsoever regarding these or other ISACA or IT Governance Institute publications assuring candidates' passage of the CISA exam.

ISACA wishes you success with the CISA exam. Your commitment to pursuing the leading certification for information systems (IS) audit, assurance, security and control professionals is exemplary, and ISACA welcomes your comments and suggestions on the use and coverage of this manual. After completion of the exam, please complete the online evaluation that corresponds to this publication *(www.isaca.org/studyaidsevaluation)*. Your observations will be invaluable as new questions, answers and explanations are prepared.

ACKNOWLEDGMENTS

This CISA® Review Questions, Answers & Explanations Manual 12th Edition is the result of the collective efforts of many volunteers. ISACA members from throughout the world participated, generously offering their talents and expertise. This international team exhibited a spirit and selflessness that has become the hallmark of contributors to this valuable manual. Their participation and insight are truly appreciated.

NEW—CISA JOB PRACTICE

Beginning in 2019, the Certified Information Systems Auditor (CISA) exam tests the new CISA job practice.

An international job practice analysis is conducted periodically to maintain the validity of the CISA certification program. A new job practice forms the basis of the CISA exam.

The primary focus of the job practice is on the current tasks performed and the knowledge used by CISAs. By gathering evidence of the current work practice of CISAs, ISACA ensures that the CISA program continues to meet the high standards for the certification of professionals throughout the world.

The findings of the CISA job practice analysis are carefully considered and directly influence the development of new test specifications to ensure that the CISA exam reflects the most current best practices.

The new job practice reflects the areas of study to be tested and is compared below to the previous job practice. The complete CISA job practice is at *www.isaca.org/cisajobpractice*.

Previous CISA Job Practice	New CISA Job Practice
Domain 1: The Process of Auditing Information Systems (21%)	**Domain 1: Information System Auditing Process (21%)**
Domain 2: Governance and Management of IT (16%)	**Domain 2: Governance and Management of IT (17%)**
Domain 3: Information Systems Acquisition, Development and Implementation (18%)	**Domain 3: Information Systems Acquisition, Development and Implementation (12%)**
Domain 4: Information Systems Operations, Maintenance and Service Management (20%)	**Domain 4: Information Systems Operations and Business Resilience (23%)**
Domain 5: Protection of Information Assets (25%)	**Domain 5: Protection of Information Assets (27%)**

Page intentionally left blank

TABLE OF CONTENTS

Page intentionally left blank

INTRODUCTION

OVERVIEW

This manual consists of 1,000 multiple-choice questions, answers and explanations. The questions are numbered **A1-1**, **A2-1**, etc.

Questions Sorted by Domain

Questions, answers and explanations are sorted by the CISA job practice domains. The CISA candidate can refer to specific domain questions to evaluate comprehension of the topics that are covered within each domain. These questions are representative of CISA exam questions, although they are not actual exam items. The questions assist the CISA candidate in understanding the materials in the *CISA® Review Manual 27th Edition* and depict the type of question format typically found on the CISA exam. The number of questions, answers and explanations provided in the five domain sections in this publication provide the CISA candidate with the maximum number of study questions.

Sample Exam

A sample exam of 150 questions is also provided in this manual. **This exam is organized according to the domain percentages specified in the CISA job practice and used on the CISA exam:**

- Domain 1—Information System Auditing Process..21 percent
- Domain 2—Governance and Management of IT..17 percent
- Domain 3—Information Systems Acquisition, Development and Implementation........12 percent
- Domain 4—Information Systems Operations and Business Resilience.........................23 percent
- Domain 5—Protection of Information Assets..27 percent

Candidates are urged to use this sample exam and the answer sheet provided to simulate an actual exam. Many candidates use this exam as a pretest to determine their strengths or weaknesses, or as a final exam. The sample exam answer sheet is provided for both uses. In addition, a CISA sample exam answer and reference key is included that cross references the exam questions to the questions in this publication, so it is convenient to refer to the explanations of the correct answers. This publication is ideal to use with the *CISA® Review Manual 27th Edition.*

The *CISA® Review Questions, Answers & Explanations Manual 12th Edition* was developed to assist CISA candidates in studying and preparing for the CISA exam. As candidates use this publication to prepare for the exam, they should note that it covers a broad spectrum of IS audit, assurance, control and security issues. Candidates should not assume that reading and working through the questions in this manual will fully prepare them for the exam. Because exam questions often relate to practical experiences, CISA candidates are cautioned to refer to their own experiences and to other publications referred to in the *CISA® Review Manual 27th Edition*. These additional references are excellent sources of further detailed information and clarification. It is recommended that candidates evaluate the job practice domains in which they feel weak or require a further understanding, and study accordingly.

This publication uses standard American English.

TYPES OF QUESTIONS ON THE CISA EXAM

CISA exam questions are developed with the intent of measuring and testing practical knowledge and applying general concepts and standards. Questions are presented in a multiple-choice format and are designed for one best answer.

The candidate is cautioned to read each question carefully. Many times, a CISA exam question will require the candidate to choose the appropriate answer that is most likely or best or choose a practice or procedure that would be performed first, related to the other answers. In every case, the candidate is required to read the question carefully, eliminate known wrong answers and then make the best choice possible. Knowing that these types of questions are asked and how to study to answer them will help candidates to answer these types of questions correctly.

Each CISA question has a stem (question) and four options (answer choices). The candidate is asked to choose the best answer from the options. The stem may be in the form of a question or incomplete statement. In some instances, a scenario or description of a problem may be included. These questions normally include a description of a situation and require the candidate to answer two or more questions based on the information provided.

When candidates prepare for the exam, they should recognize that IS audit and control is a global profession, and individual perceptions and experiences may not reflect the more global position or circumstance. Because the exam and CISA manuals are written for the international IS audit and control community, a candidate will be required to be somewhat flexible when reading an audit or control condition that may be contrary to a candidate's experience. It should be noted that CISA exam questions are written by experienced IS audit practitioners from around the world. Each question on the exam is reviewed by ISACA's CISA Test Enhancement Subcommittee and CISA Certification Working Group, which consist of international members. This geographical representation ensures that all test questions are understood equally in each country and language.

Note: ISACA review manuals are living documents. As technology advances, ISACA manuals are updated to reflect such advances. Further updates to this document before the date of the exam can be viewed at *www.isaca.org/studyaidupdates*.

Any suggestions to enhance the materials covered herein, or reference materials, should be submitted online at *support.isaca.org*.

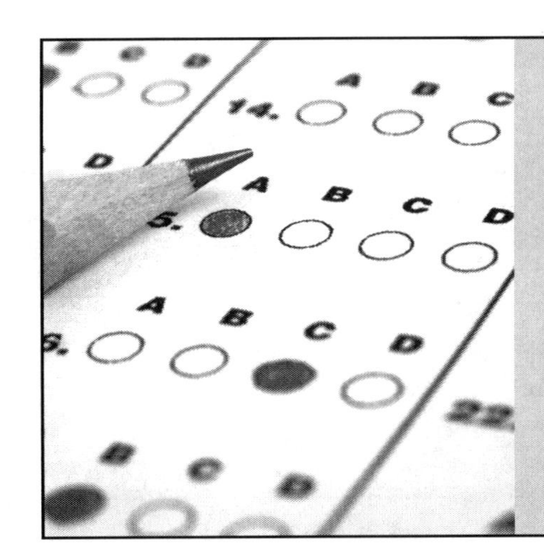

PRETEST

If you wish to take a pretest to determine strengths and weaknesses, the Sample Exam begins on page 469 and the pretest answer sheet begins on page 493. You can score your pretest with the Sample Exam Answer and Reference Key on page 491.

Page intentionally left blank

QUESTIONS, ANSWERS & EXPLANATIONS BY DOMAIN

DOMAIN 1—INFORMATION SYSTEM AUDITING PROCESS (21%)

A1-1 The internal audit department wrote some scripts that are used for continuous auditing of some information systems. The IT department asked for copies of the scripts so that they can use them for setting up a continuous monitoring process on key systems. Does sharing these scripts with IT affect the ability of the IS auditors to independently and objectively audit the IT function?

A. Sharing the scripts is not permitted because it gives IT the ability to pre-audit systems and avoid an accurate, comprehensive audit.
B. Sharing the scripts is required because IT must have the ability to review all programs and software that run on IS systems regardless of audit independence.
C. Sharing the scripts is permissible if IT recognizes that audits may still be conducted in areas not covered in the scripts.
D. Sharing the scripts is not permitted because the IS auditors who wrote the scripts would not be permitted to audit any IS systems where the scripts are being used for monitoring.

C is the correct answer.

Justification:
A. The ability of IT to continuously monitor and address any issues on IT systems does not affect the ability of IS audit to perform a comprehensive audit.
B. Sharing the scripts may be required by policy for quality assurance and configuration management, but that does not impair the ability to audit.
C. IS audit can still review all aspects of the systems. They may not be able to review the effectiveness of the scripts, but they can still audit the systems.
D. An audit of an IS system encompasses more than just the controls covered in the scripts.

A1-2 Which of the following is the **BEST** factor for determining the required extent of data collection during the planning phase of an IS compliance audit?

A. Complexity of the organization's operation
B. Findings and issues noted from the prior year
C. Purpose, objective and scope of the audit
D. Auditor's familiarity with the organization

C is the correct answer.

Justification:
A. The complexity of the organization's operation is a factor in the planning of an audit but does not directly affect the determination of how much data to collect. The extent of data collection is subject to the intensity, scope and purpose of the audit.
B. Prior findings and issues are factors in the planning of an audit but do not directly affect the determination of how much data to collect. Data must be collected outside of areas of previous findings.
C. The extent to which data will be collected during an IS audit is related directly to the purpose, objective and scope of the audit. An audit with a narrow purpose and limited objective and scope is most likely to result in less data collection than an audit with a wider purpose and scope. Statistical analysis may also determine the extent of data collection, such as sample size or means of data collection.
D. An auditor's familiarity with the organization is a factor in the planning of an audit but does not directly affect the determination of how much data to collect. The audit must be based on sufficient evidence of the monitoring of controls and not unduly influenced by the auditor's familiarity with the organization.

A1-3 An IS auditor is developing an audit plan for an environment that includes new systems. The organization's management wants the IS auditor to focus on recently implemented systems. How should the IS auditor respond?

A. Audit the new systems as requested by management.
B. Audit systems not included in last year's scope.
C. Determine the highest-risk systems and plan accordingly.
D. Audit both the systems not in last year's scope and the new systems.

C is the correct answer.

Justification:

A. Auditing the new system does not reflect a risk-based approach. Although the system can contain sensitive data and may present risk of data loss or disclosure to the organization, without a risk assessment, the decision to solely audit the newly implemented system is not a risk-based decision.
B. Auditing systems not included in the previous year's scope does not reflect a risk-based approach. In addition, management may know about problems with the new system and may be intentionally trying to steer the audit away from that vulnerable area. Although, at first, the new system may seem to be the riskiest area, an assessment must be conducted rather than relying on the judgment of the IS auditor or IT manager.
C. The best action is to conduct a risk assessment and design the audit plan to cover the areas of highest risk. ISACA IS Audit and Assurance Standard 1202 (Risk Assessment in Planning), statement 1202.1: "The IS audit and assurance function shall use an appropriate risk assessment approach and supporting methodology to develop the overall IS audit plan and determine priorities for the effective allocation of IS audit resources."
D. The creation of the audit plan should be performed in cooperation with management and based on risk. The IS auditor should not arbitrarily decide on what needs to be audited.

A1-4 An IS auditor is reviewing security controls for a critical web-based system prior to implementation. The results of the penetration test are inconclusive, and the results will not be finalized prior to implementation. Which of the following is the **BEST** option for the IS auditor?

A. Publish a report based on the available information, highlighting the potential security weaknesses and the requirement for follow-up audit testing.
B. Publish a report omitting the areas where the evidence obtained from testing was inconclusive.
C. Request a delay of the implementation date until additional security testing can be completed and evidence of appropriate controls can be obtained.
D. Inform management that audit work cannot be completed prior to implementation and recommend that the audit be postponed.

A is the correct answer.

Justification:

A. If the IS auditor cannot gain sufficient assurance for a critical system within the agreed-on time frame, this fact should be highlighted in the audit report and follow-up testing should be scheduled for a later date. Management can then determine whether any of the potential weaknesses identified were significant enough to delay the go-live date for the system.
B. It is not acceptable for the IS auditor to ignore areas of potential weakness because conclusive evidence could not be obtained within the agreed-on audit time frame. ISACA IS Audit and Assurance Standards are violated if these areas are omitted from the audit report.
C. Extending the time frame for the audit and delaying the go-live date is unlikely to be acceptable in this scenario where the system involved is business-critical. In any case, a delay to the go-live date must be the decision of business management, not the IS auditor. In this scenario, the IS auditor should present business management with all available information by the agreed-on date.
D. Failure to obtain sufficient evidence in one part of an audit engagement does not justify cancelling or postponing the audit; this violates the audit guideline concerning due professional care.

A1-5 Which of the following controls would an IS auditor look for in an environment where duties cannot be appropriately segregated?

A. Overlapping controls
B. Boundary controls
C. Access controls
D. Compensating controls

D is the correct answer.

Justification:
A. Overlapping controls are two controls addressing the same control objective or exposure. Because primary controls cannot be achieved when duties cannot or are not appropriately segregated, it is difficult to install overlapping controls.
B. Boundary controls establish the interface between the would-be user of a computer system and the computer system itself and are individual-based, not role-based, controls.
C. Access controls for resources are based on individuals and not on roles. For a lack of segregation of duties, the IS auditor expects to find that a person has higher levels of access than are ideal. The IS auditor wants to find compensating controls to address this risk.
D. Compensating controls are internal controls that are intended to reduce the risk of an existing or potential control weakness that may arise when duties cannot be appropriately segregated.

A1-6 Which of the following is the key benefit of a control self-assessment?

A. Management ownership of the internal controls supporting business objectives is reinforced.
B. Audit expenses are reduced when the assessment results are an input to external audit work.
C. Fraud detection is improved because internal business staff are engaged in testing controls.
D. Internal auditors can shift to a consultative approach by using the results of the assessment.

A is the correct answer.

Justification:
A. The objective of control self-assessment (CSA) is to have business management become more aware of the importance of internal control and their responsibility in terms of corporate governance.
B. Reducing audit expenses is not a key benefit of CSA.
C. Improved fraud detection is important but not as important as control ownership. It is not a principal objective of CSA.
D. CSA may give more insights to internal auditors, allowing them to take a more consultative role; however, this is an additional benefit, not the key benefit.

A1-7 What is the **PRIMARY** requirement that a data mining and auditing software tool should meet? The software tool should:

A. interface with various types of enterprise resource planning software and databases.
B. accurately capture data from the organization's systems without causing excessive performance problems.
C. introduce audit hooks into the organization's financial systems to support continuous auditing.
D. be customizable and support inclusion of custom programming to aid in investigative analysis.

B is the correct answer.

Justification:
A. The product must interface with the types of systems used by the organization and provide meaningful data for analysis.
B. Although all the requirements that are listed as answer choices are desirable in a software tool evaluated for auditing and data mining purposes, the most critical requirement is that the tool works effectively on the systems of the organization being audited.
C. The tool should probably work on more than just financial systems and does not necessarily require implementation of audit hooks.
D. The tool should be flexible but not necessarily customizable. It should have built-in analysis software tools.

A1-8 A long-term IT employee with a strong technical background and broad managerial experience has applied for a vacant position in the IS audit department. Determining whether to hire this individual for this position should be **PRIMARILY** based on the individual's experience and:

A. length of service, because this will help ensure technical competence.
B. age, because training in audit techniques may be impractical.
C. IT knowledge, because this will bring enhanced credibility to the audit function.
D. ability, as an IS auditor, to be independent of existing IT relationships.

D is the correct answer.

Justification:
A. Length of service does not ensure technical competency.
B. Evaluating an individual's qualifications based on the age of the individual is not a good criterion and is illegal in many parts of the world.
C. The fact that the employee has worked in IT for many years may not ensure credibility. The IS audit department's needs should be defined, and any candidate should be evaluated against those requirements.
D. Independence should be continually assessed by the auditor and management. This assessment should consider such factors as changes in personal relationships, financial interests, and prior job assignments and responsibilities.

A1-9 For a retail business with a large volume of transactions, which of the following audit techniques is the **MOST** appropriate for addressing emerging risk?

A. Use of computer-assisted audit techniques
B. Quarterly risk assessments
C. Sampling of transaction logs
D. Continuous auditing

D is the correct answer.

Justification:

A. Using software tools such as computer-assisted audit techniques to analyze transaction data can provide detailed analysis of trends and potential risk, but it is not as effective as continuous auditing, because there may be a time differential between executing the software and analyzing the results.
B. Quarterly risk assessment may be a good technique but not as responsive as continuous auditing.
C. The sampling of transaction logs is a valid audit technique; however, risk may exist that is not captured in the transaction log, and there may be a potential time lag in the analysis.
D. The implementation of continuous auditing enables a real-time feed of information to management through automated reporting processes so that management may implement corrective actions more quickly.

A1-10 An IS auditor is reviewing access to an application to determine whether recently added accounts were appropriately authorized. This is an example of:

A. variable sampling.
B. substantive testing.
C. compliance testing.
D. stop-or-go sampling.

C is the correct answer.

Justification:

A. Variable sampling is used to estimate numerical values such as dollar values.
B. Substantive testing substantiates the integrity of actual processing such as balances on financial statements. The development of substantive tests is often dependent on the outcome of compliance tests. If compliance tests indicate that there are adequate internal controls, then substantive tests can be minimized.
C. Compliance testing determines whether controls are being applied in compliance with policy. This includes tests to determine whether new accounts were appropriately authorized.
D. Stop-or-go sampling allows a test to be stopped as early as possible and is not appropriate for checking whether procedures have been followed.

A1-11 The decisions and actions of an IS auditor are **MOST** likely to affect which of the following types of risk?

A. Inherent
B. Detection
C. Control
D. Business

B is the correct answer.

Justification:

A. Inherent risk is the risk that a material error could occur, if there are no related internal controls to prevent or detect the error. Inherent risk is not usually affected by an IS auditor.
B. Detection risk is directly affected by the IS auditor's selection of audit procedures and techniques. Detection risk is the risk that a review will not detect or notice a material issue.
C. Control risk is the risk that a material error exists that would not be prevented or detected on a timely basis by the system of internal controls. Control risk can be mitigated by the actions of the organization's management.
D. Business risk is a probable situation with uncertain frequency and magnitude of loss (or gain). Business risk is usually not directly affected by an IS auditor.

A1-12 Which of the following is the **MOST** critical step when planning an IS audit?

A. Review findings from prior audits
B. Executive management's approval of the audit plan
C. Review information security policies and procedures
D. Perform a risk assessment

D is the correct answer.

Justification:

A. The findings of a previous audit are of interest to the auditor, but they are not the most critical step. The most critical step involves finding the current issues or high-risk areas, not reviewing the resolution of older issues. A review of historical audit findings could indicate that management is not resolving the items or the recommendation was ineffective.
B. Executive management is not required to approve the audit plan. It is typically approved by the audit committee or board of directors. Management could recommend areas to audit.
C. Reviewing information security policies and procedures is normally be conducted during fieldwork, not planning.
D. Of all the steps listed, performing a risk assessment is the most critical. Risk assessment is required by ISACA IS Audit and Assurance Standard 1202 (Risk Assessment in Planning), statement 1202.2: "IS audit and assurance professionals shall identify and assess risk relevant to the area under review, when planning individual engagements." In addition to the standards requirement, if a risk assessment is not performed, then high-risk areas of the auditee systems or operations may not be identified for evaluation.

A1-13 An IS auditor is reviewing a software application that is built on the principles of service-oriented architecture. What is the **INITIAL** step?

A. Understanding services and their allocation to business processes by reviewing the service repository documentation.
B. Sampling the use of service security standards as represented by the Security Assertions Markup Language.
C. Reviewing the service level agreements established for all system providers.
D. Auditing the core service and its dependencies on other systems.

A is the correct answer.

Justification:

A. A service-oriented architecture relies on the principles of a distributed environment in which services encapsulate business logic as a black box and might be deliberately combined to depict real-world business processes. Before reviewing services in detail, it is essential for the IS auditor to comprehend the mapping of business processes to services.
B. Sampling the use of service security standards as represented by the Security Assertions Markup Language is an essential follow-up step to understanding services and their allocation to business but is not the initial step.
C. Reviewing the service level agreements is an essential follow-up step to understanding services and their allocation to business but is not the initial step.
D. Auditing the core service and its dependencies with others would most likely be a part of the audit, but the IS auditor must first gain an understanding of the business processes and how the systems support those processes.

A1-14 An IS auditor conducting a review of software usage and licensing discovers that numerous PCs contain unauthorized software. Which of the following actions should the IS auditor take?

A. Delete all copies of the unauthorized software.
B. Recommend an automated process to monitor for compliance with software licensing.
C. Report the use of the unauthorized software and the need to prevent recurrence.
D. Warn the end users about the risk of using illegal software.

C is the correct answer.

Justification:

A. An IS auditor should not assume the role of the enforcing officer and take on any personal involvement in removing the unauthorized software.
B. This would detect compliance with software licensing. However, an automated solution might not be the best option in all cases.
C. The use of unauthorized or illegal software should be prohibited by an organization. An IS auditor must convince the user and management of the risk and the need to eliminate the risk. For example, software piracy can result in exposure and severe fines.
D. Auditors must report material findings to management for action. Informing the users of risk is not the primary responsibility of the IS auditor.

A1-15 An audit charter should:

A. be dynamic and change to coincide with the changing nature of technology and the audit profession.
B. clearly state audit objectives for, and the delegation of, authority to the maintenance and review of internal controls.
C. document the audit procedures designed to achieve the planned audit objectives.
D. outline the overall authority, scope and responsibilities of the audit function.

D is the correct answer.

Justification:
A. The audit charter should not be subject to changes in technology and should not significantly change over time. The charter should be approved at the highest level of management.
B. An audit charter states the authority and reporting requirements for the audit but not the details of maintenance of internal controls.
C. An audit charter is not at a detailed level and, therefore, does not include specific audit objectives or procedures.
D. An audit charter should state management's objectives for and delegation of authority to IS auditors.

A1-16 An IS auditor finds a small number of user access requests that were not authorized by managers through the normal predefined workflow steps and escalation rules. The IS auditor should:

A. perform an additional analysis.
B. report the problem to the audit committee.
C. conduct a security risk assessment.
D. recommend that the owner of the identity management system fix the workflow issues.

A is the correct answer.

Justification:
A. The IS auditor needs to perform substantive testing and additional analysis to determine why the approval and workflow processes are not working as intended. Before making any recommendation, the IS auditor should gain a good understanding of the scope of the problem and the factors that caused this incident. The IS auditor should identify whether the issue was caused by managers not following procedures, a problem with the workflow of the automated system or a combination of the two.
B. The IS auditor does not yet have enough information to report the problem.
C. Changing the scope of the IS audit or conducting a security risk assessment requires more detailed information about the processes and violations being reviewed.
D. The IS auditor must first determine the root cause and impact of the findings and does not have enough information to recommend fixing the workflow issues.

A1-17 Which of the following sampling methods is **MOST** useful when testing for compliance?

A. Attribute sampling
B. Variable sampling
C. Stratified mean-per-unit sampling
D. Difference estimation sampling

A is the correct answer.

Justification:

A. Attribute sampling is the primary sampling method used for compliance testing. Attribute sampling is a sampling model that is used to estimate the rate of occurrence of a specific quality (attribute) in a population and is used in compliance testing to confirm whether the quality exists. For example, an attribute sample may check all transactions over a certain predefined dollar amount for proper approvals.

B. Variable sampling is based on the calculation of a mean from a sample extracted from the entire population and using that to estimate the characteristics of the entire population. For example, a sample of 10 items shows an average price of US $10 per item. For the entire population of 1,000 items, the total value is estimated to be US $10,000. This is not a good way to measure compliance with a process.

C. Stratified mean sampling attempts to ensure that the entire population is represented in the sample. This is not an effective way to measure compliance.

D. Difference estimation sampling examines measure deviations and extraordinary items and is not a good way to measure compliance.

A1-18 When testing program change requests for a remote system, an IS auditor finds that the number of changes available for sampling does not provide a reasonable level of assurance. What is the **MOST** appropriate action for the IS auditor to take?

A. Develop an alternate testing procedure.
B. Report the finding to management.
C. Perform a walkthrough of the change management process.
D. Create additional sample data to test additional changes.

A is the correct answer.

Justification:

A. If a sample-size objective cannot be met with the given data, the IS auditor cannot provide assurance regarding the testing objective. In this instance, the IS auditor should develop (with audit management approval) an alternate testing procedure.

B. There is not enough evidence to report the finding as a deficiency.

C. A walkthrough should not be initiated until an analysis is performed to confirm that this could provide the required assurance.

D. It is not appropriate for an IS auditor to create sample data for the purpose of the audit.

A1-19 Which of the following situations could impair the independence of an IS auditor? The IS auditor:

A. implemented specific functionality during the development of an application.
B. designed an embedded audit module for auditing an application.
C. participated as a member of an application project team and did not have operational responsibilities.
D. provided consulting advice concerning application good practices.

A is the correct answer.

Justification:

A. Independence may be impaired if an IS auditor is, or has been, actively involved in the development, acquisition and implementation of the application system.
B. Designing an embedded audit module does not impair an IS auditor's independence.
C. IS auditors should not audit work that they have done, but just participating as a member of the application system project team does not impair an IS auditor's independence.
D. An IS auditor's independence is not impaired by providing advice on known good practices.

A1-20 The **PRIMARY** advantage of a continuous audit approach is that it:

A. does not require an IS auditor to collect evidence on system reliability while processing is taking place.
B. allows the IS auditor to review and follow up on audit issues in a timely manner.
C. places the responsibility for enforcement and monitoring of controls on the security department instead of audit.
D. simplifies the extraction and correlation of data from multiple and complex systems.

B is the correct answer.

Justification:

A. The continuous audit approach often requires an IS auditor to collect evidence on system reliability while processing is taking place.
B. Continuous audit allows audit and response to audit issues in a timely manner because audit findings are gathered in near real time.
C. Responsibility for enforcement and monitoring of controls is primarily the responsibility of management.
D. The use of continuous audit is not based on the complexity or number of systems being monitored.

A1-21 Which of the following would impair the independence of a quality assurance team?

A. Ensuring compliance with development methods
B. Checking the test assumptions
C. Correcting coding errors during the testing process
D. Checking the code to ensure proper documentation

C is the correct answer.

Justification:

A. Ensuring compliance with development methods is a valid quality assurance function.
B. Checking the test assumptions is a valid quality assurance function.
C. Correction of code should not be a responsibility of the quality assurance team, because it would not ensure segregation of duties and would impair the team's independence.
D. Checking the code to ensure proper documentation is a valid quality assurance function.

A1-22 In planning an IS audit, the **MOST** critical step is the identification of the:

A. areas of significant risk
B. skill sets of the audit staff
C. test steps in the audit
D. time allotted for the audit

A is the correct answer.

Justification:

A. When designing a risk-based audit plan, it is important to identify the areas of highest risk to determine the areas to be audited.
B. The skill sets of the audit staff should have been considered before deciding and selecting the audit. Where the skills are inadequate, the organization should consider using external resources.
C. Test steps for the audit are not as critical during the audit planning process as identifying the areas of risk that should be audited.
D. The time allotted for an audit is determined during the planning process based on the areas to be audited and is primarily based on the requirement for conducting an appropriate audit.

A1-23 The **MOST** effective audit practice to determine whether the operational effectiveness of controls is properly applied to transaction processing is:

A. control design testing.
B. substantive testing.
C. inspection of relevant documentation.
D. perform tests on risk prevention.

B is the correct answer.

Justification:

A. Testing of control design assesses whether the control is structured to meet a specific control objective. It does not help determine whether the control is operating effectively.
B. Among other methods, such as document review or walkthrough, tests of controls are the most effective procedures to assess whether controls accurately support operational effectiveness.
C. Control documents may not always describe the actual process in an accurate manner. Therefore, auditors relying on document review have limited assurance that the control is operating as intended.
D. Performing tests on risk prevention is considered compliance testing. This type of testing is used to determine whether policies are adhered to.

A1-24 The extent to which data will be collected during an IS audit should be determined based on the:

A. Availability of critical and required information.
B. Auditor's familiarity with the circumstances.
C. Auditee's ability to find relevant evidence.
D. Purpose and scope of the audit being done.

D is the correct answer.

Justification:

A. The extent to which data will be collected during an IS audit should be based on the scope, purpose and requirements of the audit and not be constrained by the ease of obtaining the information or by the IS auditor's familiarity with the area being audited.
B. An IS auditor must be objective and thorough and not subject to audit risk through preconceived expected results based on familiarity with the area being audited.
C. Collecting all the required evidence is a required element of an IS audit, and the scope of the audit should not be limited by the auditee's ability to find relevant evidence. If evidence is not readily available, the auditor must ensure that other forms of audit are considered to ensure compliance in the area that is subject to audit.
D. The extent to which data will be collected during an IS audit should be related directly to the scope and purpose of the audit. An IS audit with a narrow purpose and scope, or just a high-level review, will most likely require less data collection than an audit with a wider purpose and scope.

A1-25 While planning an IS audit, an assessment of risk should be made to provide:

A. reasonable assurance that the audit will cover material items.
B. definite assurance that material items will be covered during the audit work.
C. reasonable assurance that all items will be covered by the audit.
D. sufficient assurance that all items will be covered during the audit work.

A is the correct answer.

Justification:

A. ISACA IS Audit and Assurance Guideline 2202 (Risk Assessment and Audit Planning) states that the applied risk assessment approach should help with the prioritization and scheduling process of the IS audit and assurance work. The risk assessment should support the selection process of areas and items of audit interest and the decision process to design and conduct particular IS audit engagements.
B. Definite assurance that material items will be covered during the audit work is an impractical proposition.
C. Reasonable assurance that all items will be covered during the audit work is not the correct answer, because primarily material items need to be covered, not all items.
D. Sufficient assurance that all items will be covered is not as important as ensuring that the audit will cover all material items.

A1-26 The **MOST** appropriate action for an IS auditor to take when shared user accounts are discovered is to:

A. inform the audit committee of the potential issue.
B. review audit logs for the IDs in question.
C. document the finding and explain the risk of using shared IDs.
D. request that the IDs be removed from the system.

C is the correct answer.

Justification:

A. It is not appropriate for an IS auditor to report findings to the audit committee before conducting a more detailed review and presenting them to management for a response.
B. Review of audit logs would not be useful because shared IDs do not provide for individual accountability.
C. An IS auditor's role is to detect and document findings and control deficiencies. Part of the audit report is to explain the reasoning behind the findings. The use of shared IDs is not recommended because it does not allow for accountability of transactions. An IS auditor defers to management to decide how to respond to the findings presented.
D. It is not the role of an IS auditor to request the removal of IDs from the system.

A1-27 An IS auditor is conducting a compliance test to determine whether controls support management policies and procedures. The test will assist the IS auditor to determine:

A. that the control is operating efficiently.
B. that the control is operating as designed.
C. the integrity of data controls.
D. the reasonableness of financial reporting controls.

B is the correct answer.

Justification:

A. It is important that controls operate efficiently, but in this case the intent is to ensure that the controls support management policies and procedures. Therefore, the important issue is whether the controls are operating correctly and thereby meeting the control objective.
B. Compliance tests can be used to test the existence and effectiveness of a defined process. Understanding the objective of a compliance test is important. IS auditors want reasonable assurance that the controls they are relying on are effective. An effective control is one that meets management expectations and objectives.
C. Substantive tests, not compliance tests, are associated with data integrity.
D. Determining the reasonableness of financial reporting controls is a very narrow answer in that it is limited to financial reporting. It meets the objective of determining whether the controls are reasonable but does not ensure that the control is working correctly and thereby supporting management expectations and objectives.

A1-28 The vice president of human resources has requested an IS audit to identify payroll overpayments for the previous year. Which would be the **BEST** audit technique to use in this situation?

A. Generate sample test data
B. Generalized audit software
C. Integrated test facility
D. Embedded audit module

B is the correct answer.

Justification:

A. Test data tests for the existence of controls that might prevent overpayments, but it does not detect specific, previous miscalculations.

B. Generalized audit software features include mathematical computations, stratification, statistical analysis, sequence checking, duplicate checking and re-computations. An IS auditor, using generalized audit software, can design appropriate tests to recompute the payroll, thereby determining whether there were overpayments and to whom they were made.

C. An integrated test facility helps to identify a problem as it occurs but does not detect errors for a previous period.

D. An embedded audit module can enable the IS auditor to evaluate a process and gather audit evidence, but it does not detect errors for a previous period.

A1-29 During a security audit of IT processes, an IS auditor finds that documented security procedures do not exist. The IS auditor should:

A. Create the procedures document based on the practices.
B. Issue an opinion of the current state and end the audit.
C. Conduct compliance testing on available data.
D. Identify and evaluate existing practices.

D is the correct answer.

Justification:

A. IS auditors should not prepare documentation because the process may not be compliant with management objectives and doing so could jeopardize their independence.

B. Ending the audit and issuing an opinion will not address identification of potential risk. The auditor should evaluate the practices in place. The recommendation may still be for the organization to develop written procedures. Terminating the audit may prevent achieving one of the basic audit objectives—identification of potential risk.

C. Because there are no documented procedures, there is no basis against which to test compliance.

D. One of the main objectives of an audit is to identify potential risk; therefore, the most proactive approach is to identify and evaluate the existing security practices being followed by the organization and submit the findings and risk to management, with recommendations to document the current controls or enforce the documented procedures.

A1-30 During a risk analysis, an IS auditor identifies threats and potential impacts. Next, the IS auditor should:

A. Ensure the risk assessment is aligned to management's risk assessment process.
B. Identify information assets and the underlying systems.
C. Disclose the threats and impacts to management.
D. Identify and evaluate the existing controls.

D is the correct answer.

Justification:

A. An audit risk assessment is conducted for purposes that are different from management's risk assessment process purposes.
B. It is impossible to determine impact without first identifying the assets affected; therefore, this must already have been completed.
C. Upon completion of a risk assessment, an IS auditor should describe and discuss with management the threats and potential impacts on the assets, and recommendations for addressing the risk. However, this action cannot be done until the controls are identified and the likelihood of the threat is calculated.
D. It is important for an IS auditor to identify and evaluate the existence and effectiveness of existing and planned controls so that the risk level can be calculated after the potential threats and possible impacts are identified.

A1-31 Which of the following would normally be the **MOST** reliable evidence for an IS auditor?

A. A confirmation letter received from a third party verifying an account balance
B. Assurance from line management that an application is working as designed
C. Trend data obtained from Internet sources
D. Ratio analysis developed by the IS auditor from reports supplied by line management

A is the correct answer.

Justification:

A. Evidence obtained from independent third parties is almost always considered to be more reliable than assurance provided by local management.
B. Because management is not objective and may not understand the risk and control environment, and they are only providing evidence that the application is working correctly (not the controls), their assurance is not an acceptable level of trust for audit evidence.
C. Data collected from the Internet is not always trustworthy or independently validated.
D. Ratio analysis can identify trends and deviations from a baseline but is not reliable evidence.

A1-32 When evaluating the collective effect of preventive, detective and corrective controls within a process, an IS auditor should be aware of which of the following?

A. The point at which controls are exercised as data flow through the system.
B. Only preventive and detective controls are relevant.
C. Corrective controls are regarded as compensating.
D. Classification allows an IS auditor to determine the controls that are missing.

A is the correct answer.

Justification:

A. An IS auditor should focus on when controls are exercised as data flow through a computer system.
B. Corrective controls may also be relevant because they allow an error or problem to be corrected.
C. Corrective controls remove or reduce the effects of errors or irregularities and are not exclusively regarded as compensating controls.
D. The existence and function of controls are important but not the classification.

A1-33 Which audit technique provides the **BEST** evidence of the segregation of duties in an IT department?

A. Discussion with management
B. Review of the organization chart
C. Observation and interviews
D. Testing of user access rights

C is the correct answer.

Justification:

A. Management may not be aware of the detailed functions of each employee in the IT department and whether the controls are being followed. Therefore, discussion with the management provides only limited information regarding segregation of duties.

B. An organization chart does not provide details of the functions of the employees or whether the controls are working correctly.

C. Based on the observations and interviews, the IS auditor can evaluate the segregation of duties. By observing the IT staff performing their tasks, an IS auditor can identify whether they are performing any incompatible operations. By interviewing the IT staff, the auditor can get an overview of the tasks performed.

D. Testing of user rights provides information about the rights users have within the IS systems but does not provide complete information about the functions they perform. Observation is a better option because user rights can be changed between audits.

A1-34 After reviewing the disaster recovery planning process of an organization, an IS auditor requests a meeting with organization management to discuss the findings. Which of the following **BEST** describes the main goal of this meeting?

A. Obtain management approval of the corrective action plan.
B. Confirm factual accuracy of the findings.
C. Assist management in the implementation of corrective actions.
D. Prioritize the resolution of the items.

B is the correct answer.

Justification:

A. Management approval of the corrective action plan is not required. Management can elect to implement another corrective action plan to address the risk.

B. The goal of the meeting is to confirm the factual accuracy of the audit findings and present an opportunity for management to agree on or respond to recommendations for corrective action.

C. Implementation of corrective actions should be done after the factual accuracy of findings is established, but the work of implementing corrective action is not typically assigned to the IS auditor, because this impairs the auditor's independence.

D. Rating the audit findings provides guidance to management for allocating resources to the high-risk items first.

A1-35 An IS auditor should ensure that review of online electronic funds transfer reconciliation procedures include:

A. Vouching.
B. Authorizations.
C. Corrections.
D. Tracing.

D is the correct answer.

Justification:

A. Vouching is usually performed during the funds transfer, not during the reconciliation effort.

B. In online processing, authorizations are normally done automatically by the system, not during the reconciliation.

C. Correction entries should be reviewed during a reconciliation; however, they are normally done by an individual other than the person entrusted to do reconciliations and are not as important as tracing.

D. Tracing is a transaction reconciliation effort that involves following the transaction from the original source to its final destination. In electronic funds transfer transactions, the direction on tracing may start from the customer-printed copy of the receipt, proceed to checking the system audit trails and logs, and end with checking the master file records for daily transactions.

A1-36 An IS auditor is carrying out a system configuration review. Which of the following is the **BEST** evidence in support of the current system configuration settings?

A. System configuration values that are imported to a spreadsheet by the system administrator
B. Standard report with configuration values that are retrieved from the system by the IS auditor
C. Dated screenshot of the system configuration settings that are made available by the system administrator
D. Annual review of approved system configuration values by the business owner

B is the correct answer.

Justification:

A. Evidence that is not system-generated information can be modified before it is presented to an IS auditor. Therefore, it may not be as reliable as evidence that is obtained by the IS auditor. For example, a system administrator can change the settings or modify the graphic image before taking a screenshot.

B. Evidence that is obtained directly from the source by an IS auditor is more reliable than information that is provided by a system administrator or a business owner, because the IS auditor does not have a vested interest in the outcome of the audit.

C. The rules may be modified by the administrator prior to taking the screenshot; therefore, this is not the best evidence.

D. The annual review provided by a business owner may not reflect current information.

A1-37 The purpose of a checksum on an amount field in an electronic data interchange communication of financial transactions is to ensure:

A. Integrity.
B. Authenticity.
C. Authorization.
D. Nonrepudiation.

A is the correct answer.

Justification:

A. A checksum that is calculated on an amount field and included in the electronic data interchange communication can be used to identify unauthorized modifications.
B. Authenticity cannot be established by a checksum alone and needs other controls.
C. Authorization cannot be established by a checksum alone and needs other controls.
D. Nonrepudiation can be ensured by using digital signatures.

A1-38 Which of the following forms of evidence would an IS auditor consider the **MOST** reliable?

A. An oral statement from the auditee
B. The results of a test that is performed by an external IS auditor
C. An internally generated computer accounting report
D. A confirmation letter that is received from an outside source

B is the correct answer.

Justification:

A. An oral statement from the auditee is audit evidence but not as reliable as the results of a test that is performed by an external IS auditor.
B. An independent test that is performed by an IS auditor should always be considered a more reliable source of evidence than a confirmation letter from a third party, because the letter is the result of an analysis of the process and may not be based on authoritative audit techniques. An audit should consist of a combination of inspection, observation and inquiry by an IS auditor as determined by risk. This provides a standard methodology and reasonable assurance that the controls and test results are accurate.
C. An internally generated computer accounting report is audit evidence, but is not as reliable as the results of a test performed by an external IS auditor.
D. An independent test performed by an IS auditor should always be considered a more reliable source of evidence than a confirmation letter from a third party, because a letter is subjective and may not have been generated as a part of an authoritative audit or conform to audit standards.

A1-39 An IS auditor who has discovered unauthorized transactions during a review of electronic data interchange (EDI) transactions is likely to recommend improving the:

A. EDI trading partner agreements.
B. Physical controls for terminals.
C. Authentication techniques for sending and receiving messages.
D. Program change control procedures.

C is the correct answer.

Justification:

A. The electronic data interchange trading partner agreements minimize exposure to legal issues but do not resolve the problem of unauthorized transactions.
B. Physical control is important and may provide protection from unauthorized people accessing the system but does not provide protection from unauthorized transactions by authorized users.
C. Authentication techniques for sending and receiving messages play a key role in minimizing exposure to unauthorized transactions.
D. Change control procedures do not resolve the issue of unauthorized transactions.

A1-40 An IS auditor is validating a control that involves a review of system-generated exception reports. Which of the following is the **BEST** evidence of the effectiveness of the control?

A. Walk-through with the reviewer of the operation of the control
B. System-generated exception reports for the review period with the reviewer's sign-off
C. A sample system-generated exception report for the review period, with follow-up action items noted by the reviewer
D. Management's confirmation of the effectiveness of the control for the review period

C is the correct answer.

Justification:

A. A walk-through highlights how a control is designed to work, but it seldom highlights the effectiveness of the control, or exceptions or constraints in the process.
B. Reviewer sign-off does not demonstrate the effectiveness of the control if the reviewer does not note follow-up actions for the exceptions identified.
C. A sample of a system-generated report with evidence that the reviewer followed up on the exception represents the best possible evidence of the effective operation of the control, because there is documented evidence that the reviewer reviewed the exception report and took actions based on the exception report.
D. Management's confirmation of effectiveness of the control suffers from lack of independence—management might be biased toward the effectiveness of the controls put in place.

A1-41 A company has recently upgraded its purchase system to incorporate electronic data interchange (EDI) transmissions. Which of the following controls should be implemented in the EDI interface to provide for efficient data mapping?

A. Key verification
B. One-for-one checking
C. Manual recalculations
D. Functional acknowledgments

D is the correct answer.

Justification:
A. Key verification is used for encryption and protection of data but not for data mapping.
B. One-for-one checking validates that transactions are accurate and complete but does not map data.
C. Manual recalculations are used to verify that the processing is correct but do not map data.
D. Acting as an audit trail for electronic data interchange transactions, functional acknowledgments are one of the main controls used in data mapping.

A1-42 Which of the following sampling methods would be the **MOST** effective to determine whether purchase orders issued to vendors have been authorized as per the authorization matrix?

A. Variable sampling
B. Stratified mean per unit
C. Attribute sampling
D. Unstratified mean per unit

C is the correct answer.

Justification:
A. Variable sampling is the method used for substantive testing, which involves testing transactions for quantitative aspects such as monetary values.
B. Stratified mean per unit is used in variable sampling.
C. Attribute sampling is the method used for compliance testing. In this scenario, the operation of a control is being evaluated, and therefore, the attribute of whether each purchase order was correctly authorized would be used to determine compliance with the control.
D. Unstratified mean per unit is used in variable sampling.

A1-43 The **BEST** method of confirming the accuracy of a system tax calculation is by:

A. review and analysis of the source code of the calculation programs.
B. recreating program logic using generalized audit software to calculate monthly totals.
C. preparing simulated transactions for processing and comparing the results to predetermined results.
D. automatic flowcharting and analysis of the source code of the calculation programs.

C is the correct answer.

Justification:
A. A review of source code is not an effective method of ensuring that the calculation is being computed correctly.
B. Recreating program logic may lead to errors, and monthly totals are not accurate enough to ensure correct computations.
C. Preparing simulated transactions for processing and comparing the results to predetermined results is the best method for confirming the accuracy of a tax calculation.
D. Flowcharting and analysis of source code are not effective methods to address the accuracy of individual tax calculations.

A1-44 An IS auditor performing a review of application controls would evaluate the:

A. efficiency of the application in meeting the business processes.
B. impact of any exposures discovered.
C. business processes served by the application.
D. application's optimization.

B is the correct answer.

Justification:

A. The IS auditor is reviewing the effectiveness of the controls, not the suitability of the application to meet business needs.
B. An application control review involves the evaluation of the application's automated controls and an assessment of any exposures resulting from the control weaknesses.
C. The other choices may be objectives of an application audit but are not part of an audit restricted to a review of the application controls.
D. One area to be reviewed may be the efficiency and optimization of the application, but this is not the area being reviewed in this audit.

A1-45 Corrective action has been taken by an auditee immediately after the identification of a reportable finding. The IS auditor should:

A. include the finding in the final report, because the IS auditor is responsible for an accurate report of all findings.
B. not include the finding in the final report because management resolved the item.
C. not include the finding in the final report, because corrective action can be verified by the IS auditor during the audit.
D. include the finding in the closing meeting for discussion purposes only.

A is the correct answer.

Justification:

A. Including the finding in the final report is a generally accepted audit practice. If an action is taken after the audit started and before it ended, the audit report should identify the finding and describe the corrective action taken. An audit report should reflect the situation, as it existed at the start of the audit. All corrective actions taken by the auditee should be reported in writing.
B. The audit report should contain all relevant findings and the response from management even if the finding has been resolved. This would mean that subsequent audits may test for the continued resolution of the control.
C. The audit report should contain the finding so that it is documented and the removal of the control subsequent to the audit would be noticed.
D. The audit report should contain the finding and resolution, and this can be mentioned in the final meeting. The audit report should list all relevant findings and the response from management.

A1-46 The internal IS audit team is auditing controls over sales returns and is concerned about fraud. Which of the following sampling methods will **BEST** assist the IS auditors?

A. Stop-or-go
B. Classical variable
C. Discovery
D. Probability-proportional-to-size

C is the correct answer.

Justification:

A. Stop-or-go is a sampling method that helps limit the size of a sample and allows the test to be stopped at the earliest possible moment.
B. Classical variable sampling is associated with dollar amounts and has a sample based on a representative sample of the population but is not focused on fraud.
C. Discovery sampling is used when an IS auditor is trying to determine whether a type of event has occurred. Therefore, it is suited to assess the risk of fraud and to identify whether a single occurrence has taken place.
D. Probability-proportional-to-size sampling is typically associated with cluster sampling when there are groups within a sample. The question does not indicate that an IS auditor is searching for a threshold of fraud.

A1-47 When developing a risk-based audit strategy, an IS auditor should conduct a risk assessment to ensure that:

A. Controls needed to mitigate risk are in place.
B. Vulnerabilities and threats are identified.
C. Audit risk is considered.
D. A gap analysis is appropriate.

B is the correct answer.

Justification:

A. Understanding whether appropriate controls that are required to mitigate risk are in place is a resultant effect of an audit.
B. While developing a risk-based audit strategy, it is critical that the risk and vulnerabilities are understood. They determine the areas to be audited and the extent of coverage.
C. Audit risk is an inherent aspect of auditing, directly related to the audit process and not relevant to the risk analysis of the environment to be audited.
D. A gap analysis is normally done to compare the actual state to an expected or desirable state.

A1-48 During an exit interview, in cases where there is disagreement regarding the impact of a finding, an IS auditor should:

A. Ask the auditee to sign a release form accepting full legal responsibility.
B. Elaborate on the significance of the finding and the risk of not correcting it.
C. Report the disagreement to the audit committee for resolution.
D. Accept the auditee's position because they are the process owners.

B is the correct answer.

Justification:

A. Management is always responsible and liable for risk. The role of the IS auditor is to inform management of the findings and associated risk discovered in an audit.

B. If the auditee disagrees with the impact of a finding, it is important for an IS auditor to elaborate and clarify the risk and exposures because the auditee may not fully appreciate the magnitude of the exposure. The goal should be to enlighten the auditee or uncover new information of which an IS auditor may not have been aware. Anything that appears to threaten the auditee lessens effective communications and sets up an adversarial relationship, but an IS auditor should not automatically agree just because the auditee expresses an alternate point of view.

C. The audit report contains the finding from the IS auditor and the response from management. It is the responsibility of management to accept risk or mitigate it appropriately. The role of the auditor is to inform management clearly and thoroughly so that the best decision can be made.

D. The IS auditor must be professional, competent and independent. They must not just accept an explanation or argument from management, unless the process used to generate the finding was flawed.

A1-49 To ensure that audit resources deliver the best value to the organization, the **FIRST** step in an audit project is to:

A. Schedule the audits and monitor the time spent on each audit.
B. Train the IS audit staff on current technology used in the organization.
C. Develop the audit plan based on a detailed risk assessment.
D. Monitor progress of audits and initiate cost control measures.

C is the correct answer.

Justification:

A. Monitoring the audits and the time spent on audits is not effective if the wrong areas are being audited. It is most important to develop a risk-based audit plan to ensure effective use of audit resources.

B. The IS auditor may have specialties, or the audit team may rely on outside experts to conduct very specialized audits. It is not necessary for each IS auditor to be trained on all new technology.

C. Although monitoring the time and audit programs, and adequate training improve the IS audit staff's productivity (efficiency and performance), ensuring that the resources and efforts being dedicated to audit are focused on higher-risk areas delivers value to the organization.

D. Monitoring audits and initiating cost controls does not ensure the effective use of audit resources.

A1-50 Which of the following should be the **FIRST** action of an IS auditor during a dispute with a department manager over audit findings?

A. Retest the control to validate the finding.
B. Engage a third party to validate the finding.
C. Include the finding in the report with the department manager's comments.
D. Revalidate the supporting evidence for the finding.

D is the correct answer.

Justification:
A. Retesting the control normally occurs after the evidence has been revalidated.
B. Although there are cases where a third party may be needed to perform specialized audit procedures, an IS auditor should first revalidate the supporting evidence to determine whether there is a need to engage a third party.
C. Before putting a disputed finding or management response in the audit report, the IS auditor should take care to review the evidence that is used in the finding to ensure audit accuracy.
D. Conclusions drawn by an IS auditor should be adequately supported by evidence, and any compensating controls or corrections that are pointed out by a department manager should be taken into consideration. Therefore, the first step is to revalidate the evidence for the finding. If, after revalidating and retesting, there are unsettled disagreements, those issues should be included in the report.

A1-51 An IS auditor should use statistical sampling, and not judgment (nonstatistical) sampling, when:

A. The probability of error must be objectively quantified.
B. The auditor wants to avoid sampling risk.
C. Generalized audit software is unavailable.
D. The tolerable error rate cannot be determined.

A is the correct answer.

Justification:
A. Given an expected error rate and confidence level, statistical sampling is an objective method of sampling, which helps an IS auditor determine the sample size and quantify the probability of error (confidence coefficient).
B. Sampling risk is the risk of a sample not being representative of the population. This risk exists for judgment and statistical samples.
C. Statistical sampling can use generalized audit software, but it is not required.
D. The tolerable error rate must be predetermined for both judgment and statistical sampling.

A1-52 What is the **BEST** action for an IS auditor to take when an outsourced monitoring process for remote access is inadequate and management disagrees because intrusion detection system (IDS) and firewall controls are in place?

A. Revise the finding in the audit report per management's feedback.
B. Retract the finding because the IDS controls are in place.
C. Retract the finding because the firewall rules are monitored.
D. Document the identified finding in the audit report.

D is the correct answer.

Justification:

A. The IS auditor may include the management response in the report, but that will not affect the requirement to report the finding.
B. The finding remains valid and the management response is documented; however, the audit may indicate a need to review the validity of the management response.
C. The finding remains valid and the management response is documented; however, the audit may indicate a need to review the validity of the management response.
D. IS auditor independence dictates that the additional information provided by the auditee is taken into consideration. Normally, an IS auditor does not automatically retract or revise the finding.

A1-53 An organization uses a bank to process its weekly payroll. Time sheets and payroll adjustment forms (e.g., hourly rate changes and terminations) are completed and delivered to the bank, which prepares the checks and reports for distribution. To **BEST** ensure payroll data accuracy:

A. Payroll reports should be compared to input forms.
B. Gross payroll should be recalculated manually.
C. Checks should be compared to input forms.
D. Checks should be reconciled with output reports.

A is the correct answer.

Justification:

A. The best way to confirm data accuracy, when input is provided by the organization and output is generated by the bank, is to verify the data input (input forms) with the results of the payroll reports.
B. Recalculating gross payroll manually only verifies whether the processing is correct and not the data accuracy of inputs.
C. Comparing checks to input forms is not feasible because checks contain the processed information and input forms contain the input data.
D. Reconciling checks with output reports only confirms that checks were issued as stated on output reports.

A1-54 Which of the following represents the **GREATEST** potential risk in an electronic data interchange (EDI) environment?

A. Lack of transaction authorizations
B. Loss or duplication of EDI transmissions
C. Transmission delay
D. Deletion or manipulation of transactions prior to, or after, establishment of application controls

A is the correct answer.

Justification:
A. Because the interaction between parties is electronic, there is no inherent authentication occurring; therefore, lack of transaction authorization is the greatest risk.
B. Loss or duplication of electronic data interchange transmissions is an example of risk, but because all transactions should be logged, the impact is not as great as that of unauthorized transactions.
C. Transmission delays may terminate the process or hold the line until the normal time for processing has elapsed; however, there will be no loss of data.
D. Deletion or manipulation of transactions prior to, or after, establishment of application controls is an example of risk. Logging detects any alteration to the data, and the impact is not as great as that of unauthorized transactions.

A1-55 During the planning stage of an IS audit, the **PRIMARY** goal of an IS auditor is to:

A. Address audit objectives.
B. Collect sufficient evidence.
C. Specify appropriate tests.
D. Minimize audit resources.

A is the correct answer.

Justification:
A. ISACA IS Audit and Assurance Standards require that an IS auditor plan the audit work to address the audit objectives. The activities described in the other options are all undertaken to address audit objectives and, thus, are secondary.
B. The IS auditor does not collect evidence in the planning stage of an audit.
C. Specifying appropriate tests is not the primary goal of audit planning.
D. Effective use of audit resources is a goal of audit planning, not minimizing audit resources.

A1-56 When selecting audit procedures, an IS auditor should use professional judgment to ensure that:

A. Sufficient evidence will be collected.
B. Significant deficiencies will be corrected within a reasonable period.
C. All material weaknesses will be identified.
D. Audit costs will be kept at a minimum level.

A is the correct answer.

Justification:

A. Procedures are processes that an IS auditor may follow in an audit engagement. In determining the appropriateness of any specific procedure, an IS auditor should use professional judgment that is appropriate to the specific circumstances. Professional judgment involves a subjective and often qualitative evaluation of conditions arising during an audit. Judgment addresses a grey area where binary (yes/no) decisions are not appropriate, and the IS auditor's past experience plays a key role in making a judgment. The IS auditor should use judgment in assessing the sufficiency of evidence to be collected. ISACA's guidelines provide information on how to meet the standards when performing IS audit work.

B. The correction of deficiencies is the responsibility of management and is not a part of the audit procedure selection process.

C. Identifying material weaknesses is the result of appropriate competence, experience and thoroughness in planning and executing the audit, and not of professional judgment. Professional judgment is not a primary input to the financial aspects of the audit. Audit procedures and use of professional judgment cannot ensure that all deficiencies/weaknesses will be identified and corrected.

D. Professional judgment ensures that audit resources and costs are used wisely, but this is not the primary objective of the auditor when selecting audit procedures.

A1-57 A substantive test to verify that tape library inventory records are accurate is:

A. Determining whether bar code readers are installed.
B. Determining whether the movement of tapes is authorized.
C. Conducting a physical count of the tape inventory.
D. Checking whether receipts and issues of tapes are accurately recorded.

C is the correct answer.

Justification:

A. Determining whether bar code readers are installed is a compliance test.

B. Determining whether the movement of tapes is authorized is a compliance test.

C. A substantive test includes gathering evidence to evaluate the integrity (i.e., the completeness, accuracy and validity) of individual transactions, data or other information. Conducting a physical count of the tape inventory is a substantive test.

D. Checking whether receipts and issues of tapes are accurately recorded is a compliance test.

A1-58 An appropriate control for ensuring the authenticity of orders received in an electronic data interchange system application is to:

A. Acknowledge receipt of electronic orders with a confirmation message.
B. Perform reasonableness checks on quantities ordered before filling orders.
C. Verify the identity of senders and determine if orders correspond to contract terms.
D. Encrypt electronic orders.

C is the correct answer.

Justification:

A. Acknowledging the receipt of electronic orders with a confirming message is good practice but will not authenticate orders from customers.
B. Performing reasonableness checks on quantities ordered before placing orders is a control for ensuring the correctness of the organization's orders, not the authenticity of its customers' orders.
C. An electronic data interchange system is subject not only to the usual risk exposures of computer systems but also to those arising from the potential ineffectiveness of controls on the part of the trading partner and the third-party service provider, making authentication of users and messages a major security concern.
D. Encrypting sensitive messages is an appropriate step but does not prove authenticity of messages received.

A1-59 An IS auditor finds that the answers received during an interview with a payroll clerk do not support job descriptions and documented procedures. Under these circumstances, the IS auditor should:

A. conclude that the controls are inadequate.
B. expand the scope to include substantive testing.
C. place greater reliance on previous audits.
D. suspend the audit.

B is the correct answer.

Justification:

A. Based solely on the interview with the payroll clerk, the IS auditor will not be able to collect evidence to conclude on the adequacy of existing controls.
B. If the answers provided to an IS auditor's questions are not confirmed by documented procedures or job descriptions, the IS auditor should expand the scope of testing the controls and include additional substantive tests.
C. Placing greater reliance on previous audits is an inappropriate action, because it provides no current knowledge of the adequacy of the existing controls.
D. Suspending the audit is an inappropriate action, because it provides no current knowledge of the adequacy of the existing controls.

A1-60 An external IS auditor issues an audit report pointing out the lack of firewall protection features at the perimeter network gateway and recommending a specific vendor product to address this vulnerability. The IS auditor has failed to exercise:

A. Professional independence.
B. Organizational independence.
C. Technical competence.
D. Professional competence.

A is the correct answer.

Justification:

A. When an IS auditor recommends a specific vendor, the auditor's professional independence is compromised.
B. Organizational independence has no relevance to the content of an audit report and should be considered at the time of accepting the engagement.
C. Technical competence is not relevant to the requirement of independence.
D. Professional competence is not relevant to the requirement of independence.

A1-61 The **PRIMARY** reason an IS auditor performs a functional walk-through during the preliminary phase of an audit assignment is to:

A. Understand the business process.
B. Comply with auditing standards.
C. Identify control weakness.
D. Develop the risk assessment.

A is the correct answer.

Justification:

A. Understanding the business process is the first step an IS auditor needs to perform.
B. ISACA IS Audit and Assurance Standards encourage adoption of the audit procedures/processes required to assist the IS auditor in performing IS audits more effectively. However, standards do not require an IS auditor to perform a process walk-through at the commencement of an audit engagement.
C. Identifying control weaknesses is not the primary reason for the walk-through and typically occurs at a later stage in the audit.
D. The main reason is to understand the business process. The risk assessment is developed after the business process is understood.

A1-62 In the process of evaluating program change controls, an IS auditor uses source code comparison software to:

A. Examine source program changes without information from IS personnel.
B. Detect a source program change made between acquiring a copy of the source and the comparison run.
C. Confirm that the control copy is the current version of the production program.
D. Ensure that all changes made in the current source copy are tested.

A is the correct answer.

Justification:

A. When an IS auditor uses a source code comparison to examine source program changes without information from IS personnel, the IS auditor has an objective, independent and relatively complete assurance of program changes, because the source code comparison identifies the changes.
B. The changes detected by the source code comparison are between two versions of the software. This does not detect changes made since the acquisition of the copy of the software.
C. This is a function of library management, not source code comparison. An IS auditor gains this assurance separately.
D. Source code comparison detects all changes between an original and a changed program; however, the comparison will not ensure that the changes have been adequately tested.

A1-63 The **PRIMARY** purpose for meeting with auditees prior to formally closing a review is to:

A. Confirm that the auditors did not overlook any important issues.
B. Gain agreement on the findings.
C. Receive feedback on the adequacy of the audit procedures.
D. Test the structure of the final presentation.

B is the correct answer.

Justification:

A. The closing meeting identifies any misunderstandings or errors in the audit but does not identify any important issues overlooked in the audit.
B. The primary purpose for meeting with auditees prior to formally closing a review is to gain agreement on the findings and responses from management.
C. The closing meeting may obtain comments from management on the conduct of the audit but is not intended to be a formal review of the adequacy of the audit procedures.
D. The structure of an audit report and the presentation follows accepted standards and practices. The closing meeting may indicate errors in the audit or presentation but is not intended to test the structure of the presentation.

A1-64 Which of the following audit techniques **BEST** helps an IS auditor in determining whether there have been unauthorized program changes since the last authorized program update?

A. Test data run
B. Code review
C. Automated code comparison
D. Review of code migration procedures

C is the correct answer.

Justification:
A. Test data runs permit the auditor to verify the processing of preselected transactions but provide no evidence about unauthorized changes or unexercised portions of a program.
B. Code review is the process of reading program source code listings to determine whether the code follows coding standards or contains potential errors or inefficient statements. A code review can be used as a means of code comparison, but it is inefficient and unlikely to detect any changes in the code, especially in a large program.
C. An automated code comparison is the process of comparing two versions of the same program to determine whether the two correspond. It is an efficient technique because it is an automated procedure.
D. The review of code migration procedures does not detect unauthorized program changes.

A1-65 When preparing an audit report, the IS auditor should ensure that the results are supported by:

A. Statements from IS management.
B. Work papers of other auditors.
C. An organizational control self-assessment.
D. Sufficient and appropriate audit evidence.

D is the correct answer.

Justification:
A. Statements from IS management may be included in the audit analysis but these statement alone are not considered a sufficient basis for issuing a report.
B. Work papers from other auditors may be used to substantiate and validate a finding but should not be used without the additional evidence of the work papers from the IS auditor who is preparing the report.
C. The results of a control self-assessment may assist the IS auditor in determining risk and compliance but on its own is not enough to support the audit report.
D. ISACA's IS Audit and Assurance Standard on reporting requires that the IS auditor has sufficient and appropriate audit evidence to support the reported results. Statements from IS management provide a basis for obtaining concurrence on matters that cannot be verified with empirical evidence. The report should be based on evidence that is collected during the review even though the IS auditor may have access to the work papers of other auditors. The results of an organizational control self-assessment can supplement the audit findings.

A1-66 While evaluating software development practices in an organization, an IS auditor notes that the quality assurance (QA) function reports to project management. The **MOST** important concern for an IS auditor is the:

A. Effectiveness of the QA function because it should interact between project management and user management.
B. Efficiency of the QA function because it should interact with the project implementation team.
C. Effectiveness of the project manager because the project manager should interact with the QA function.
D. Efficiency of the project manager because the QA function needs to communicate with the project implementation team.

A is the correct answer.

Justification:

A. To be effective, the quality assurance (QA) function should be independent of project management. If it is not, project management may put pressure on the QA function to approve an inadequate product.
B. The efficiency of the QA function is not impacted by interacting with the project implementation team. The QA team does not release a product for implementation until it meets QA requirements.
C. The project manager responds to the issues raised by the QA team. This does not impact the effectiveness of the project manager.
D. The QA function's interaction with the project implementation team should not impact the efficiency of the project manager.

A1-67 The final decision to include a material finding in an audit report should be made by the:

A. audit committee.
B. auditee's manager.
C. IS auditor.
D. chief executive officer.

C is the correct answer.

Justification:

A. The audit committee should not impair the independence, professionalism and objectivity of the IS auditor by influencing what is included in the audit report.
B. The IS auditor's manager may recommend what should or should not be included in an audit report, but the auditee's manager should not influence the content of the report.
C. The IS auditor should make the final decision about what to include or exclude from the audit report.
D. The chief executive officer must not provide influence over the content of an audit report because that would be a breach of the independence of the audit function.

A1-68 While reviewing sensitive electronic work papers, the IS auditor noticed that they were not encrypted. This could compromise the:

A. Audit trail of the versioning of the work papers.
B. Approval of the audit phases.
C. Access rights to the work papers.
D. Confidentiality of the work papers.

D is the correct answer.

Justification:

A. Audit trails do not, by themselves, affect the confidentiality, but are part of the reason for requiring encryption.
B. Audit phase approvals do not, by themselves, affect the confidentiality of the work papers, but are part of the reason for requiring encryption.
C. Access to the work papers should be limited by need to know; however, a lack of encryption breaches the confidentiality of the work papers, not the access rights to the papers.
D. Encryption provides confidentiality for the electronic work papers.

A1-69 The **MOST** important reason for an IS auditor to obtain sufficient and appropriate audit evidence is to:

A. Comply with regulatory requirements.
B. Provide a basis for drawing reasonable conclusions.
C. Ensure complete audit coverage.
D. Perform the audit according to the defined scope.

B is the correct answer.

Justification:

A. Complying with regulatory requirements is relevant to an audit but is not the most important reason why sufficient and relevant evidence is required.
B. The scope of an IS audit is defined by its objectives. This involves identifying control weaknesses relevant to the scope of the audit. Obtaining sufficient and appropriate evidence assists the auditor in not only identifying control weaknesses but also documenting and validating them.
C. Ensuring coverage is relevant to conducting an IS audit but is not the most important reason why sufficient and relevant evidence is required. The reason for obtaining evidence is to ensure that the audit conclusions are factual and accurate.
D. The execution of an audit to meet its defined scope is relevant to an audit but is not the reason why sufficient and relevant evidence is required.

A1-70 After initial investigation, an IS auditor has reasons to believe that fraud may be present. The IS auditor should:

A. Expand activities to determine whether an investigation is warranted.
B. Report the matter to the audit committee.
C. Report the possibility of fraud to management.
D. Consult with external legal counsel to determine the course of action to be taken.

A is the correct answer.

Justification:

A. An IS auditor's responsibilities for detecting fraud include evaluating fraud indicators and deciding whether any additional action is necessary or whether an investigation should be recommended.
B. The IS auditor should notify the appropriate authorities within the organization only if it has determined that the indicators of fraud are sufficient to recommend an investigation.
C. The IS auditor should report the possibility of fraud to top management only after there is sufficient evidence to launch an investigation. This may be affected by whether management may be involved in the fraud.
D. Normally, the IS auditor does not have authority to consult with external legal counsel.

A1-71 An IS auditor notes that failed login attempts to a core financial system are automatically logged and the logs are retained for a year by the organization. This logging is:

A. An effective preventive control.
B. A valid detective control.
C. Not an adequate control.
D. A corrective control.

C is the correct answer.

Justification:

A. Generation of an activity log is not a preventive control because it cannot prevent inappropriate access.
B. Generation of an activity log is not a detective control because it does not help in detecting inappropriate access unless it is reviewed by appropriate personnel.
C. Generation of an activity log is not a control by itself. It is the review of such a log that makes the activity a control (i.e., generation plus review equals control).
D. Generation of an activity log is not a corrective control because it does not correct the effect of inappropriate access.

A1-72 An organization's IS audit charter should specify the:

A. plans for IS audit engagements.
B. objectives and scope of IS audit engagements.
C. detailed training plan for the IS audit staff.
D. role of the IS audit function.

D is the correct answer.

Justification:

A. Planning is the responsibility of audit management.
B. The objectives and scope of each IS audit should be agreed on in an engagement letter. The charter would specify the objectives and scope of the audit function but not of individual engagements.
C. A training plan that is based on the audit plan should be developed by audit management.
D. An IS audit charter establishes the role of the information systems audit function. The charter should describe the overall authority, scope and responsibilities of the audit function. It should be approved by the highest level of management and, if available, by the audit committee.

A1-73 Which of the following should an IS auditor use to detect duplicate invoice records within an invoice master file?

A. Attribute sampling
B. Computer-assisted audit techniques
C. Compliance testing
D. Integrated test facility

B is the correct answer.

Justification:

A. Attribute sampling aids in identifying records meeting specific conditions but does not compare one record to another to identify duplicates. To detect duplicate invoice records, the IS auditor should check all items that meet the criteria and not just a sample of the items.
B. Computer-assisted audit techniques (CAATs) enable the IS auditor to review the entire invoice file to look for those items that meet the selection criteria.
C. Compliance testing determines whether controls procedures are adhered to. Using CAATs is the better option because it is most likely more efficient to search for duplicates.
D. An integrated test facility allows the IS auditor to test transactions through the production system but does not compare records to identify duplicates.

A1-74 When developing a risk management program, what is the **FIRST** activity to be performed?

A. Threat assessment
B. Classification of data
C. Inventory of assets
D. Criticality analysis

C is the correct answer.

Justification:

A. The assets need to be identified first. A listing of the threats that can affect the assets is a later step in the process.
B. Data classification is required for defining access controls and in criticality analysis, but the assets (including data) need be identified before doing classification.
C. Identification of the assets to be protected is the first step in the development of a risk management program.
D. Criticality analysis is a later step in the process after the assets have been identified.

A1-75 When evaluating the controls of an electronic data interchange (EDI) application, an IS auditor should **PRIMARILY** be concerned with the risk of:

A. Excessive transaction turnaround time.
B. Application interface failure.
C. Improper transaction authorization.
D. Nonvalidated batch totals.

C is the correct answer.

Justification:
A. An excessive turnaround time is an inconvenience, but not a serious risk.
B. The failure of the application interface is a risk, but not the most serious issue. Usually such a problem is temporary and easily fixed.
C. Foremost among the risk associated with electronic data interchange (EDI) is improper transaction authorization. Because the interaction with the parties is electronic, there is no inherent authentication. Improper authentication poses a serious risk of financial loss.
D. The integrity of EDI transactions is important, but not as significant as the risk of unauthorized transactions

A1-76 Which of the following would be **MOST** useful for an IS auditor for accessing and analyzing digital data to collect relevant audit evidence from diverse software environments?

A. Structured Query Language
B. Application software reports
C. Data analytics controls
D. Computer-assisted auditing techniques

D is the correct answer.

Justification:
A. Structured Query Language provides options for auditors to query specific tables of a database according to audit objectives. However, skills are required to query specific databases, and a user must be able to understand the record structure to access the data.
B. Reports from application software may be useful, but they are not be as beneficial as computer-assisted auditing techniques (CAATs).
C. Data analytics controls might be a good technique to use for control testing, but they are not as comprehensive as CAATs.
D. CAATs are tools used for accessing data in an electronic form from diverse software environments, record formats, etc. CAATs serve as useful tools for collecting and evaluating audit evidence according to audit objectives and can create efficiencies for collecting this evidence.

A1-77 Which of the following sampling methods is the **MOST** appropriate for testing automated invoice authorization controls to ensure that exceptions are not made for specific users?

A. Variable sampling
B. Judgmental sampling
C. Stratified random sampling
D. Systematic sampling

C is the correct answer.

Justification:

A. Variable sampling is used for substantive testing to determine the monetary or volumetric impact of characteristics of a population. This is not the most appropriate in this case.

B. In judgmental sampling, professionals place a bias on the sample (e.g., all sampling units over a certain value, all for a specific type of exception or all negatives). It should be noted that a judgmental sample is not statistically based, and results should not be extrapolated over the population because the sample is unlikely to be representative of the population.

C. Stratification is the process of dividing a population into subpopulations with similar characteristics explicitly defined, so that each sampling unit can belong to only one stratum. This method of sampling ensures that all sampling units in each subgroup have a known, nonzero chance of selection. It is the most appropriate in this case.

D. Systematic sampling involves selecting sampling units using a fixed interval between selections with the first interval having a random start. This is not the most appropriate in this case.

A1-78 An IS auditor who was involved in designing an organization's business continuity plan (BCP) has been assigned to audit the plan. The IS auditor should:

A. decline the assignment.
B. inform management of the possible conflict of interest after completing the audit assignment.
C. inform the BCP team of the possible conflict of interest prior to beginning the assignment.
D. communicate the possibility of conflict of interest to audit management prior to starting the assignment.

D is the correct answer.

Justification:

A. Declining the assignment could be acceptable only after obtaining management approval or it is appropriately disclosed to management, audit management and other stakeholders.

B. Approval should be obtained prior to commencement and not after the completion of the assignment.

C. Informing the BCP team of the possible conflict of interest prior to starting the assignment is not the correct answer because the BCP team does not have the authority to decide on this issue.

D. A possible conflict of interest, likely to affect the IS auditor's independence, should be brought to the attention of management prior to starting the assignment.

A1-79 The **PRIMARY** purpose of an IT forensic audit is:

A. To participate in investigations related to corporate fraud.
B. The systematic collection and analysis of evidence after a system irregularity.
C. To assess the correctness of an organization's financial statements.
D. To preserve evidence of criminal activity.

B is the correct answer.

Justification:
A. Forensic audits are not limited to corporate fraud.
B. The systematic collection and analysis of evidence after a system irregularity best describes a forensic audit. The evidence collected can then be analyzed and used in judicial proceedings.
C. Assessing the correctness of an organization's financial statements is not the primary purpose of most forensic audits.
D. Forensics is the investigation of evidence related to a crime or misbehavior. Preserving evidence is the forensic process, but not the primary purpose.

A1-80 An IS auditor reviews one day of logs for a remotely managed server and finds one case where logging failed, and the backup restarts cannot be confirmed. What should the IS auditor do?

A. Issue an audit finding.
B. Seek an explanation from IS management.
C. Review the classifications of data held on the server.
D. Expand the sample of logs reviewed.

D is the correct answer.

Justification:
A. At this stage it is too preliminary to issue an audit finding. Seeking an explanation from management is advisable, but it is better to gather additional evidence to properly evaluate the seriousness of the situation.
B. Without gathering more information on the incident and the frequency of the incident, it is difficult to obtain a meaningful explanation from management.
C. A backup failure, which has not been established at this point, will be serious if it involves critical data. However, the issue is not the importance of the data on the server, where a problem has been detected, but whether a systematic control failure that impacts other servers exists.
D. IS Audit and Assurance Standards require that an IS auditor gather sufficient and appropriate audit evidence. The IS auditor has found a potential problem and now needs to determine whether this is an isolated incident or a systematic control failure.

A1-81 In a small organization, the function of release manager and application programmer are performed by the same employee. What is the **BEST** compensating control in this scenario?

A. Hiring additional staff to provide segregation of duties
B. Preventing the release manager from making program modifications
C. Logging of changes to development libraries
D. Verifying that only approved program changes are implemented

D is the correct answer.

Justification:

A. Establishing segregation of duties is not a compensating control; it is a preventive control. In a small organization, it may not be feasible to hire new staff, which is why a compensating control may be necessary.
B. Since the release manager is performing dual roles, preventing them from making program modifications is not feasible, and, in a small organization, segregation of duties may not be possible.
C. Logging changes to development libraries does not detect changes to production libraries.
D. Compensating controls are used to mitigate risk when proper controls are not feasible or practical. In a small organization, it may not be feasible to hire new staff, which is why a compensating control may be necessary. Verifying program changes has roughly the same effect as intended by full segregation of duties.

A1-82 Which of the following is the **FIRST** step in an IT risk assessment for a risk-based audit?

A. Identify all IT systems and controls that are relevant to audit objectives.
B. List all controls from the audit program to select ones matching with audit objectives.
C. Review the results of a risk self-assessment.
D. Understand the business, its operating model and key processes.

D is the correct answer.

Justification:

A. Understanding the business environment comes first; this is followed by understanding the IT environment.
B. Listing controls and matching them to audit objectives is not the first step of risk assessment. This step follows understanding the business environment and the IT systems.
C. A risk self-assessment is optional and applicable for some types of audit engagements.
D. Risk-based auditing must be based on the understanding of the business, operating model and environment. This is the first step in an IT risk assessment for a risk-based audit.

A1-83 An IS auditor discovers that devices connected to the network are not included in a network diagram that had been used to develop the scope of the audit. The chief information officer explains that the diagram is being updated and awaiting final approval. The IS auditor should **FIRST**:

A. expand the scope of the IS audit to include the devices that are not on the network diagram.
B. evaluate the impact of the undocumented devices on the audit scope.
C. note a control deficiency because the network diagram has not been approved.
D. plan follow-up audits of the undocumented devices.

B is the correct answer.

Justification:

A. It is important that the IS auditor does not immediately assume that everything on the network diagram provides information about the risk affecting a network/system. There is a process in place for documenting and updating the network diagram.

B. In a risk-based approach to an IS audit, the scope is determined by the impact that the devices will have on the audit. If the undocumented devices do not impact the audit scope, then they may be excluded from the current audit engagement. The information provided on a network diagram can vary depending on what is being illustrated—for example, the network layer and cross connections.

C. In this case, there is simply a mismatch in timing between the completion of the approval process and when the IS audit began. There is no control deficiency to be reported.

D. Planning for follow-up audits of the undocumented devices is contingent on the risk that the undocumented devices have on the ability of the entity to meet the audit scope.

A1-84 An IS auditor is testing employee access to a large financial system, and the IS auditor selected a sample from the current employee list provided by the auditee. Which of the following evidence is the **MOST** reliable to support the testing?

A. A spreadsheet provided by the system administrator
B. Human resources access documents signed by employees' managers
C. A list of accounts with access levels generated by the system
D. Observations performed onsite in the presence of a system administrator

C is the correct answer.

Justification:

A. A spreadsheet supplied by the system administrator may not be complete or may be inaccurate. Documentary evidence should be collected to support the auditee's spreadsheet.

B. The human resources access documents signed by managers are good evidence; however, they are not as objective as the system-generated access list, because access may have changed, or the documents may have been incorrect when they were signed.

C. The access list generated by the system is the most reliable, because it is the most objective evidence to perform a comparison against the samples selected. The evidence is objective, because it was generated by the system rather than by an individual.

D. The observations are good evidence to understand the internal control structure; however, observations are not efficient for many users. Observations are not objective enough for substantive tests.

A1-85 During a compliance audit of a small bank, the IS auditor notes that the IT and accounting functions are being performed by the same user of the financial system. Which of the following reviews that are conducted by the user's supervisor represents the **BEST** compensating control?

A. Audit trails that show the date and time of the transaction
B. A daily report with the total numbers and dollar amounts of each transaction
C. User account administration
D. Computer log files that show individual transactions

D is the correct answer.

Justification:

A. An audit trail of only the date and time of the transaction is not sufficient to compensate for the risk of multiple functions being performed by the same individual.
B. Review of the summary financial reports does not compensate for the segregation of duties issue.
C. Supervisor review of user account administration can be a good control; however, it may not detect inappropriate activities where a person fills multiple roles.
D. Computer logs record the activities of individuals during their access to a computer system or data file and record any abnormal activities, such as the modification or deletion of financial data.

A1-86 A system developer transfers to the audit department to serve as an IT auditor. When production systems are to be reviewed by this employee, which of the following will become the **MOST** significant concern?

A. The work may be construed as a self-audit.
B. Audit points may largely shift to technical aspects.
C. The employee may not have sufficient control assessment skills.
D. The employee's knowledge of business risk may be limited.

A is the correct answer.

Justification:

A. Because the employee had been a developer, it is recommended that the audit coverage should exclude the systems developed by this employee to avoid any conflicts of interests.
B. Because the employee has a technical background, it is possible that the audit findings tend to focus on technical matters. However, this is normally corrected in the review process before it is carried out in production.
C. Because auditing is a new role for this employee, they may not have adequate control assessment skills. However, this can be addressed by on-the-job training and is not be as big of a concern as a potential conflict of interest.
D. Because this employee was previously employed in the organization's IT department, it is possible to build upon the employee's current understanding of the business to address any gaps in knowledge.

A1-87 Which of the following **BEST** describes the objective of an IS auditor discussing the audit findings with the auditee?

A. Communicate results to the auditee.
B. Develop time lines for the implementation of suggested recommendations.
C. Confirm the findings and propose a course of corrective action.
D. Identify compensating controls to the identified risk.

C is the correct answer.

Justification:

A. Based on this discussion, the IS auditor will finalize the report and present the report to relevant levels of senior management after the findings are confirmed. This discussion should, however, also address a timetable for remediation of the audit findings.
B. This discussion informs management of the findings of the audit, and, based on these discussions, management may agree to develop an implementation plan for the suggested recommendations, along with the time lines.
C. Before communicating the results of an audit to senior management, the IS auditor should discuss the findings with the auditee. The goal of this discussion is to confirm the accuracy of the findings and to propose or recommend a course of corrective action.
D. At the draft report stage, the IS auditor may recommend various controls to mitigate the risk, but the purpose of the meeting is to validate the findings of the audit with management.

A1-88 Which of the following responsibilities would **MOST** likely compromise the independence of an IS auditor when reviewing the risk management process?

A. Participating in the design of the risk management framework
B. Advising on different implementation techniques
C. Facilitating risk awareness training
D. Performing a due diligence review of the risk management processes

A is the correct answer.

Justification:

A. Participating in the design of the risk management framework involves designing controls, which compromises the independence of the IS auditor to audit the risk management process.
B. Advising on different implementation techniques does not compromise the IS auditor's independence because the IS auditor will not be involved in the decision-making process.
C. Facilitating awareness training does not hamper the IS auditor's independence because the auditor will not be involved in the decision-making process.
D. Due diligence reviews are a type of audit generally related to mergers and acquisitions.

A1-89 Which of the following would be the **GREATEST** concern if audit objectives are not established during the initial phase of an audit program?

A. Key stakeholders are incorrectly identified.
B. Control costs will exceed planned budget.
C. Important business risk may be overlooked.
D. Previously audited areas may be inadvertently included.

C is the correct answer.

Justification:

A. In certain cases, it may be more difficult to discuss findings when incorrect stakeholders are identified, thus delaying the communication of audit findings. However, this is not as concerning as important business risk not being included in audit scope.
B. Many factors determine the cost of controls. Therefore, it is difficult to state that only audit objectives will determine the control cost. However, this is not as important if key risk is not identified.
C. Without an audit scope, the appropriate risk assessment has not been performed, and therefore, the auditor might not audit those areas of highest risk for the organization.
D. Auditing previously audited areas is not an efficient use of resources; however, this is not as big of a concern as key risk not being identified.

A1-90 An IS auditor wants to analyze audit trails on critical servers to discover potential anomalies in user or system behavior. Which of the following is the **MOST** suitable for performing that task?

A. Computer-aided software engineering tools
B. Embedded data collection tools
C. Trend/variance detection tools
D. Heuristic scanning tools

C is the correct answer.

Justification:

A. Computer-aided software engineering tools are used to assist in software development.
B. Embedded (audit) data collection software, such as systems control audit review file or systems audit review file, is used to provide sampling and production statistics, but not to conduct an audit log analysis.
C. Trend/variance detection tools look for anomalies in user or system behavior, such as invoices with increasing invoice numbers.
D. Heuristic scanning tools are a type of virus scanning used to indicate possible infected traffic.

A1-91 While performing an audit of an accounting application's internal data integrity controls, an IS auditor identifies a major control deficiency in the change management software supporting the accounting application. The **MOST** appropriate action for the IS auditor to take is to:

A. Continue to test the accounting application controls and inform the IT manager about the control deficiency and recommend possible solutions.
B. Complete the audit and not report the control deficiency because it is not part of the audit scope.
C. Continue to test the accounting application controls and include the deficiency in the final report.
D. Cease all audit activity until the control deficiency is resolved.

C is the correct answer.

Justification:

A. The IS auditor should not assume that the IT manager will follow through on a verbal notification to resolve the change management control deficiency, and it is inappropriate to offer consulting services on issues discovered during an audit.
B. Although not technically within the audit scope, it is the responsibility of the IS auditor to report findings discovered during an audit that can have a material impact on the effectiveness of controls.
C. It is the responsibility of the IS auditor to report on findings that can have a material impact on the effectiveness of controls—whether or not they are within the scope of the audit.
D. It is not the role of the IS auditor to demand that IT work be completed before performing or completing an audit.

A1-92 Which of the following will **MOST** successfully identify overlapping key controls in business application systems?

A. Reviewing system functionalities that are attached to complex business processes
B. Submitting test transactions through an integrated test facility
C. Replacing manual monitoring with an automated auditing solution
D. Testing controls to validate that they are effective

C is the correct answer.

Justification:

A. In general, highly complex business processes may have more key controls than business areas with less complexity; however, finding, with certainty, unnecessary controls in complex areas is not always possible. If a well-thought-out key control structure was established from the beginning, finding any overlap in key controls will not be possible.
B. An integrated test facility is an audit technique to test the accuracy of the processes in the application system. It may find control flaws in the application system, but it would be difficult to find the overlap in key controls.
C. As part of the effort to realize continuous audit management, there are cases for introducing an automated monitoring and auditing solution. All key controls need to be clearly aligned for systematic implementation; thus, analysts can discover unnecessary or overlapping key controls in existing systems.
D. By testing controls to validate whether they are effective, the IS auditor can identify whether there are overlapping controls; however, the process of implementing an automated auditing solution would better identify overlapping controls.

A1-93 When performing a risk analysis, the IS auditor should **FIRST**:

A. Review the data classification program.
B. Identify the organization's information assets.
C. Identify the inherent risk of the system.
D. Perform a cost-benefit analysis for controls.

B is the correct answer.

Justification:
A. After the business objectives and the underlying systems are identified, the greatest degree of risk management effort should be focused towards those assets containing data considered most sensitive to the organization. The data classification program assists the IS auditor in identifying these assets.
B. The first step of the risk assessment process is to identify the systems and processes that support the business objectives because risk to those processes impacts the achievement of business goals.
C. Inherent risk is the exposure without considering the actions that management has taken or might take. The purpose of a risk assessment is to identify vulnerabilities so that mitigating controls can be established. However, one must first understand the business and its supporting systems to best identify systems requiring the most risk assessment effort.
D. Designing and implementing controls to mitigate inherent risk of critical systems can only be performed after the above steps have been taken.

A1-94 After identifying the findings, the IS auditor should **FIRST**:

A. Gain agreement on the findings.
B. Determine mitigation measures for the findings.
C. Inform senior management of the findings.
D. Obtain remediation deadlines to close the findings.

A is the correct answer.

Justification:
A. If findings are not agreed upon and confirmed by both parties, then there may be an issue during sign-off on the final audit report or while discussing findings with management. When agreement is obtained with the auditee, it implies the finding is understood and a clear plan of action can be determined.
B. Although the auditor may recommend mitigation measures, the organization ultimately decides and implements the mitigation strategies as a function of risk management.
C. Before senior management is informed, it is imperative that the auditor informs the auditee and gains agreement on the audit findings to correctly communicate the risk.
D. Obtaining remediation deadlines to close the findings is not the first step in communicating the audit findings.

A1-95 A **PRIMARY** benefit derived for an organization employing control self-assessment techniques is that it:

A. Can identify high-risk areas that might need a detailed review later.
B. Allows IS auditors to independently assess risk.
C. Can be used as a replacement for traditional audits.
D. Allows management to relinquish responsibility for control.

A is the correct answer.

Justification:

A. Control self-assessment (CSA) is predicated on the review of high-risk areas that either need immediate attention or may require a more thorough review later.
B. CSA requires the involvement of IS auditors and line management. The internal audit function shifts some of the control monitoring responsibilities to the functional areas.
C. CSA is not a replacement for traditional audits. CSA is not intended to replace audit's responsibilities, but to enhance them.
D. CSA does not allow management to relinquish its responsibility for control.

A1-96 Which of the following is the **FIRST** step performed prior to creating a risk ranking for the annual internal IS audit plan?

A. Prioritize the identified risk.
B. Define the audit universe.
C. Identify the critical controls.
D. Determine the testing approach.

B is the correct answer.

Justification:

A. After the audit universe is defined, the IS auditor can prioritize risk based on its overall impact on different operational areas of the organization covered under the audit universe.
B. In a risk-based audit approach, the IS auditor identifies risk to the organization based on the nature of the business. To plan an annual audit cycle, the types of risk must be ranked. To rank the types of risk, the auditor must first define the audit universe by considering the IT strategic plan, organizational structure and authorization matrix.
C. The controls that help in mitigating high-risk areas are generally critical controls and their effectiveness provides assurance on mitigation of risk. However, this cannot be done unless the types of risk are ranked.
D. The testing approach is based on the risk ranking.

A1-97 Which of the following is **MOST** likely be considered a conflict of interest for an IS auditor who is reviewing a cybersecurity implementation?

A. Delivering cybersecurity awareness training
B. Designing the cybersecurity controls
C. Advising on the cybersecurity framework
D. Conducting the vulnerability assessment

B is the correct answer.

Justification:

A. Delivering cybersecurity awareness training is typically an operational responsibility, but it is not nearly as strong as a conflict of interest as the auditor designing controls and then reviewing them.
B. If an auditor designs the controls, a conflict of interest arises in the neutrality of the auditor to address deficiencies during an audit. This is in violation of the ISACA Code of Ethics.
C. Part of the role of an IS auditor can be to advise on a cybersecurity framework, provided that such advice does not rise to the level of designing specific controls that the auditor would later review.
D. Conducting a vulnerability assessment can be the responsibility of the IS auditor and does not present a conflict of interest.

A1-98 An IS auditor identified a business process to be audited. The IS auditor should **NEXT** identify the:

A. Most valuable information assets.
B. IS audit resources to be deployed.
C. Auditee personnel to be interviewed.
D. Control objectives and activities.

D is the correct answer.

Justification:

A. All assets need to be identified, not just information assets. To determine the key information assets to be audited, the IS auditor should first determine which control objectives and key control activities should be validated. Only information assets that are related to the control objectives and key control activities are relevant for scoping the audit.
B. Only after determining which controls and related relevant information assets are to be validated can the IS auditor decide on the key IS audit resources (with the relevant skill sets) that should be deployed for the audit.
C. Only after determining the key control activities to be validated can the IS auditor identify the relevant process personnel who should be interviewed.
D. After the business process is identified, the IS auditor should first identify the control objectives and activities associated with the business process that should be validated in the audit.

A1-99 The effect of which of the following should have priority in planning the scope and objectives of an IS audit?

A. Applicable statutory requirements
B. Applicable corporate standards
C. Applicable industry good practices
D. Organizational policies and procedures

A is the correct answer.

Justification:

A. The effect of applicable statutory requirements must be factored in while planning an IS audit—the IS auditor has no options regarding statutory requirements because there can be no limitation of scope relating to statutory requirements.
B. Statutory requirements always take priority over corporate standards.
C. Industry good practices help plan an audit; however, good practices are not mandatory and can be deviated from, to meet organization objectives.
D. Organizational policies and procedures are important, but statutory requirements always take priority. Organizational policies must be in alignment with statutory requirements.

A1-100 An external IS auditor discovers that systems in the scope of the audit were implemented by an associate. In such a circumstance, IS audit management should:

A. Remove the IS auditor from the engagement.
B. Cancel the engagement.
C. Disclose the issue to the client.
D. Take steps to restore the IS auditor's independence.

C is the correct answer.

Justification:

A. It is not necessary to withdraw the IS auditor unless there is a statutory limitation, which exists in certain countries.
B. Canceling the engagement is not required if properly disclosed and accepted.
C. In circumstances in which the IS auditor's independence is impaired and the IS auditor continues to be associated with the audit, the facts surrounding the issue of the IS auditor's independence should be disclosed to the appropriate management and in the report.
D. This is not a feasible solution. The independence of the IS auditor cannot be restored while continuing to conduct the audit.

A1-101 An IS auditor is planning to evaluate the control design effectiveness that is related to an automated billing process. Which of the following is the **MOST** effective approach for the auditor to adopt?

A. Interview
B. Inquiry
C. Reperformance
D. Walk-through

D is the correct answer.

Justification:

A. An interview is not as strong an evidence as an observation or walk-throughs. In addition, personnel might add some bias to interviews if they know they are being interviewed for an audit.
B. Inquiry can be used to understand the controls in a process only if it is accompanied by verification of evidence. However, interviewees might be biased if they know they are being audited.
C. Reperformance is used to evaluate the operating effectiveness of the control rather than the design of the control.
D. Walk-throughs involve a combination of inquiry and inspection of evidence with respect to business process controls. This is the most effective basis for evaluation of the design of the control, because it actually exists.

A1-102 Which of the following is the **MAIN** reason to perform a risk assessment in the planning phase of an IS audit?

A. To ensure management's concerns are addressed
B. To provide reasonable assurance material items will be addressed
C. To ensure the audit team will perform audits within budget
D. To develop audit program and procedures needed to perform the audit

B is the correct answer.

Justification:

A. Management concerns have no bearing on the risk assessment process. If management has concerns and wants the auditor to focus on a certain area, the auditor should ensure adequate time is allocated to address the concerns.
B. A risk assessment helps to focus the audit procedures on the highest risk areas included in the scope of the audit. The concept of reasonable assurance is also important.
C. A risk assessment is performed to determine where to place time and personnel resources, while budget constraints are limited to time resources.
D. A risk assessment is not used in the development of the audit program and procedures. However, the risk assessment is used to allocate resources to audits.

A1-103 Which of the following is **MOST** important to ensure before communicating the audit findings to top management during the closing meeting?

A. Risk statement includes an explanation of a business impact.
B. Findings are clearly tracked back to evidence.
C. Recommendations address root causes of findings.
D. Remediation plans are provided by responsible parties.

B is the correct answer.

Justification:
A. It is important to have a well-elaborated risk statement; however, it might not be relevant if findings are not accurate.
B. Without adequate evidence, the findings hold no ground; therefore, this must be verified before communicating the findings.
C. It is important to address the root causes of findings, and it may be not included in the report. However, it might not be relevant if findings are not accurate.
D. In some cases, top-management might expect to see remediation plans during debriefing of the findings; however, the accuracy of findings should be proved first.

A1-104 The **MAIN** advantage of an IS auditor directly extracting data from a general ledger systems is:

A. Reduction of human resources needed to support the audit
B. Reduction in the time to have access to the information
C. Greater flexibility for the audit department
D. Greater assurance of data validity

D is the correct answer.

Justification:
A. Although the burden on human resources to support the audit may decrease if the IS auditor directly extracts the dates, this advantage is not as significant as the increased data validity.
B. This will not necessarily reduce the time to have access to the information because time will need to be scheduled for training and granting access.
C. There may be more flexibility for the IS auditor to adjust the data extracts to meet various audit requirements; however, this is not the main advantage.
D. If the IS auditor executes the data extraction, there is greater assurance that the extraction criteria will not interfere with the required completeness, and, therefore, all required data will be collected. Asking IT to extract the data may expose the risk of filtering out exceptions that should be seen by the auditor. Also, if the IS auditor collects the data, all internal references correlating the various data tables/elements will be understood, and this knowledge may reveal vital elements to the completeness and correctness of the overall audit activity.

A1-105 An IS auditor wants to determine the number of purchase orders that are not appropriately approved. Which of the following sampling techniques should an IS auditor use to make such a conclusion?

A. Attribute
B. Variable
C. Stop-or-go
D. Judgment

A is the correct answer.

Justification:

A. Attribute sampling is used to test compliance of transactions to controls—in this instance, the existence of appropriate approval.

B. Variable sampling is used in substantive testing situations and deals with population characteristics that vary, such as monetary values and weights.

C. Stop-or-go sampling is used when the expected occurrence rate is extremely low.

D. Judgment sampling is not relevant here. It refers to a subjective approach of determining sample size and selection criteria of elements of the sample.

A1-106 An IS auditor uses computer-assisted audit techniques (CAATs) to collect and analyze data. Which of the following attributes of evidence is **MOST** affected by using CAATs?

A. Usefulness
B. Reliability
C. Relevance
D. Adequacy

B is the correct answer.

Justification:

A. Usefulness of audit evidence pulled by computer-assisted audit techniques (CAATs) is determined by the audit objective, and the use of CAATs does not have as direct of an impact on usefulness as reliability.

B. Because the data are directly collected by the IS auditor, the audit findings can be reported with an emphasis on the reliability of the records that are produced and maintained in the system. The reliability of the source of information used provides reassurance on the generated findings.

C. Relevance of audit evidence pulled by CAATs is determined by the audit objective, and the use of CAATs does not have as direct of an impact on relevance as reliability.

D. Adequacy of audit evidence pulled by CAATs is determined by the processes and personnel who author the data, and the use of CAATs does not have any impact on competence.

A1-107 An internal IS audit function is planning a general IS audit. Which of the following activities takes place during the **FIRST** step of the planning phase?

A. Development of an audit program
B. Define the audit scope
C. Identification of key information owners
D. Development of a risk assessment

D is the correct answer.

Justification:
A. The results of the risk assessment are used for the input for the audit program.
B. The output of the risk assessment helps define the scope.
C. A risk assessment must be performed prior to identifying key information owners. Key information owners are generally not directly involved during the planning process of an audit.
D. A risk assessment should be performed to determine how internal audit resources should be allocated to ensure that all material items will be addressed.

A1-108 Which of the following is the **MOST** important skill that an IS auditor should develop to understand the constraints of conducting an audit?

A. Managing audit staff
B. Allocating resources
C. Project management
D. Attention to detail

C is the correct answer.

Justification:
A. Managing audit staff is not the only aspect of conducting an audit.
B. Allocating resources, such as time and personnel, are needed for overall project management skills.
C. Audits often involve resource management, deliverables, scheduling and deadlines that are similar to project management good practices.
D. Attention to detail is needed, but it is not a constraint of conducting audits.

A1-109 What is the **MAJOR** benefit of conducting a control self-assessment over a traditional audit?

A. It detects risk sooner.
B. It replaces the internal audit function.
C. It reduces the audit workload.
D. It reduces audit resource requirements.

A is the correct answer.

Justification:
A. Control self-assessments (CSAs) require employees to assess the control stature of their own function. CSAs help to increase the understanding of business risk and internal controls. Because they are conducted more frequently than audits, CSAs help to identify risk in a timelier manner.
B. CSAs do not replace the internal audit function; an audit must still be performed to ensure that controls are present.
C. CSAs may not reduce the audit function's workload and are not a major difference between the two approaches.
D. CSAs do not affect the need for audit resources. Although the results of the CSA may serve as a reference point for the audit process, they do not affect the scope or depth of audit work that needs to be performed.

A1-110 An IS auditor is reviewing a project risk assessment and notices that the overall residual risk level is high due to confidentiality requirements. Which of the following types of risk is normally high due to the number of unauthorized users the project may affect?

A. Control risk
B. Compliance risk
C. Inherent risk
D. Residual risk

C is the correct answer.

Justification:

A. Control risk can be high, but it is not due to internal controls not being identified, evaluated or tested, and is not due to the number of users or business areas affected.

B. Compliance risk is the penalty applied to current and future earnings for nonconformance to laws and regulations and may not be impacted by the number of users and business areas affected.

C. Inherent risk is normally high due to the number of users and business areas that may be affected. Inherent risk is the risk level or exposure without considering the actions that management has taken or might take.

D. Residual risk is the remaining risk after management has implemented a risk response and is not based on the number of users or business areas affected.

A1-111 An IS auditor discovers a potential material finding. The **BEST** course of action is to:

A. report the potential finding to business management.
B. discuss the potential finding with the audit committee.
C. increase the scope of the audit.
D. perform additional testing.

D is the correct answer.

Justification:

A. The item should be confirmed through additional testing before it is reported to management.

B. The item should be confirmed through additional testing before it is discussed with the audit committee.

C. Additional testing to confirm the potential finding should be within the scope of the engagement. Increasing the scope could demand more needed audit resources and could be subject to risk creep.

D. The IS auditor should perform additional testing to ensure that it is a finding. An auditor can quickly lose credibility if it is later discovered the finding was not justified or accurate.

A1-112 Which of the following is in the **BEST** position to approve changes to the audit charter?

A. Board of directors
B. Audit committee
C. Executive management
D. Director of internal audit

B is the correct answer.

Justification:

A. The board of directors does not need to approve the charter; it is best presented to the audit committee for approval.

B. The audit committee is a subgroup of the board of directors. The audit department should report to the audit committee and the audit charter should be approved by the committee.

C. Executive management is not required to approve the audit charter and will not have the independence to approve the charter. The audit committee is in the best position to approve the charter because it is an independent and senior group.

D. While the director of internal audit may draft the charter and make changes, the audit committee should have the final approval of the charter.

A1-113 An IS auditor reviewing the process of log monitoring wants to evaluate the organization's manual review process. Which of the following audit techniques would the auditor **MOST** likely employ to fulfill this purpose?

A. Inspection
B. Inquiry
C. Walk-through
D. Reperformance

C is the correct answer.

Justification:

A. Inspection is just one component of a walk-through and by itself does not supply enough information to provide a full understanding of the overall process and identify potential control weaknesses.

B. Inquiry provides only general information on how the control is executed. It does not necessarily enable the IS auditor to determine whether the control performer has an in-depth understanding of the control.

C. Walk-through procedures usually include a combination of inquiry, observation, inspection of relevant documentation and reperformance of controls. A walk-through of the manual log review process follows the manual log review process from start to finish to gain a thorough understanding of the overall process and identify potential control weaknesses.

D. Reperformance of the control is carried out by the IS auditor and does not provide assurance of the competency of the auditee.

A1-114 An IS auditor is comparing equipment in production with inventory records. This type of testing is an example of:

A. substantive testing.
B. compliance testing.
C. analytical testing.
D. control testing.

A is the correct answer.

Justification:

A. Substantive testing obtains audit evidence on the completeness, accuracy or existence of activities or transactions during the audit period.
B. Compliance testing is evidence gathering for the purpose of testing an enterprise's compliance with control procedures. This differs from substantive testing in which evidence is gathered to evaluate the integrity of individual transactions, data or other information.
C. Analytical testing evaluates the relationship of two sets of data and discerns inconsistencies in the relationship.
D. Control testing is the same as compliance testing.

A1-115 Which of the following does a lack of adequate controls represent?

A. An impact
B. A vulnerability
C. An asset
D. A threat

B is the correct answer.

Justification:

A. Impact is the measure of the consequence (including financial loss, reputational damage, loss of customer confidence) that a threat event may have.
B. The lack of adequate controls represents a vulnerability, exposing sensitive information and data to the risk of malicious damage, attack or unauthorized access by hackers, employee error, environmental threat or equipment failure. This could result in a loss of sensitive information, financial loss, legal penalties or other losses.
C. An asset is something of either tangible or intangible value worth protecting, including people, systems, infrastructure, finances and reputation.
D. A threat is a potential cause of an unwanted incident.

A1-116 An IS auditor notes daily reconciliation of visitor access card inventory is not aligned with the organization's procedures. Which of the following is the auditor's **BEST** course of action?

A. Do not report the lack of reconciliation.
B. Recommend regular physical inventory counts.
C. Report the lack of daily reconciliations.
D. Recommend the implementation of a more secure access system.

C is the correct answer.

Justification:
A. Absence of discrepancy in physical count only confirms absence of any impact but cannot be a reason to overlook failure of operation of the control. The issue should be reported because the control was not followed.
B. While the IS auditor may in some cases recommend a change in procedures, the primary goal is to observe and report when the current process is deficient.
C. The IS auditor should report the lack of daily reconciliation as an exception, because a physical inventory count gives assurance only at a point in time and the practice is not in compliance with management's mandated activity.
D. While the IS auditor may in some cases recommend a more secure solution, the primary goal is to observe and report when the current process is deficient.

A1-117 During an audit, the IS auditor notes the application developer also performs quality assurance testing on another application. Which of the following is the **MOST** important course of action for the auditor?

A. Recommend compensating controls.
B. Review the code created by the developer.
C. Analyze the quality assurance dashboards.
D. Report the identified condition.

D is the correct answer.

Justification:
A. Although compensating controls may be a good idea, the primary response in this case should be to report the condition, because the risk associated with this should be reported to the users of the audit report
B. Evaluating the code created by the application developer is not the appropriate response in this case. The IS auditor may evaluate a sample of changes to determine whether the developer tested his/her own code, but the primary response should be to report the condition.
C. Analyzing the quality assurance dashboards can help evaluate the actual impact of the lack of segregation of duties but does not address the underlying risk. The primary response should be to report the condition.
D. The software quality assurance role should be independent and separate from development and development activities. The same person should not hold both roles because this would cause a segregation of duties concern. The IS auditor should report this condition when identified.

A1-118 An IS auditor is reviewing risk and controls of a bank's wire transfer system. To ensure that the bank's financial risk is properly addressed, the IS auditor will most likely review which of the following?

A. Privileged access to the wire transfer system
B. Wire transfer procedures
C. Fraud monitoring controls
D. Employee background checks

B is the correct answer.

Justification:

A. Privileged access, such as administrator access, is necessary to manage user account privileges and should not be granted to end users. The wire transfer procedures are a better control to review to ensure that there is segregation of duties of the end users to help prevent fraud.
B. Wire transfer procedures include segregation of duties controls. This helps prevent internal fraud by not allowing one person to initiate, approve and send a wire. Therefore, the IS auditor should review the procedures as they relate to the wire system.
C. Fraud monitoring is a detective control and does not prevent financial loss. Segregation of duties is a preventive control which is part of the wire transfer procedures.
D. Although controls related to background checks are important, the controls related to segregation of duties as found in the wire transfer procedures are more critical.

A1-119 An IS auditor is determining the appropriate sample size for testing the existence of program change approvals. Previous audits did not indicate any exceptions, and management has confirmed that no exceptions have been reported for the review period. In this context, the IS auditor can adopt a:

A. lower confidence coefficient, resulting in a smaller sample size.
B. higher confidence coefficient, resulting in a smaller sample size.
C. higher confidence coefficient, resulting in a larger sample size.
D. lower confidence coefficient, resulting in a larger sample size.

A is the correct answer.

Justification:

A. When internal controls are strong, a lower confidence coefficient can be adopted, which will enable the use of a smaller sample size.
B. A higher confidence coefficient will result in the use of a larger sample size.
C. A higher confidence coefficient need not be adopted in this situation because internal controls are strong.
D. A lower confidence coefficient will result in the use of a smaller sample size.

A1-120 Why does an audit manager review the staff's audit papers, even when the IS auditors have many years of experience?

A. Internal quality requirements
B. The audit guidelines
C. The audit methodology
D. Professional standards

D is the correct answer.

Justification:

A. Internal quality requirements may exist but are superseded by the requirement of supervision to comply with professional standards.

B. Audit guidelines exist to provide guidance on how to achieve compliance with professional standards. For example, they may provide insights on the purpose of supervision and examples of how supervisory duties are to be performed to achieve compliance with professional standards.

C. An audit methodology is a well-configured process/procedure to achieve audit objectives. While an audit methodology is a meaningful tool, supervision is generally driven by compliance with professional standards.

D. Professional standards from ISACA, The Institute of Internal Auditors and the International Federation of Accountants require supervision of audit staff to accomplish audit objectives and comply with competence, professional proficiency and documentation requirements, and more.

A1-121 Which technique will **BEST** test for the existence of dual control when auditing the wire transfer systems of a bank?

A. Analysis of transaction logs
B. Reperformance
C. Observation
D. Interviewing personnel

C is the correct answer.

Justification:

A. Analysis of transaction logs would help to show that dual control is in place but does not necessarily guarantee that this process is being followed consistently. Therefore, observation is the better test technique.

B. Although reperformance could provide assurance that dual control was in effect, reperforming wire transfers at a bank would not be an option for an IS auditor.

C. Dual control requires that two people carry out an operation. The observation technique helps to ascertain whether two individuals do get involved in execution of the operation and an element of oversight exists. It is obvious if one individual is masquerading and filling in the role of the second person.

D. Interviewing personnel is useful to determine the level of awareness and understanding of the personnel carrying out the operations. However, it does provide direct evidence confirming the existence of dual control, because the information provided may not accurately reflect the process being performed.

A1-122 In a risk-based IS audit, where both inherent and control risk have been assessed as high, an IS auditor would **MOST** likely compensate for this scenario by performing additional:

A. Stop-or-go sampling.
B. Substantive testing.
C. Compliance testing.
D. Discovery sampling.

B is the correct answer.

Justification:

A. Stop-or-go sampling is used when an IS auditor believes few errors will be found in the population, and, thus, is not the best type of testing to perform in this case.
B. Because both the inherent and control risk are high in this case, additional testing is required. Substantive testing obtains audit evidence on the completeness, accuracy or existence of activities or transactions during the audit period.
C. Compliance testing is evidence gathering for the purpose of testing an enterprise's compliance with control procedures. Although performing compliance testing is important, performing additional substantive testing is more appropriate in this case.
D. Discovery sampling is a form of attribute sampling that is used to determine a specified probability of finding at least one example of an occurrence (attribute) in a population, typically used to test for fraud or other irregularities. In this case, additional substantive testing is the better option.

A1-123 The **PRIMARY** objective of the audit initiation meeting with an IS audit client is to:

A. Discuss the scope of the audit.
B. Identify resource requirements of the audit.
C. Select the methodology of the audit.
D. Collect audit evidence.

A is the correct answer.

Justification:

A. The primary objective of the initiation meeting with an audit client is to help define the scope of the audit.
B. Determining the resource requirements of the IS audit is typically done by IS audit management during the early planning phase of the project rather than at the initiation meeting.
C. Selecting the methodology of the audit is not normally an objective of the initiation meeting.
D. For most audits, collecting audit evidence is performed during the course of the engagement and is not normally collected during the initiation meeting.

A1-124 The **PRIMARY** purpose of the IS audit charter is to:

A. Establish the organizational structure of the audit department.
B. Illustrate the reporting responsibilities of the is audit function.
C. Detail the resource requirements needed for the audit function.
D. Outline the responsibility and authority of the is audit function.

D is the correct answer.

Justification:
A. The IS audit charter does not set forth the organizational structure of the IS audit department. The charter serves as a directive to create the IS audit function.
B. The IS audit charter does not dictate the reporting requirements of the IS audit department. The charter sets forth the purpose, responsibility, authority and accountability of the information systems audit function.
C. Resources are determined by the audit and not the charter.
D. The primary purpose of the IS audit charter is to set forth the purpose, responsibility, authority and accountability of the IS audit function. The charter document grants authority to the audit function on behalf of the board of directors and organization stakeholders.

A1-125 Which of the following is **MOST** important for an IS auditor to understand when auditing an ecommerce environment?

A. The technology architecture of the ecommerce environment
B. The policies, procedures and practices forming the control environment
C. The nature and criticality of the business processes supported by the application
D. Continuous monitoring of control measures for system availability and reliability

C is the correct answer.

Justification:
A. Understanding the technology architecture of the ecommerce environment is important; however, it is vital that the nature and criticality of the business process supported by the ecommerce application are well understood.
B. Although the policies, procedure and practices that form the internal control environment need to be in alignment with the ecommerce environment, this is not the most important element that the IS auditor needs to understand.
C. The ecommerce application enables the execution of business transactions. Therefore, it is important to understand the nature and criticality of the business process supported by the ecommerce application to identify specific controls to review.
D. The availability of the ecommerce environment is important, but this is only one of the aspects to be considered with respect to business processes that are supported by the ecommerce application.

A1-126 During an IS audit, which is the **BEST** method for an IS auditor to evaluate the implementation of segregation of duties within an IT department?

A. Discuss with the IT managers.
B. Review the IT job descriptions.
C. Research past IT audit reports.
D. Evaluate the organizational structure.

A is the correct answer.

Justification:

A. Discussing the implementation of segregation of duties with the IT managers is the best way to determine how responsibilities are assigned within the department.
B. Job descriptions may not be the best source of information because they can be outdated or what is documented in the job descriptions may be different from what is actually performed.
C. Past IS audit reports are not the best source of information because they may not accurately describe how IT responsibilities are assigned.
D. Evaluating the organizational structure may give a limited view on the allocation of IT responsibilities. The responsibilities also may have changed over time.

A1-127 A financial institution with multiple branch offices has an automated control that requires the branch manager to approve transactions more than a certain amount. What type of audit control is this?

A. Detective
B. Preventive
C. Corrective
D. Directive

B is the correct answer.

Justification:

A. Detective controls identify events after they have happened. In this case, the action of the branch manager would prevent an event from occurring.
B. Having a manager approve transactions more than a certain amount is considered a preventive control.
C. A corrective control serves to remedy problems discovered by detective controls. In this case, the action of the branch manager is a preventive control.
D. A directive control is a manual control that typically consists of a policy or procedure that specifies what actions are to be performed. In this case, there is an automated control that prevents an event from occurring.

A1-128 During an application software review, an IS auditor identified minor weaknesses in a relevant database environment that is out of scope for the audit. The **BEST** option is to:

A. Include a review of the database controls in the scope.
B. Document for future review.
C. Work with database administrators to correct the issue.
D. Report the weaknesses as observed.

D is the correct answer.

Justification:
A. Executing audits and reviews outside the scope is not advisable. In this case, the weakness identified is considered to be a minor issue, and it is sufficient to report the issue and address it at a later time.
B. In this case, the weakness identified is considered to be a minor issue. The IS auditor should formally report the weaknesses as an observation rather than documenting it to address during a future audit.
C. It is not appropriate for the IS auditor to work with database administrators to correct the issue.
D. Any weakness noticed should be reported, even if it is outside the scope of the current audit. Weaknesses identified during an application software review need to be reported to management.

A1-129 A centralized antivirus system determines whether each personal computer has the latest signature files and installs the latest signature files before allowing a PC to connect to the network. This is an example of a:

A. directive control.
B. corrective control.
C. compensating control.
D. detective control.

B is the correct answer.

Justification:
A. Directive controls, such as IT policies and procedures, do not apply in this case because this is an automated control.
B. Corrective controls are designed to correct errors, omissions and unauthorized uses and intrusions, when they are detected. This provides a mechanism to detect when malicious events have happened and correct the situation.
C. A compensating control is used where other controls are not sufficient to protect the system. In this case, the corrective control in place will effectively protect the system from access via an unpatched device.
D. Detective controls exist to detect and report when errors, omissions and unauthorized uses or entries occur.

A1-130 Due to unexpected resource constraints of the IS audit team, the audit plan, as originally approved, cannot be completed. Assuming the situation is communicated in the audit report, which course of action is **MOST** acceptable?

A. Test the adequacy of the control design.
B. Test the operational effectiveness of controls.
C. Focus on auditing high-risk areas.
D. Rely on management testing of controls.

C is the correct answer.

Justification:

A. Testing the adequacy of control design is not the best course of action because this does not ensure that controls operate effectively as designed.
B. Testing control operating effectiveness does not ensure that the audit plan is focused on areas of greatest risk.
C. Reducing the scope and focusing on auditing high-risk areas is the best course of action.
D. The reliance on management testing of controls does not provide an objective verification of the control environment.

A1-131 Which of the following **BEST** ensures the effectiveness of controls related to interest calculation for an accounting system?

A. Reperformance
B. Process walk-through
C. Observation
D. Documentation review

A is the correct answer.

Justification:

A. To ensure the effectiveness of controls, it is most effective to conduct reperformance. When the same result is obtained after the performance by an independent person, this provides the strongest assurance.
B. Process walk-through may help the auditor understand the controls better; however, it may not be as useful as conducting reperformance for a sample of transactions.
C. Observation is a valid audit method to verify that operators are using the system appropriately; however, conducting reperformance is a better method.
D. Documentation review may be of some value for understanding the control environment; however, conducting reperformance is a better method.

A1-132 Which of the following choices would be the **BEST** source of information when developing a risk-based audit plan?

A. Process owners identify key controls.
B. System custodians identify vulnerabilities.
C. Peer auditors understand previous audit results.
D. Senior management identify key business processes.

D is the correct answer.

Justification:

A. Although process owners should be consulted to identify key controls, senior management is a better source to identify business processes, which are more important.
B. System custodians is a good source to better understand the risk and controls as they apply to specific applications; however, senior management is a better source to identify business processes, which are more important.
C. The review of previous audit results is one input into the audit planning process; however, if previous audits focused on a limited or a restricted scope or if the key business processes have changed and/or new business processes have been introduced, then this does contribute to the development of a risk-based audit plan.
D. Developing a risk-based audit plan must start with the identification of key business processes, which determine and identify the risk that needs to be addressed.

A1-133 While auditing a third-party IT service provider, an IS auditor discovered that access reviews were not being performed as required by the contract. The IS auditor should:

A. Report the issue to IT management.
B. Discuss the issue with the service provider.
C. Perform a risk assessment.
D. Perform an access review.

A is the correct answer.

Justification:

A. During an audit, if there are material issues that are of concern, they need to be reported to management in the audit report.
B. The IS auditor may discuss the issue with the service provider; however, the appropriate response is to report the issue to IT management because they are ultimately responsible.
C. This issue can serve as an input for a future risk assessment, but the issue of noncompliance should be reported to management regardless of whether the IS auditor believes there is a significant risk.
D. The IS auditor could perform an access review as part of the audit to determine if there are errors, but not on behalf of the third-party IT service provider. It is more important to report the issue in the audit report to management.

A1-134 Which of the following is the **PRIMARY** requirement for reporting IS audit results? The report is:

A. Prepared according to a predefined and standard template.
B. Backed by sufficient and appropriate audit evidence.
C. Comprehensive in coverage of enterprise processes.
D. Reviewed and approved by audit management.

B is the correct answer.

Justification:

A. Preparation of the IS audit report according to a predefined and standard template may be useful in ensuring that all key aspects are provided in a uniform structure, but this does not demonstrate that audit findings are based on evidence that can be proven, if required.
B. ISACA IS audit standards require that reports should be backed by sufficient and appropriate audit evidence so that they demonstrate the application of the minimum standard of performance, and the findings and recommendations can be validated, if required.
C. The scope and coverage of IS audit is defined by a risk assessment process, which may not always provide comprehensive coverage of processes of the enterprise.
D. While from an operational standpoint an audit report should be reviewed and approved by audit management, the more critical consideration is that all conclusions are backed by sufficient and appropriate audit evidence.

A1-135 An IS auditor performing an audit of the risk assessment process should **FIRST** confirm that:

A. Reasonable threats to the information assets are identified.
B. Technical and organizational vulnerabilities have been analyzed.
C. Assets have been identified and ranked.
D. The effects of potential security breaches have been evaluated.

C is the correct answer.

Justification:

A. The threats facing each of the organization's assets should be analyzed according to their value to the organization. This occurs after identifying and ranking assets.
B. Analyzing how these weaknesses, in the absence of mitigating controls, will impact the organization's information assets occurs after the assets and weaknesses have been identified.
C. Identification and ranking of information assets (e.g., data criticality, sensitivity, locations of assets) will set the tone or scope of how to assess risk in relation to the organizational value of the asset.
D. The effect of security breaches is dependent on the value of the assets and the threats, vulnerabilities and effectiveness of mitigating controls. The impact of an attack against a weakness should be identified so that controls can be evaluated to determine if they effectively mitigate the weaknesses.

A1-136 Which of the following represents an example of a preventive control with respect to IT personnel?

A. A security guard stationed at the server room door
B. An intrusion detection system
C. Implementation of a badge entry system for the IT facility
D. A fire suppression system in the server room

C is the correct answer.

Justification:
A. A security guard is a deterrent control.
B. An intrusion detection system is a detective control.
C. Preventive controls are used to reduce the probability of an adverse event. A badge entry system prevents unauthorized entry to the facility.
D. A fire suppression system is a corrective control.

A1-137 Which of the following is an attribute of the control self-assessment approach?

A. Broad stakeholder involvement
B. Auditors are the primary control analysts
C. Limited employee participation
D. Policy driven

A is the correct answer.

Justification:
A. The control self-assessment (CSA) approach emphasizes management of and accountability for developing and monitoring the controls of an organization's business processes. The attributes of CSA include empowered employees, continuous improvement, extensive employee participation and training—all of which are representations of broad stakeholder involvement.
B. IS auditors are the primary control analysts in a traditional audit approach. CSA involves many stakeholders, not just auditors.
C. Limited employee participation is an attribute of a traditional audit approach.
D. Policy-driven is an attribute of a traditional audit approach.

A1-138 An IS auditor conducting a review of disaster recovery planning (DRP) at a financial processing organization discovered the following:

- The existing DRP was compiled two years earlier by a systems analyst in the organization's IT department using transaction flow projections from the operations department.
- The DRP was presented to the deputy chief executive officer (CEO) for approval and formal issue, but it is still awaiting attention.
- The DRP has never been updated, tested or circulated to key management and staff, although interviews show that each would know what action to take for its area if a disruptive incident occurred.

The IS auditor's report should recommend that:

A. The deputy chief executive officer (CEO) is censured for failure to approve the plan.
B. A board of senior managers is set up to review the existing plan.
C. The existing plan is approved and circulated to all key management and staff.
D. A manager coordinates the creation of a new or revised plan within a defined time limit.

D is the correct answer.

Justification:

A. Censuring the deputy CEO will not improve the current situation and is generally not within the scope of an IS auditor to recommend.
B. Establishing a board to review the disaster recovery plan (DRP), which is two years out of date, may achieve an updated DRP but is not likely to be a speedy operation; issuing the existing DRP would be imprudent without first ensuring that it is workable.
C. The current DRP may be unacceptable or ineffective and recommending the approval of the DRP may be unwise. The best way to develop a DRP in a short time is to make an experienced manager responsible for coordinating the knowledge of other managers into a single, formal document within a defined time limit.
D. The primary concern is to establish a workable DRP that reflects current processing volumes to protect the organization from any disruptive incident.

A1-139 An IS auditor finds that a disaster recovery plan (DRP) for critical business functions does not cover all systems. Which of the following is the **MOST** appropriate course of action for the IS auditor?

A. Alert management and evaluate the impact of not covering all systems.
B. Cancel the audit.
C. Complete the audit of the systems covered by the existing DRP.
D. Postpone the audit until the systems are added to the DRP.

A is the correct answer.

Justification:

A. An IS auditor should make management aware that some systems are omitted from the disaster recovery plan (DRP). An IS auditor should continue the audit and include an evaluation of the impact of not including all systems in the DRP.
B. Canceling the audit is an inappropriate action.
C. Ignoring the fact that some systems are not covered would violate audit standards that require reporting all material findings and is an inappropriate action.
D. Postponing the audit is an inappropriate action. The audit should be completed according to the initial scope with identification to management of the risk of systems not being covered.

A1-140 Which of the following is **MOST** effective for monitoring transactions exceeding predetermined thresholds?

A. Generalized audit software
B. An integrated test facility
C. Regression tests
D. Transaction snapshots

A is the correct answer.

Justification:

A. **Generalized audit software (GAS) is a data analytic tool that can be used to filter large amounts of data.**
B. Integrated test facilities test the processing of the data and cannot be used to monitor real-time transactions.
C. Regression tests are used to test new versions of software to ensure that previous changes and functionality are not inadvertently overwritten or disabled by the new changes.
D. Gathering information through snapshots alone is not sufficient. GAS will assist with an analysis of the data.

A1-141 Which of the following is **MOST** important to ensure that effective application controls are maintained?

A. Exception reporting
B. Manager oversight
C. Control self-assessment
D. Peer reviews

C is the correct answer.

Justification:

A. Exception reporting only looks at errors or problems but will not ensure controls are still working.
B. Manager oversight is important but may not be a consistent or well-defined process compared to control self-assessment.
C. **CSA is the review of business objectives and internal controls in a formal and documented collaborative process. It includes testing the design of automated application controls.**
D. Peer reviews lack the direct involvement of audit specialists and management.

A1-142 The success of a control self-assessment depends highly on:

A. Line managers assuming a portion of the responsibility for control monitoring
B. Assigning staff managers, the responsibility for building controls
C. The implementation of a stringent control policy and rule-driven controls
D. The implementation of supervision and monitoring of controls of assigned duties

A is the correct answer.

Justification:

A. **The primary objective of a control self-assessment (CSA) program is to leverage the internal audit function by shifting some of the control monitoring responsibilities to the functional area line managers. The success of a CSA program depends on the degree to which line managers assume responsibility for controls. This enables line managers to detect and respond to control errors promptly.**
B. CSA requires managers to participate in the monitoring of controls.
C. The implementation of stringent controls will not ensure controls are working correctly.
D. Better supervision is a compensating and detective control and may assist in ensuring control effectiveness but would work best when used in a formal process such as CSA.

A1-143 Which of the following is evaluated as a preventive control by an IS auditor performing an audit?

A. Transaction logs
B. Before and after image reporting
C. Table lookups
D. Tracing and tagging

C is the correct answer.

Justification:
A. Transaction logs are a detective control and provide audit trails.
B. Before and after image reporting makes it possible to trace the impact that transactions have on computer records. This is a detective control.
C. Table lookups are preventive controls; input data are checked against predefined tables, which prevent any undefined data to be entered.
D. Tracing and tagging is used to test application systems and controls but is not a preventive control in itself.

A1-144 Which of the following is a **PRIMARY** objective of embedding an audit module while developing online application systems?

A. To collect evidence while transactions are processed
B. To reduce requirements for periodic internal audits
C. To identify and report fraudulent transactions
D. To increase efficiency of the audit function

A is the correct answer.

Justification:
A. Embedding a module for continuous auditing within an application processing a large number of transactions provides timely collection of audit evidence during processing and is the primary objective. The continuous auditing approach allows the IS auditor to monitor system reliability on a continuous basis and to gather selective audit evidence through the computer.
B. An embedded audit module enhances the effectiveness of internal audit by ensuring timely availability of required evidence. It may not reduce the requirements for periodic internal audits, but it will increase their efficiency. Also, the question pertains to the development process for new application systems, and not to subsequent internal audits.
C. An audit module collects data on transactions that may help identify fraudulent transactions, but it does not identify fraudulent transactions inherently.
D. Although increased efficiency may be an added benefit of an embedded audit module, it is not the primary objective.

A1-145 An IS audit department considers implementing continuous auditing techniques for a multinational retail enterprise that requires high availability of its key systems. A **PRIMARY** benefit of continuous auditing is that:

A. Effective preventive controls are enforced.
B. System integrity is ensured.
C. Errors can be corrected in a timely fashion.
D. Fraud can be detected more quickly.

D is the correct answer.

Justification:

A. Continuous monitoring is detective in nature and, therefore, does not necessarily assist the IS auditor in monitoring for preventive controls. The approach will detect and monitor for errors that have already occurred. In addition, continuous monitoring will benefit the internal audit function in reducing the use of auditing resources and in the timely reporting of errors or inconsistencies.
B. System integrity is typically associated with preventive controls such as input controls and quality assurance reviews. These controls do not typically benefit an internal auditing function implementing continuous monitoring. Continuous monitoring benefits the internal audit function because it reduces the use of auditing resources.
C. Continuous audit will detect errors but not correct them. Correcting errors is the function of the organization's management and not the internal audit function. Continuous auditing benefits the internal audit function because it reduces the use of auditing resources to create a more efficient auditing function.
D. Continuous auditing techniques assist the auditing function in reducing the use of auditing resources through continuous collection of evidence. This approach assists the IS auditors in identifying fraud in a timely fashion and allows the auditors to focus on relevant data.

A1-146 An IS auditor wants to determine the effectiveness of managing user access to a server room. Which of the following is the **BEST** evidence of effectiveness?

A. Observation of a logged event
B. Review of the procedure manual
C. Interview with management
D. Interview with security personnel

A is the correct answer.

Justification:

A. Observation of the process to reset an employee's security access to the server room and the subsequent logging of this event provide the best evidence of the adequacy of the physical security control.
B. Although reviewing the procedure manual can be helpful in gaining an overall understanding of a process, it is not evidence of the effectiveness of the execution of a control.
C. Although interviewing management can be helpful in gaining an overall understanding of a process, it is not evidence of the effectiveness of the execution of a control.
D. Although interviewing security personnel can be helpful in gaining an overall understanding of a process, it is not evidence of the effectiveness of the execution of a control.

A1-147 As part of audit planning, an IS auditor is designing various data validation tests to effectively detect transposition and transcription errors. Which of the following will **BEST** help in detecting these errors?

A. Range check
B. Validity check
C. Duplicate check
D. Check digit

D is the correct answer.

Justification:

A. Range checks can only ensure that data fall within a predetermined range but cannot detect transposition errors.
B. Validity checks are generally programmed checking of data validity in accordance with predetermined criteria.
C. Duplicate check analysis is used to test defined or selected primary keys for duplicate primary key values.
D. A check digit is a numeric value that has been calculated mathematically and is added to data to ensure that original data have not been altered or that an incorrect, but valid, match has occurred. The check digit control is effective in detecting transposition and transcription errors.

A1-148 The **MAIN** purpose of the annual IS audit plan is to:

A. Allocate resources for audits.
B. Reduce the impact of audit risk.
C. Develop a training plan for auditors.
D. Minimize the audit costs.

A is the correct answer.

Justification:

A. Because IS audit assignments need to be accomplished with limited time and human resources, audits are scheduled and prioritized as determined by IS audit management.
B. Audit risk is inherent to all audits, and the schedule has no bearing on the impact to audit risk.
C. Developing a training plan for auditors is important, but it is not the main purpose of an IS audit plan.
D. Minimizing the audit costs could be one of the objectives of annual IS audit plan. However, this would be a result of ensuring audit resources are used effectively.

A1-149 Which of the following would be expected to approve the audit charter?

A. Chief financial officer
B. Chief executive officer
C. Audit steering committee
D. Audit committee

D is the correct answer.

Justification:

A. The chief financial officer (CFO) does not approve the audit charter but may be responsible for allocating funds in support of the audit charter. The CFO may also be a part of the audit committee or audit steering committee but would not approve the charter on their own.
B. The chief executive officer (CEO) does not approve the audit charter. The CEO may be informed, but they are independent of the audit committee.
C. The steering committee would most likely be composed of various members of senior management whose purpose is to work under the framework of the audit charter and would not approve the charter itself.
D. One of the primary functions of the audit committee is to create and approve the audit charter.

A1-150 Which of the following is the **PRIMARY** purpose of a risk-based audit?

A. High-impact areas are addressed first.
B. Audit resources are allocated efficiently.
C. Material areas are addressed first.
D. Management concerns are prioritized.

C is the correct answer.

Justification:
A. High-impact does not necessarily indicate high risk. Risk also takes into consideration probability.
B. Although a risk-based audit approach does address allocation of resources, that is not the primary function of a risk-based audit approach.
C. Material risk is audited according to the risk ranking, thus enabling the audit team to concentrate on high-risk areas first.
D. Management concerns may not be aligned with high-risk areas.

A1-151 An auditee disagrees with an audit finding. Which of the following is the **BEST** course of action for the IT auditor to take?

A. Discuss the finding with the IT auditor's manager.
B. Retest the control to confirm the finding.
C. Elevate the risk associated with the control.
D. Discuss the finding with the auditee's manager.

A is the correct answer.

Justification:
A. Discussing the disagreement with the auditor's manager is the best course of action because other actions can weaken relationships with the auditee and auditor.
B. This may unnecessarily expend human and time resources. The audit manager should determine if controls need to be retested.
C. Elevating the risk will not address the disagreement.
D. It is usually best to consult the audit manager prior to escalating the issue the auditee's manager. This could prove to be an adversarial action.

DOMAIN 2—GOVERNANCE AND MANAGEMENT OF IT (17%)

A2-1 Organizations requiring employees to take a mandatory vacation each year **PRIMARILY** want to ensure:

A. adequate cross-training exists between functions.
B. an effective internal control environment is in place by increasing morale.
C. potential irregularities in processing are identified by a temporary replacement.
D. the risk of processing errors is reduced.

C is the correct answer.

Justification:

A. Cross-training is a good practice to follow but can be achieved without the requirement for mandatory vacation.
B. Good employee morale and high levels of employee satisfaction are worthwhile objectives, but they should not be considered a means to achieve an effective internal control system.
C. Employees who perform critical and sensitive functions within an organization should be required to take some time off to help ensure that irregularities and fraud are detected.
D. Although rotating employees could contribute to fewer processing errors, this is not typically a reason to require a mandatory vacation policy.

A2-2 An IS auditor is verifying IT policies and finds that some of the policies have not been approved by management (as required by policy), but the employees strictly follow the policies. What should the IS auditor do **FIRST**?

A. Ignore the absence of management approval because employees follow the policies.
B. Recommend immediate management approval of the policies.
C. Emphasize the importance of approval to management.
D. Report the absence of documented approval.

D is the correct answer.

Justification:

A. Absence of management approval is an important (material) finding and, although it is not currently an issue with relation to compliance because the employees are following the policy without approval, it may be a problem at a later time and should be resolved.
B. Although the IS auditor would likely recommend that the policies should be approved as soon as possible and may also remind management of the critical nature of this issue, the first step is to report this issue to the relevant stakeholders.
C. The first step is to report the finding and provide recommendations later.
D. The IS auditor must report the finding. Unapproved policies may present a potential risk to the organization, even if they are being followed, because this technicality may prevent management from enforcing the policies in some cases and may present legal issues. For example, if an employee was terminated as a result of violating an organization policy, and it was discovered that the policies had not been approved, the organization may face an expensive lawsuit.

A2-3 What is the **PRIMARY** consideration for an IS auditor reviewing the prioritization and coordination of IT projects and program management?

A. Projects are aligned with the organization's strategy.
B. Identified project risk is monitored and mitigated.
C. Controls related to project planning and budgeting are appropriate.
D. IT project metrics are reported accurately.

A is the correct answer.

Justification:

A. The primary goal of IT projects is to add value to the business, so they must be aligned with the business strategy to achieve the intended results. Therefore, the IS auditor should first focus on ensuring this alignment.
B. An adequate process for monitoring and mitigating identified project risk is important; however, strategic alignment helps in assessing identified risk in business terms.
C. Completion of projects within a predefined time and budget is important; however, the focus of project management should be on achieving the desired outcome of the project, which is aligned with the business strategy.
D. Adequate reporting of project status is important but may or may not help in providing the strategic perspective of project deliverables.

A2-4 In a review of the human resources policies and procedures within an organization, an IS auditor is **MOST** concerned with the absence of a:

A. requirement for periodic job rotations.
B. process for formalized exit interviews.
C. termination checklist.
D. requirement for new employees to sign a nondisclosure agreement.

C is the correct answer.

Justification:

A. Job rotation is a valuable control to ensure continuity of operations, but not the most serious human resources policy risk.
B. Holding an exit interview is desirable when possible to gain feedback but is not a serious risk.
C. A termination checklist is critical to ensure the logical and physical security of an enterprise. In addition to preventing the loss of enterprise property that was issued to the employee, there is the risk of unauthorized access, intellectual property theft and even sabotage by a disgruntled former employee.
D. Signing a nondisclosure agreement (NDA) is a recommended human resources practice, but a lack of an NDA is not the most serious risk listed.

A2-5 Which of the following factors is **MOST** critical when evaluating the effectiveness of an IT governance implementation?

A. Ensure that assurance objectives are defined.
B. Determine stakeholder requirements and involvement.
C. Identify relevant risk and related opportunities.
D. Determine relevant enablers and their applicability.

B is the correct answer.

Justification:
A. Stakeholders' needs and their involvement form the basis for scoping the IT governance implementation. This will be used to define assurance objectives.
B. The most critical factor to be considered in auditing an IT governance implementation is to determine stakeholder requirements and involvement. This drives the success of the project. Based on this, the assurance scope and objectives are determined.
C. The relevant risk and related opportunities are identified and driven by the assurance objectives.
D. The relevant enablers and their applicability for the IT governance implementation are considered based on assurance objectives.

A2-6 Which of the following is the **BEST** reason to implement a policy that places conditions on secondary employment for IT employees?

A. To prevent the misuse of corporate resources
B. To prevent conflicts of interest
C. To prevent employee performance issues
D. To prevent theft of IT assets

B is the correct answer.

Justification:
A. The misuse of corporate resources is an issue that must be addressed but is not necessarily related to secondary employment.
B. The best reason to implement and enforce a policy governing secondary employment is to prevent conflicts of interest. Policies should be in place to control IT employees seeking secondary employment from releasing sensitive information or working for a competing otganization. Conflicts of interest can result in serious risk such as fraud, theft of intellectual property or other improprieties.
C. Employee performance can certainly be an issue if an employee is overworked or has insufficient time off, but that should be dealt with as a management function and not the primary reason to have a policy on secondary employment.
D. Theft of assets is a problem but not necessarily related to secondary employment.

A2-7 An IS auditor has been assigned to review an organization's information security policy. Which of the following issues represents the **HIGHEST** potential risk?

A. The policy has not been updated in more than one year.
B. The policy includes no revision history.
C. The policy is approved by the security administrator.
D. The company does not have an information security policy committee.

C is the correct answer.

Justification:

A. Although the information security policy should be updated on a regular basis, the specific time period may vary based on the organization. Although reviewing policies annually is a good practice, the policy may be updated less frequently and still be relevant and effective. An outdated policy is still enforceable, whereas a policy without proper approval is not enforceable.
B. The lack of a revision history with respect to the IS policy document is an issue but not as significant as not having it approved by management. A new policy, for example, may not have been subject to any revisions yet.
C. The information security policy should have an owner who has management responsibility for the development, review, approval and evaluation of the security policy. The position of security administrator is typically a staff-level position (not management), and therefore does not have the authority to approve the policy. In addition, an individual in a more independent position should also review the policy. Without proper management approval, enforcing the policy may be problematic, leading to compliance or security issues.
D. Although a policy committee drawn from across the company is a good practice and may help write better policies, a good policy can be written by a single person, and the lack of a committee is not a problem by itself.

A2-8 When performing a review of a business process reengineering (BPR) effort, which of the following is of **PRIMARY** concern?

A. Controls are eliminated as part of the streamlining BPR effort.
B. Resources are not adequate to support the BPR process.
C. The audit department does not have a consulting role in the BPR effort.
D. The BPR effort includes employees with limited knowledge of the process area.

A is the correct answer.

Justification:

A. A primary risk of business process reengineering (BPR) is that controls are eliminated as part of the reengineering effort. This is the primary concern.
B. The BPR process can be a resource-intensive initiative; however, the more important issue is whether critical controls are eliminated as a result of the BPR effort.
C. Although BPR efforts often involve many different business functions, it is not a significant concern if audit is not involved, and, in most cases, it is not appropriate for audit to be involved in such an effort.
D. A recommended good practice for BPR is to include individuals from all parts of the enterprise, even those with limited knowledge of the process area. Therefore, this is not a concern.

A2-9 When auditing the IT governance framework and IT risk management practices existing within an organization, the IS auditor identified some undefined responsibilities regarding IT management and governance roles. Which of the following recommendations is the **MOST** appropriate?

A. Review the strategic alignment of IT with the business.
B. Implement accountability rules within the organization.
C. Ensure that independent IS audits are conducted periodically.
D. Create a chief risk officer role in the organization.

B is the correct answer.

Justification:

A. While the strategic alignment of IT with the business is important, it is not directly related to the gap identified in this scenario.

B. IT risk is managed by embedding accountability into the enterprise. The IS auditor should recommend the implementation of accountability rules to ensure that all responsibilities are defined within the organization. Note that this question asks for the best recommendation—not about the finding itself.

C. Performing more frequent IS audits is not helpful if the accountability rules are not clearly defined and implemented.

D. Recommending the creation of a new role (e.g., chief risk officer) is not helpful if the accountability rules are not clearly defined and implemented.

A2-10 An IS auditor is performing a review of the software quality management process in an organization. The **FIRST** step should be to:

A. Verify how the organization complies the standards.
B. Identify and report the existing controls.
C. Review the metrics for quality evaluation.
D. Request all standards adopted by the organization.

D is the correct answer.

Justification:

A. The auditor needs to know what standards the organization has adopted and then measure compliance with those standards. Determining how the organization follows the standards is secondary to knowing what the standards are. The other items listed—verifying how well standards are being followed, identifying relevant controls and reviewing the quality metrics—are secondary to the identification of standards.

B. The first step is to know the standards and what policies and procedures are mandated for the organization, then to document the controls and measure compliance.

C. The metrics cannot be reviewed until the auditor has a copy of the standards that describe or require the metrics.

D. Because an audit measures compliance with the standards of the organization, the first step of the review of the software quality management process should be to determine the evaluation criteria in the form of standards adopted by the organization. The evaluation of how well the organization follows their own standards cannot be performed until the IS auditor has determined what standards exist.

A2-11 An IS auditor found that the enterprise architecture (EA) recently adopted by an organization has an adequate current-state representation. However, the organization has started a separate project to develop a future-state representation. The IS auditor should:

A. Recommend that this separate project be completed as soon as possible.
B. Report this issue as a finding in the audit report.
C. Recommend the adoption of the Zachmann framework.
D. Rescope the audit to include the separate project as part of the current audit.

B is the correct answer.

Justification:

A. The IS auditor does not ordinarily provide input on the timing of projects, but rather provides an assessment of the current environment. The most critical issue in this scenario is that the enterprise architecture (EA) is undergoing change, so the IS auditor should be most concerned with reporting this issue.

B. It is critical for the EA to include the future state because the gap between the current state and the future state will determine IT strategic and tactical plans. If the EA does not include a future-state representation, it is not complete, and this issue should be reported as a finding.

C. The organization is free to choose any EA framework, and the IS auditor should not recommend a specific framework.

D. Changing the scope of an audit to include the secondary project is not required, although a follow-up audit may be desired.

A2-12 An IS auditor is evaluating management's risk assessment of information systems. The IS auditor should **FIRST** review:

A. Controls in place.
B. Effectiveness of the controls.
C. Mechanism for monitoring the risk.
D. Threats/vulnerabilities affecting the assets.

D is the correct answer.

Justification:

A. The controls are irrelevant until the IS auditor knows the threats and risk that the controls are intended to address.

B. The effectiveness of the controls must be measured in relation to the risk (based on assets, threats and vulnerabilities) that the controls are intended to address.

C. The first step must be to determine the risk that is being managed before reviewing the mechanism of monitoring risk.

D. One of the key factors to be considered while assessing the information systems risk is the value of the systems (the assets) and the threats and vulnerabilities affecting the assets. The risk related to the use of information assets should be evaluated in isolation from the installed controls.

A2-13 The **PRIMARY** benefit of an enterprise architecture initiative is to:

A. Enable the organization to invest in the most appropriate technology.
B. Ensure security controls are implemented on critical platforms.
C. Allow development teams to be more responsive to business requirements.
D. Provide business units with greater autonomy to select it solutions that fit their needs.

A is the correct answer.

Justification:

A. The primary focus of the enterprise architecture (EA) is to ensure that technology investments are consistent with the platform, data and development standards of the IT organization; therefore, the goal of the EA is to help the organization to implement the technology that is most effective.

B. Ensuring that security controls are implemented on critical platforms is important, but this is not the function of the EA. The EA may be concerned with the design of security controls; however, the EA would not help to ensure that they were implemented. The primary focus of the EA is to ensure that technology investments are consistent with the platform, data and development standards of the IT organization.

C. While the EA process may enable development teams to be more efficient, because they are creating solutions based on standard platforms using standard programming languages and methods, the more critical benefit of the EA is to provide guidance for IT investments of all types, which encompasses much more than software development.

D. A primary focus of the EA is to define standard platforms, databases and interfaces. Business units that invest in technology would need to select IT solutions that meet their business needs and are compatible with the EA of the enterprise. There may be instances when a proposed solution works better for a business unit but is not at all consistent with the EA of the enterprise, so there would be a need to compromise to ensure that the application can be supported by IT. Overall, the EA would restrict the ability of business units in terms of the potential IT systems that they may wish to implement. The support requirements would not be affected in this case.

A2-14 Which of the following situations is addressed by a software escrow agreement?

A. The system administrator requires access to software to recover from a disaster.
B. A user requests to have software reloaded onto a replacement hard drive.
C. The vendor of custom-written software goes out of business.
D. An IS auditor requires access to software code written by the organization.

C is the correct answer.

Justification:

A. Access to software should be managed by an internally managed software library. Escrow refers to the storage of software with a third party—not the internal libraries.

B. Providing the user with a backup copy of software is not escrow. Escrow requires that a copy be kept with a trusted third party.

C. A software escrow is a legal agreement between a software vendor and a customer to guarantee access to source code. The application source code is held by a trusted third party, according to the contract. This agreement is necessary in the event that the software vendor goes out of business, there is a contractual dispute with the customer or the software vendor fails to maintain an update of the software as promised in the software license agreement.

D. Software escrow is used to protect the intellectual property of software developed by one organization and sold to another organization. This is not used for software being reviewed by an auditor of the organization that wrote the software.

A2-15 An IS auditor reviews an organizational chart **PRIMARILY** for:

A. Understanding of the complexity of the organizational structure.
B. Investigating various communication channels.
C. Understanding the responsibilities and authority of individuals.
D. Investigating the network connected to different employees.

C is the correct answer.

Justification:
A. Understanding the complexity of the organizational structure is not the primary reason to review an organizational chart because the chart will not necessarily depict the complexity.
B. The organizational chart is a key tool for an auditor to understand roles and responsibilities and reporting lines but is not used for examining communications channels.
C. An organizational chart provides information about the responsibilities and authority of individuals in the organization. This helps an IS auditor to know if there is a proper segregation of functions.
D. A network diagram will provide information about the usage of various communication channels and will indicate the connection of users to the network.

A2-16 Sharing risk is a key factor in which of the following methods of managing risk?

A. Transferring risk
B. Tolerating risk
C. Terminating risk
D. Treating risk

A is the correct answer.

Justification:
A. Transferring risk (e.g., by taking an insurance policy) is a way to share risk.
B. Tolerating risk means that the risk is accepted, but not shared.
C. Terminating risk would not involve sharing the risk because the organization has chosen to terminate the process associated with the risk.
D. There are several ways of treating or controlling the risk, which may involve reducing or sharing the risk, but this is not as precise an answer as transferring the risk.

A2-17 A team conducting a risk analysis is having difficulty projecting the financial losses that could result from a risk. To evaluate the potential impact, the team should:

A. Compute the amortization of the related assets.
B. Calculate a return on investment.
C. Apply a qualitative approach.
D. Spend the time needed to define the loss amount exactly.

C is the correct answer.

Justification:
A. Amortization is used in a profit and loss statement, not in computing potential losses.
B. A return on investment (ROI) is computed when there is predictable savings or revenues that can be compared to the investment needed to realize the revenues.
C. The common practice when it is difficult to calculate the financial losses is to take a qualitative approach, in which the manager affected by the risk defines the impact in terms of a weighted factor (e.g., one is a very low impact to the business and five is a very high impact).
D. Spending the time needed to define exactly the total amount is normally a wrong approach. If it has been difficult to estimate potential losses (e.g., losses derived from erosion of public image due to a hack attack), that situation is not likely to change, and the result will be a not well-supported evaluation.

A2-18 While reviewing a quality management system, the IS auditor should **PRIMARILY** focus on collecting evidence to show that:

A. Quality management systems comply with good practices.
B. Continuous improvement targets are being monitored.
C. Standard operating procedures of it are updated annually.
D. Key performance indicators are defined.

B is the correct answer.

Justification:

A. Generally, good practices are adopted according to business requirements. Therefore, conforming to good practices may or may not be a requirement of the business.
B. Continuous and measurable improvement of quality is the primary requirement to achieve the business objective for the quality management system (QMS).
C. Updating operating procedures is part of implementing the QMS; however, it must be part of change management and not an annual activity.
D. Key performance indicators may be defined in a QMS, but they are of little value if they are not being monitored.

A2-19 An IS auditor discovers several IT-based projects were implemented and not approved by the steering committee. What is the **GREATEST** concern for the IS auditor?

A. The IT department's projects will not be adequately funded.
B. IT projects are not following the system development life cycle process.
C. IT projects are not consistently formally approved.
D. The IT department may not be working toward a common goal.

D is the correct answer.

Justification:

A. Funding for the projects may be addressed through various budgets and may not require steering committee approval. The primary concern would be to ensure that the project is working toward meeting the goals of the company.
B. Although requiring steering committee approval may be part of the system development life cycle process, the greater concern would be whether the projects are working toward the corporate goals. Without steering committee approval, it would be difficult to determine whether these projects are following the direction of the corporate goals.
C. Although having a formal approval process is important, the greatest concern would be for the steering committee to provide corporate direction for the projects.
D. The steering committee provides direction and control over projects to ensure that the company is making appropriate investments. Without approval, the project may or may not be working toward the company's goals.

A2-20 Value delivery from IT to the business is **MOST** effectively achieved by:

A. Aligning the IT strategy with the enterprise strategy
B. Embedding accountability in the enterprise
C. Providing a positive return on investment
D. Establishing an enterprisewide risk management process

A is the correct answer.

Justification:
A. IT's value delivery to the business is driven by aligning IT with the enterprise's strategy.
B. Embedding accountability in the enterprise promotes risk management (another element of corporate governance).
C. While return on investment is important, it is not the only criterion by which the value of IT is assessed.
D. Enterprisewide risk management is critical to IT governance; however, by itself, it will not guarantee that IT delivers value to the business unless the IT strategy is aligned with the enterprise strategy.

A2-21 During a feasibility study regarding outsourcing IT processing, the relevance for the IS auditor of reviewing the vendor's business continuity plan is to:

A. Evaluate the adequacy of the service levels that the vendor can provide in a contingency.
B. Evaluate the financial stability of the service bureau and its ability to fulfill the contract.
C. Review the experience of the vendor´s staff.
D. Test the business continuity plan.

A is the correct answer.

Justification:
A. A key factor in a successful outsourcing environment is the capability of the vendor to face a contingency and continue to support the organization's processing requirements.
B. Financial stability is not related to the vendor's business continuity plan (BCP).
C. Experience of the vendor's staff is not related to the vendor's BCP.
D. The review of the vendor's BCP during a feasibility study is not a way to test the vendor's BCP.

A2-22 An IS auditor is evaluating a newly developed IT policy for an organization. Which of the following factors does the IS auditor consider **MOST** important to facilitate compliance with the policy upon its implementation?

A. Existing IT mechanisms enabling compliance
B. Alignment of the policy to the business strategy
C. Current and future technology initiatives
D. Regulatory compliance objectives defined in the policy

A is the correct answer.

Justification:
A. The organization should be able to comply with a policy when it is implemented. The most important consideration when evaluating the new policy should be the existing mechanisms in place that enable the organization and its employees to comply with the policy.
B. Policies should be aligned with the business strategy, but this does not affect an organization's ability to comply with the policy upon implementation.
C. Current and future technology initiatives should be driven by the needs of the business and would not affect an organization's ability to comply with the policy.
D. Regulatory compliance objectives may be defined in the IT policy, but that would not facilitate compliance with the policy. Defining objectives would only result in the organization knowing the desired state and would not aid in achieving compliance.

A2-23 The **MOST** likely effect of the lack of senior management commitment to IT strategic planning is:

A. Lack of investment in technology
B. Lack of a methodology for systems development
C. Technology not aligning with organization objectives
D. Absence of control over technology contracts

C is the correct answer.

Justification:

A. Lack of management commitment will almost certainly affect investment, but the primary loss will be the lack of alignment of IT strategy with the strategy of the business.
B. Systems development methodology is a process-related function and not a key concern of management.
C. A steering committee should exist to ensure that the IT strategies support the organization's goals. The absence of an information technology committee or a committee not composed of senior managers is an indication of a lack of top-level management commitment. This condition increases the risk that IT is aligned with organization strategy.
D. Approval for contracts is a business process and would be controlled through financial process controls. This is not applicable here.

A2-24 Which of the following is a function of an IT steering committee?

A. Monitoring vendor-controlled change control and testing
B. Ensuring a separation of duties within the information's processing environment
C. Approving and monitoring the status of IT plans and budgets
D. Liaising between the IT department and end users

C is the correct answer.

Justification:

A. Vendor change control is a sourcing issue and should be monitored by IT management.
B. Ensuring a separation of duties within the information's processing environment is an IT management responsibility.
C. The IT steering committee typically serves as a general review board for major IT projects and should not become involved in routine operations; therefore, one of its functions is to approve and monitor major projects, such as the status of IT plans and budgets.
D. Liaising between the IT department and end users is a function of the individual parties and not a committee responsibility.

A2-25 An IS auditor is performing a review of an organization's governance model. Which of the following should be of **MOST** concern to the auditor?

A. The information security policy is not periodically reviewed by senior management.
B. A policy ensuring systems are patched in a timely manner does not exist.
C. The audit committee did not review the organization's mission statement.
D. An organizational policy related to information asset protection does not exist.

A is the correct answer.

Justification:
A. Data security policies should be reviewed/refreshed once every year to reflect changes in the organization's environment. Policies are fundamental to the organization's governance structure, and, therefore, this is the greatest concern.
B. While it is a concern that there is no policy related to system patching, the greater concern is that the information security policy is not reviewed periodically by senior management.
C. Mission statements tend to be long term because they are strategic in nature and are established by the board of directors and management. This is not the IS auditor's greatest concern because proper governance oversight could lead to meeting the objectives of the organization's mission statement.
D. While it is a concern that there is no policy related to the protection of information assets, the greater concern is that the security policy is not reviewed periodically by senior management because top level support is fundamental to information security governance.

A2-26 Involvement of senior management is **MOST** important in the development of:

A. Strategic plans.
B. IT policies.
C. IT procedures.
D. Standards and guidelines.

A is the correct answer.

Justification:
A. Strategic plans provide the basis for ensuring that the enterprise meets its goals and objectives. Involvement of senior management is critical to ensuring that the plan adequately addresses the established goals and objectives.
B. IT policies are created and enforced by IT management and information security. They are structured to support the overall strategic plan.
C. IT procedures are developed to support IT policies. Senior management is not involved in the development of procedures.
D. Standards and guidelines are developed to support IT policies. Senior management is not involved in the development of standards, baselines and guidelines.

A2-27 Effective IT governance ensures that the IT plan is consistent with the organization's:

A. Business plan.
B. Audit plan.
C. Security plan.
D. Investment plan.

A is the correct answer.

Justification:

A. To govern IT effectively, IT and business should be moving in the same direction, requiring that the IT plans are aligned with an organization's business plans.
B. The audit plan is not part of the IT plan.
C. The security plan is not a responsibility of IT and does not need to be consistent with the IT plan.
D. The investment plan is not part of the IT plan.

A2-28 Establishing the level of acceptable risk is the responsibility of:

A. Quality assurance management.
B. Senior business management.
C. The chief information officer.
D. The chief security officer.

B is the correct answer.

Justification:

A. Quality assurance (QA) is concerned with reliability and consistency of processes. The QA team is not responsible for determining an acceptable risk level.
B. Senior management should establish the acceptable risk level because they have the ultimate or final responsibility for the effective and efficient operation of the organization as a senior manager of the business process. The person can be the QA, chief information officer (CIO), or the chief security officer (CSO), but the responsibility rests with the business manager.
C. The establishment of acceptable risk levels is a senior business management responsibility. The CIO is the most senior official of the enterprise who is accountable for IT advocacy; aligning IT and business strategies; and planning, resourcing and managing the delivery of IT services, information and the deployment of associated human resources. The CIO is rarely the person that determines acceptable risk levels because this could be a conflict of interest unless the CIO is the senior business process owner.
D. The establishment of acceptable risk levels is a senior business management responsibility. The CSO is responsible for enforcing the decisions of the senior management team unless the CIO is the business process manager.

A2-29 IT governance is **PRIMARILY** the responsibility of the:

A. chief executive officer.
B. board of directors.
C. IT steering committee.
D. audit committee.

B is the correct answer.

Justification:

A. The chief executive officer is instrumental in implementing IT governance according to the directions of the board of directors.
B. IT governance is primarily the responsibility of the executives and shareholders (as represented by the board of directors).
C. The IT steering committee monitors and facilitates deployment of IT resources for specific projects "in support of business plans. The IT steering committee enforces governance on behalf of the board of directors.
D. The audit committee reports to the board of directors and executes governance-related audits. The audit committee should monitor the implementation of audit recommendations.

A2-30 From a control perspective, the key element in job descriptions is that they:

A. Provide instructions on how to do the job and define authority.
B. Are current, documented and readily available to the employee.
C. Communicate management's specific job performance expectations.
D. Establish responsibility and accountability for the employee's actions.

D is the correct answer.

Justification:

A. Providing instructions on how to do the job and defining authority addresses the managerial and procedural aspects of the job and is a management responsibility. Job descriptions, which are a human resources (HR)-related function, are primarily used to establish job requirements and accountability.
B. It is important that job descriptions are current, documented and readily available to the employee, but this, in itself, is not the key element of the job description. Job descriptions, which are an HR-related function, are primarily used to establish job requirements and accountability.
C. Communication of management's specific expectations for job performance would not necessarily be included in job descriptions.
D. From a control perspective, a job description should establish responsibility and accountability. This aids in ensuring that users are given system access in accordance with their defined job responsibilities and are accountable for how they use that access.

A2-31 Which of the following **BEST** provides assurance of the integrity of new staff?

A. Background screening
B. References
C. Bonding
D. Qualifications listed on a resume

A is the correct answer.

Justification:

A. A background screening is the primary method for assuring the integrity of a prospective staff member. This may include criminal history checks, driver's license abstracts, financial status checks, verification of education, etc.

B. References are important and would need to be verified, but they are not as reliable as background screening because the references themselves may not be validated as trustworthy.

C. Bonding is directed at due-diligence compliance and does not ensure integrity.

D. Qualifications listed on a résumé may be used to demonstrate proficiency but will not indicate the integrity of the candidate employee.

A2-32 When an employee is terminated from service, the **MOST** important action is to:

A. hand over all of the employee's files to another designated employee.
B. complete a backup of the employee's work.
C. notify other employees of the termination.
D. disable the employee's logical access.

D is the correct answer.

Justification:

A. All the work of the terminated employee needs to be handed over to a designated employee; however, this is not as critical as removing terminated employee access.

B. All the work of the terminated employee needs to be backed up, but this is not as critical as removing terminated employee access.

C. The employees need to be notified of the termination, but this is not as critical as removing terminated employee access.

D. There is a probability that a terminated employee may misuse access rights; therefore, disabling the terminated employee's logical access is the most important and immediate action to take.

A2-33 A business unit has selected a new accounting application and did not consult with IT early in the selection process. The **PRIMARY** risk is that:

A. The security controls of the application may not meet requirements.
B. The application may not meet the requirements of the business users.
C. The application technology may be inconsistent with the enterprise architecture.
D. The application may create unanticipated support issues for IT.

C is the correct answer.

Justification:
A. Although security controls should be a requirement for any application, the primary focus of the enterprise architecture (EA) is to ensure that new applications are consistent with enterprise standards. Although the use of standard supported technology may be more secure, this is not the primary benefit of the EA.
B. When selecting an application, the business requirements and the suitability of the application for the IT environment must be considered. If the business units selected their application without IT involvement, they are more likely to choose a solution that fits their business process the best with less emphasis on how compatible and supportable the solution will be in the enterprise, and this is not a concern.
C. The primary focus of the EA is to ensure that technology investments are consistent with the platform, data and development standards of the IT organization. The EA defines both a current and future state in areas such as the use of standard platforms, databases or programming languages. If a business unit selected an application using a database or operating system that is not part of the EA for the business, this increases the cost and complexity of the solution and ultimately delivers less value to the business.
D. Although any new software implementation may create support issues, the primary benefit of the EA is ensuring that the IT solutions deliver value to the business. Decreased support costs may be a benefit of the EA, but the lack of IT involvement in this case would not affect the support requirements.

A2-34 Many organizations require an employee to take a mandatory vacation (holiday) of a week or more to:

A. Ensure that the employee maintains a good quality of life, which will lead to greater productivity.
B. Reduce the opportunity for an employee to commit an improper or illegal act.
C. Provide proper cross-training for another employee.
D. Eliminate the potential disruption caused when an employee takes vacation one day at a time.

B is the correct answer.

Justification:
A. Maintaining a good quality of life is important, but the primary reason for a mandatory vacation is to catch fraud or errors.
B. Required vacations/holidays of a week or more in duration in which someone other than the regular employee performs the job function of the employee on vacation is often mandatory for sensitive positions because this reduces the opportunity to commit improper or illegal acts. During this time off, it may be possible to discover any fraudulent activity that was taking place.
C. Providing cross-training is an important management function, but the primary reason for mandatory vacations is to detect fraud or errors.
D. Enforcing a rule that all vacations must be taken a week at a time is a management decision but not related to a mandatory vacation policy. The primary reason for mandatory vacations is to detect fraud or errors.

A2-35 A local area network (LAN) administrator normally is restricted from:

A. having end-user responsibilities.
B. reporting to the end-user manager.
C. having programming responsibilities.
D. being responsible for LAN security administration.

C is the correct answer.

Justification:
A. Although not ideal, a local area network (LAN) administrator may have end-user responsibilities.
B. The LAN administrator may report to the director of the information processing facility (IPF) or, in a decentralized operation, to the end-user manager.
C. A LAN administrator should not have programming responsibilities because that could allow modification of production programs without proper separation of duties, but the LAN administrator may have end-user responsibilities.
D. In small organizations, the LAN administrator may also be responsible for security administration over the LAN.

A2-36 A decision support system is used to help high-level management:

A. Solve highly structured problems.
B. Combine the use of decision models with predetermined criteria.
C. Make decisions based on data analysis and interactive models.
D. Support only structured decision-making tasks.

C is the correct answer.

Justification:
A. A decision support system (DSS) is aimed at solving less structured problems.
B. A DSS combines the use of models and analytic techniques with traditional data access and retrieval functions but is not limited by predetermined criteria.
C. A DSS emphasizes flexibility in the decision-making approach of management through data analysis and the use of interactive models, not fixed criteria.
D. A DSS supports semistructured decision-making tasks.

A2-37 During an audit, the IS auditor discovers that the human resources (HR) department uses a cloud-based application to manage employee records. The HR department engaged in a contract outside of the normal vendor management process and manages the application on its own. Which of the following is of **GREATEST** concern?

A. Maximum acceptable downtime metrics have not been defined in the contract.
B. The IT department does not manage the relationship with the cloud vendor.
C. The help desk call center is in a different country, with different privacy requirements.
D. Organization-defined security policies are not applied to the cloud application.

D is the correct answer.

Justification:

A. Maximum acceptable downtime is a good metric to have in the contract to ensure application availability; however, human resources (HR) applications are usually not mission-critical, and therefore, maximum acceptable downtime is not the most significant concern in this scenario.
B. The responsibility for managing the relationship with a third party should be assigned to a designated individual or service management team; however, it is not essential that the individual or team belong to the IT department.
C. An organization-defined security policy ensures that help desk personnel do not have access to personnel data, and this is covered under the security policy. The more critical issue is that the application complied with the security policy.
D. Cloud applications should adhere to the organization-defined security policies to ensure that the data in the cloud are protected in a manner consistent with internal applications. These include, but are not limited to, the password policy, user access management policy and data classification policy.

A2-38 Before implementing an IT balanced scorecard, an organization must:

A. Deliver effective and efficient services.
B. Define key performance indicators.
C. Provide business value to IT projects.
D. Control IT expenses.

B is the correct answer.

Justification:

A. A balanced scorecard (BSC) is a method of specifying and measuring the attainment of strategic results. It will measure the delivery of effective and efficient services, but an organization may not have those in place prior to using a BSC.
B. Because a BSC is a way to measure performance, a definition of key performance indicators is required before implementing an IT BSC.
C. A BSC will measure the value of IT to business, not the other way around.
D. A BSC will measure the performance of IT, but the control over IT expenses is not a key requirement for implementing a BSC.

A2-39 To support an organization's goals, an IT department should have:

A. A low-cost philosophy.
B. Long- and short-term plans.
C. Leading-edge technology.
D. Plans to acquire new hardware and software.

B is the correct answer.

Justification:

A. A low-cost philosophy is one objective, but more important is the cost-benefit and the relation of IT investment cost to business strategy.
B. To ensure its contribution to the realization of an organization's overall goals, the IT department should have long- and short-range plans that are consistent with the organization's broader and strategic plans for attaining its goals.
C. Leading-edge technology is an objective, but IT plans would be needed to ensure that those plans are aligned with organizational goals.
D. Plans to acquire new hardware and software could be a part of the overall plan but would be required only if hardware or software is needed to achieve the organizational goals.

A2-40 In reviewing the IT short-range (tactical) plan, an IS auditor should determine whether:

A. There is an integration of IT and business personnel within projects.
B. There is a clear definition of the IT mission and vision.
C. A strategic information technology planning scorecard is in place.
D. The plan correlates business objectives to IT goals and objectives.

A is the correct answer.

Justification:

A. The integration of IT and business personnel in projects is an operational issue and should be considered while reviewing the short-range plan. A strategic plan provides a framework for the IT short-range plan.
B. A clear definition of the IT mission and vision would be covered by a strategic plan.
C. A strategic information technology planning scorecard would be covered by a strategic plan.
D. Business objectives correlating to IT goals and objectives would be covered by a strategic plan.

A2-41 Which of the following does an IS auditor consider the **MOST** relevant to short-term planning for an IT department?

A. Allocating resources
B. Adapting to changing technologies
C. Conducting control self-assessments
D. Evaluating hardware needs

A is the correct answer.

Justification:

A. The IT department should specifically consider the manner in which resources are allocated in the short term. The IS auditor ensures that the resources are being managed adequately.
B. Investments in IT need to be aligned with top management strategies rather than be relevant to short-term planning and focus on technology for technology's sake.
C. Conducting control self-assessments is not as critical as allocating resources during short-term planning for the IT department.
D. Evaluating hardware needs is not as critical as allocating resources during short-term planning for the IT department.

A2-42 Which of the following goals do you expect to find in an organization's strategic plan?

A. Results of new software testing
B. An evaluation of information technology needs
C. Short-term project plans for a new planning system
D. Approved suppliers for products offered by the company

D is the correct answer.

Justification:

A. Results of a new accounting package is a tactical or short-term goal and would not be included in a strategic plan.
B. An evaluation of information technology needs is a way to measure performance, but not a goal to be found in a strategic plan.
C. Short-term project plans is project-oriented and is a method of implementing a goal but not the goal in itself. The goal would be to have better project management—the new system is how to achieve that goal.
D. Approved suppliers of choice for the product is a strategic business objective that is intended to focus the overall direction of the business and, thus, is a part of the organization's strategic plan.

A2-43 Which of the following does an IS auditor consider to be **MOST** important when evaluating an organization's IT strategy? That it:

A. Was approved by line management.
B. Does not vary from the IT department's preliminary budget.
C. Complies with procurement procedures.
D. Supports the business objectives of the organization.

D is the correct answer.

Justification:

A. A strategic plan is a senior management responsibility and would receive input from line managers but would not be approved by them.
B. The budget should not vary from the plan.
C. Procurement procedures are organizational controls, but not a part of strategic planning.
D. Strategic planning sets corporate or department objectives into motion. Both long-term and short-term strategic plans should be consistent with the organization's broader plans and business objectives for attaining these goals.

A2-44 An organization has contracted with a vendor for a turnkey solution for their electronic toll collection system (ETCS). The vendor has provided its proprietary application software as part of the solution. The contract should require that:

A. A backup server is available to run ETCS operations with up-to-date data.
B. A backup server is loaded with all relevant software and data.
C. The systems staff of the organization is trained to handle any event.
D. Source code of the ETCS application is placed in escrow.

D is the correct answer.

Justification:
A. Having a backup server with current data is critical but not as critical as ensuring the availability of the source code.
B. Having a backup server with relevant software is critical but not as critical as ensuring the availability of the source code.
C. Having staff training is critical but not as critical as ensuring the availability of the source code.
D. Whenever proprietary application software is purchased, the contract should provide for a source code escrow agreement. This agreement ensures that the purchasing organization has the opportunity to modify the software should the vendor cease to be in business.

A2-45 When reviewing the IT strategy, an IS auditor can **BEST** assess whether the strategy supports the organizations' business objectives by determining whether IT:

A. Has all the personnel and equipment it needs.
B. Plans are consistent with management strategy.
C. Uses its equipment and personnel efficiently and effectively.
D. Has sufficient excess capacity to respond to changing directions.

B is the correct answer.

Justification:
A. Having personnel and equipment is an important requirement to meet the IT strategy but will not ensure that the IT strategy supports business objectives.
B. The only way to know if IT strategy will meet business objectives is to determine if the IT plan is consistent with management strategy and that it relates IT planning to business plans.
C. Using equipment and personnel efficiently and effectively is an effective method for determining the proper management of the IT function but does not ensure that the IT strategy is aligned with business objectives.
D. Having sufficient excess capacity to respond to changing directions is important to show flexibility to meet organizational changes but is not in itself a way to ensure that IT is aligned with business goals.

A2-46 An IS auditor of a large organization is reviewing the roles and responsibilities of the IT function and finds some individuals serving multiple roles. Which one of the following combinations of roles should be of **GREATEST** concern for the IS auditor?

A. Network administrators are responsible for quality assurance.
B. System administrators are application programmers.
C. End users are security administrators for critical applications.
D. Systems analysts are database administrators.

B is the correct answer.

Justification:

A. Ideally, network administrators should not be responsible for quality assurance because they could approve their own work. However, that is not as serious as the combination of system administrator and application programmer, which would allow nearly unlimited abuse of privilege.
B. When individuals serve multiple roles, this represents a separation-of-duties problem with associated risk. System administrators should not be application programmers, due to the associated rights of both functions. A person with both system and programming rights can do almost anything on a system, including creating a back door. The other combinations of roles are valid from a separation of duties perspective.
C. In some distributed environments, especially with small staffing levels, users may also manage security.
D. While a database administrator is a very privileged position it would not be in conflict with the role of a systems analyst.

A2-47 Which of the following is the **GREATEST** risk of an inadequate policy definition for ownership of data and systems?

A. User management coordination does not exist.
B. Specific user accountability cannot be established.
C. Unauthorized users may have access to modify data.
D. Audit recommendations may not be implemented.

C is the correct answer.

Justification:

A. The greatest risk is from unauthorized users being able to modify data. User management is important but not the greatest risk.
B. User accountability is important but not as great a risk as the actions of unauthorized users.
C. Without a policy defining who has the responsibility for granting access to specific systems, there is an increased risk that individuals can gain (be given) system access when they should not have authorization. The ability of unauthorized users to modify data is greater than the risk of authorized user accounts not being controlled properly.
D. The failure to implement audit recommendations is a management problem but not as serious as the ability of unauthorized users making modifications.

A2-48 An IS audit department is planning to minimize the risk of short-term employees. Activities contributing to this objective are documented procedures, knowledge sharing, cross-training and:

A. Succession planning.
B. Staff job evaluation.
C. Responsibilities definitions.
D. Employee award programs.

A is the correct answer.

Justification:

A. Succession planning ensures that internal personnel with the potential to fill key positions in the organization are identified and developed.

B. Job evaluation is the process of determining the worth of one job in relation to that of the other jobs in a company so that a fair and equitable wage and salary system can be established.

C. Staff responsibilities definitions provide for well-defined roles and responsibilities; however, they do not minimize dependency on key individuals.

D. Employee award programs provide motivation; however, they do not minimize dependency on key individuals.

A2-49 The rate of change in technology increases the importance of:

A. Outsourcing the IT function.
B. Implementing and enforcing sound processes.
C. Hiring qualified personnel.
D. Meeting user requirement.

B is the correct answer.

Justification:

A. Outsourcing the IT function is a business decision and not directly related to the rate of technological change, nor does the rate of change increase the importance of outsourcing.

B. Change control requires that good change management processes be implemented and enforced.

C. Personnel in a typical IT department can often be trained in new technologies to meet organizational requirements.

D. Although meeting user requirements is important, it is not directly related to the rate of technological change in the IT environment.

A2-50 An IS auditor finds that not all employees are aware of the enterprise's information security policy. The IS auditor should conclude that:

A. This lack of knowledge may lead to unintentional disclosure of sensitive information.
B. Information security is not critical to all functions.
C. Is audit should provide security training to the employees.
D. The audit finding will cause management to provide continuous training to staff.

A is the correct answer.

Justification:

A. All employees should be aware of the enterprise's information security policy to prevent unintentional disclosure of sensitive information. Training is a preventive control. Security awareness programs for employees can prevent unintentional disclosure of sensitive information to outsiders.
B. Information security is everybody's business, and all staff should be trained in how to handle information correctly.
C. Providing security awareness training is not an IS audit function.
D. Management may agree to or reject an audit finding. The IS auditor cannot be assured that management will act upon an audit finding unless they are aware of its impact; therefore, the auditor must report the risk associated with lack of security awareness.

A2-51 Which of the following is responsible for the approval of an information security policy?

A. IT department
B. Security committee
C. Security administrator
D. Board of directors

D is the correct answer.

Justification:

A. The IT department is responsible for the execution of the policy, having no authority in framing the policy.
B. The security committee also functions within the broad security policy framed by the board of directors.
C. The security administrator is responsible for implementing, monitoring and enforcing the security rules that management has established and authorized.
D. Normally, the approval of an information systems security policy is the responsibility of top management or the board of directors.

A2-52 While reviewing the IT governance processes of an organization, an IS auditor discovers the firm has recently implemented an IT balanced scorecard (BSC). The implementation is complete; however, the IS auditor notices that performance indicators are not objectively measurable. What is the **PRIMARY** risk presented by this situation?

A. Key performance indicators are not reported to management and management cannot determine the effectiveness of the BSC.
B. IT projects could suffer from cost overruns.
C. Misleading indications of IT performance may be presented to management.
D. IT service level agreements may not be accurate.

C is the correct answer.

Justification:
A. If the performance indicators are not objectively measurable, the most significant risk would be the presentation of misleading performance results to management. This could result in a false sense of assurance and, as a result, IT resources may be misallocated, or strategic decisions may be based on incorrect information. Whether or not the performance indicators are correctly defined, the results would be reported to management.
B. Although project management issues could arise from performance indicators that were not correctly defined, the presentation of misleading performance to management is a much more significant risk.
C. The IT balanced scorecard is designed to measure IT performance. To measure performance, a sufficient number of performance drivers (key performance indicators [KPIs]) must be defined and measured over time. Failure to have objective KPIs may result in arbitrary, subjective measures that may be misleading and lead to unsound decisions.
D. Although performance management issues related to service level agreements could arise from performance indicators that were not correctly defined, the presentation of misleading performance to management is a much more significant risk.

A2-53 Which of the following should be included in an organization's information security policy?

A. A list of key IT resources to be secured
B. The basis for access control authorization
C. Identity of sensitive security assets
D. Relevant software security features

B is the correct answer.

Justification:
A. A list of key IT resources to be secured is more detailed than that which should be included in a policy.
B. The security policy provides the broad framework of security as laid down and approved by senior management. It includes a definition of those authorized to grant access and the basis for granting the access.
C. The identity of sensitive security assets is more detailed than that which should be included in a policy.
D. A list of the relevant software security features is more detailed than that which should be included in a policy.

A2-54 Which of the following is the initial step in creating a firewall policy?

A. A cost-benefit analysis of methods for securing the applications
B. Identification of network applications to be externally accessed
C. Identification of vulnerabilities associated with network applications to be externally accessed
D. Creation of an application traffic matrix showing protection methods

B is the correct answer.

Justification:

A. Identifying methods to protect against identified vulnerabilities and their comparative cost-benefit analysis is the third step.
B. Identification of the applications required across the network should be the initial step. After identification, depending on the physical location of these applications in the network and the network model, the person in charge will be able to understand the need for, and possible methods of, controlling access to these applications.
C. Having identified the externally accessed applications, the second step is to identify vulnerabilities (weaknesses) associated with the network applications.
D. The fourth step is to analyze the application traffic and create a matrix showing how each type of traffic will be protected.

A2-55 Which of the following is an implementation risk within the process of decision support systems?

A. Management control
B. Semistructured dimensions
C. Inability to specify purpose and usage patterns
D. Changes in decision processes

C is the correct answer.

Justification:

A. Management control is not a type of risk, but a characteristic of a decision support system (DSS).
B. Semistructured dimensions is not a type of risk, but a characteristic of a DSS.
C. The inability to specify purpose and usage patterns is a risk that developers need to anticipate while implementing a DSS.
D. Changes in decision processes are not a type of risk, but a characteristic of a DSS.

A2-56 Which of the following is **MOST** critical for the successful implementation and maintenance of a security policy?

A. Assimilation of the framework and intent of a written security policy by all appropriate parties
B. Management support and approval for the implementation and maintenance of a security policy
C. Enforcement of security rules by providing punitive actions for any violation of security rules
D. Stringent implementation, monitoring and enforcing of rules by the security officer through access control software

A is the correct answer.

Justification:

A. Assimilation of the framework and intent of a written security policy by all levels of management and users of the system is critical to the successful implementation and maintenance of the security policy. If a policy is not assimilated into daily actions, it will not be effective.
B. Management support and commitment is, no doubt, important, but for successful implementation and maintenance of a security policy, educating the users on the importance of security is paramount.
C. Punitive actions are needed to enforce the policy but are not the key to successful implementation.
D. The stringent implementation, monitoring and enforcing of rules by the security officer through access control software, and provision for punitive actions for violation of security rules is important, but it is dependent on the support and education of management and users on the importance of security.

A2-57 A comprehensive and effective email policy should address the issues of email structure, policy enforcement, monitoring and:

A. recovery.
B. retention.
C. rebuilding.
D. reuse.

B is the correct answer.

Justification:

A. Email policy should address the business and legal requirements of email retention. Addressing the retention issue in the email policy would facilitate recovery.
B. Besides being a good practice, laws and regulations may require an organization to keep information that has an impact on the financial statements. The prevalence of lawsuits in which email communication is held in the same regard as the official form of classic paper makes the retention policy of corporate email a necessity. All email generated on an organization's hardware is the property of the organization, and an email policy should address the retention of messages, considering both known and unforeseen litigation. The policy should also address the destruction of emails after a specified time to protect the nature and confidentiality of the messages themselves.
C. Email policy should address the business and legal requirements of email retention. Addressing the retention issue in the email policy would facilitate rebuilding.
D. Email policy should address the business and legal requirements of email retention. Reuse of email is not a policy matter.

A2-58 An organization is considering making a major investment to upgrade technology. Which of the following choices is the **MOST** important to consider?

A. A cost analysis
B. The security risk of the current technology
C. Compatibility with existing systems
D. A risk analysis

D is the correct answer.

Justification:
A. The information system solution should be cost-effective, but this is not the most important aspect.
B. The security risk of the current technology is one of the components of the risk analysis, and alone is not the most important factor.
C. Compatibility with existing systems is one consideration; however, the new system may be a major upgrade that is not compatible with existing systems, so this is not the most important consideration.
D. Prior to implementing new technology, an organization should perform a risk assessment, which is then presented to business unit management for review and acceptance.

A2-59 Which of the following choices is the **PRIMARY** benefit of requiring a steering committee to oversee IT investment?

A. To conduct a feasibility study to demonstrate IT value
B. To ensure that investments are made according to business requirements
C. To ensure that proper security controls are enforced
D. To ensure that a standard development methodology is implemented

B is the correct answer.

Justification:
A. A steering committee may use a feasibility study in its reviews; however, it is not responsible for performing/conducting the study.
B. A steering committee consists of representatives from the business and IT and ensures that IT investment is based on business objectives rather than on IT priorities.
C. The steering committee is not responsible for enforcing security controls.
D. The steering committee is not responsible for implementing development methodologies.

A2-60 IS control objectives are useful to IS auditors because they provide the basis for understanding the:

A. Desired result or purpose of implementing specific control procedures
B. Best IS security control practices relevant to a specific entity.
C. Techniques for securing information
D. Security policy

A is the correct answer.

Justification:
A. An IS control objective is defined as the statement of the desired result or purpose to be achieved by implementing control procedures in a particular IS activity.
B. Control objectives provide the actual objectives for implementing controls and may or may not be based on good practices.
C. Techniques are the means of achieving an objective, but it is more important to know the reason and objective for the control than to understand the technique itself.
D. A security policy mandates the use of IS controls, but the controls are not used to understand policy.

A2-61 The initial step in establishing an information security program is the:

A. Development and implementation of an information security standards manual
B. Performance of a comprehensive security control review by the IS auditor
C. Adoption of a corporate information security policy statement
D. Purchase of security access control software

C is the correct answer.

Justification:

A. The security program is driven by policy and the standards are driven by the program. The initial step is to have a policy and ensure that the program is based on the policy.
B. Audit and monitoring of controls related to the program can only come after the program is set up.
C. A policy statement reflects the intent and support provided by executive management for proper security and establishes a starting point for developing the security program.
D. Access control software is an important security control but only after the policy and program are defined.

A2-62 Which of the following is the **MOST** important function to be performed by IT management when a service has been outsourced?

A. Ensuring that invoices are paid to the provider
B. Participating in systems design with the provider
C. Renegotiating the provider's fees
D. Monitoring the outsourcing provider's performance

D is the correct answer.

Justification:

A. Payment of invoices is a finance function, which would be completed per contractual requirements.
B. Participating in systems design is a by-product of monitoring the outsourcing provider's performance.
C. Renegotiating fees is usually a one-time activity and is not as important as monitoring the vendor's performance.
D. In an outsourcing environment, the enterprise is dependent on the performance of the service provider. Therefore, it is critical that the outsourcing provider's performance bis monitored to ensure that services are delivered to the enterprise as required.

A2-63 An organization purchased a third-party application and made significant modifications. While auditing the development process for this critical, customer-facing application, the IS auditor noted that the vendor has been in business for only one year. Which of the following helps to mitigate the risk relating to continued application support?

A. A viability study on the vendor
B. A software escrow agreement
C. Financial evaluation of the vendor
D. A contractual agreement for future enhancements

B is the correct answer.

Justification:

A. Although a viability study on the vendor may provide some assurance on the long-term availability of the vendor's services to the entity, in this case, it is more important that the company has the rights to the source code.

B. Considering that the vendor has been in the business for only one year, the biggest concern is financial stability or viability of the vendor and the risk of the vendor going out of business. The best way that this risk can be addressed is to have a software escrow agreement for the source code of the application, which provides the entity access to the source code if the vendor goes out of business.

C. Considering that the vendor has been in business for only one year, financial evaluation of the vendor would not be of much value and cannot provide assurance on the long-term availability of the vendor's services to the entity. In this case, it is more important that the company has rights to the source code.

D. A contractual agreement, while binding, is not enforceable or only has limited value in the event of bankruptcy.

A2-64 An IS auditor reviewing an outsourcing contract of IT facilities expects it to define the:

A. Hardware configuration.
B. Access control software.
C. Ownership of intellectual property.
D. Application development methodology.

C is the correct answer.

Justification:

A. The hardware configuration is generally irrelevant as long as the functionality, availability and security can be affected, which are specific contractual obligations.

B. The access control software is generally irrelevant as long as the functionality, availability and security can be affected, which are specific contractual obligations.

C. The contract must specify who owns the intellectual property (i.e., information being processed and application programs). Ownership of intellectual property is a significant cost and is a key aspect to be defined in an outsourcing contract.

D. The development methodology should be of no real concern in an outsourcing contract.

A2-65 While conducting an audit of a service provider, an IS auditor observes that the service provider has outsourced a part of the work to another provider. Because the work involves confidential information, the IS auditor's **PRIMARY** concern should be that the:

A. Requirement for securely protecting of information can be compromised.
B. Contract may be terminated because prior permission from the outsourcer was not obtained.
C. Other service provider to whom work has been outsourced is not subject to audit.
D. Outsourcer will approach the other service provider directly for further work.

A is the correct answer.

Justification:

A. Many countries have enacted regulations to protect the confidentiality of information maintained in their countries and/or exchanged with other countries. When a service provider outsources part of its services to another service provider, there is a potential risk that the confidentiality of the information will be compromised.
B. Terminating the contract for a violation of the terms of the contract could be a concern but is not related to ensuring the security of information.
C. The outsourcer not being subject to an audit could be a concern but is not related to ensuring the security of information.
D. There is no reason why an IS auditor should be concerned with the outsourcer approaching the other service providers directly for further work.

A2-66 A benefit of open system architecture is that it:

A. Facilitates interoperability within different systems.
B. Facilitates the integration of proprietary components.
C. Will be a basis for volume discounts from equipment vendors.
D. Allows for the achievement of more economies of scale for equipment.

A is the correct answer.

Justification:

A. Open systems are those for which suppliers provide components whose interfaces are defined by public standards, thus facilitating interoperability between systems made by different vendors.
B. Closed system components are built to proprietary standards so that other suppliers' systems cannot or will not interface with existing systems.
C. The ability to obtain volume discounts is achieved through the use of bulk purchasing or a primary vendor, not through open system architecture.
D. Open systems may be less expensive than proprietary systems depending on the supplier, but the primary benefit of open system architecture is its interoperability between vendors.

A2-67 The risk associated with electronic evidence gathering is **MOST** likely reduced by an email:

A. Destruction policy.
B. Security policy.
C. Archive policy.
D. Audit policy.

C is the correct answer.

Justification:
A. The email retention policy would include the destruction or deletion of emails. This must be compliant with legal requirements to retain emails.
B. A security policy is too high level and would not address the risk of inadequate retention of emails or the ability to provide access to emails when required.
C. With a policy of well-archived email records, access to or retrieval of specific email records to comply with legal requirements is possible.
D. An audit policy would not address the legal requirement to provide emails as electronic evidence.

A2-68 The output of the risk management process is an input for making:

A. Business plans.
B. Audit charters.
C. Security policy decisions.
D. Software design decisions.

C is the correct answer.

Justification:
A. Making a business plan is not the ultimate goal of the risk management process.
B. Risk management can help create the audit plan, but not the audit charter.
C. The risk management process is about making specific, security-related decisions, such as the level of acceptable risk.
D. Risk management will drive the design of security controls in software but influencing security policy is more important.

A2-69 An IS auditor was hired to review e-business security. The IS auditor's first task was to examine each existing e-business application, looking for vulnerabilities. What would be the next task?

A. Immediately report the risk to the chief information officer and chief executive officer.
B. Examine the e-business application in development.
C. Identify threats and the likelihood of occurrence.
D. Check the budget available for risk management.

C is the correct answer.

Justification:
A. The risk can only be determined after the threats, likelihood and vulnerabilities are all documented.
B. The first step is to identify the risk levels to existing applications and then to apply those to applications in development. Risk can only be identified after the threats and likelihood have also been determined.
C. To determine the risk associated with e-business, an IS auditor must identify the assets, look for vulnerabilities, and then identify the threats and the likelihood of occurrence.
D. The budget available for risk management is not relevant at this point because the risk has not yet been determined.

A2-70 An IS auditor reviewing the IT organization is **MOST** concerned if the IT steering committee:

A. Is responsible for project approval and prioritization.
B. Is responsible for developing the long-term it plan.
C. Reports the status of IT projects to the board of directors.
D. Is responsible for determining business goals.

D is the correct answer.

Justification:

A. The IT steering committee is responsible for project approval and prioritization.
B. The IT steering committee is responsible for oversight of the development of the long-term IT plan.
C. The IT steering committee advises the board of directors on the status of developments in IT.
D. Determining the business goals is the responsibility of senior management and not of the IT steering committee. IT should support business goals and be driven by the business—not the other way around.

A2-71 An IS auditor was asked to review a contract for a vendor being considered to provide data center services. Which is the **BEST** way to determine whether the terms of the contract are adhered to after the contract is signed?

A. Require the vendor to provide monthly status reports.
B. Have periodic meetings with the client IT manager.
C. Conduct periodic audit reviews of the vendor.
D. Require that performance parameters be stated within the contract.

C is the correct answer.

Justification:

A. Although providing monthly status reports may show that the vendor is meeting contract terms, without independent verification these data may not be reliable.
B. Having periodic meetings with the client IT manager will assist with understanding the current relationship with the vendor, but meetings may not include vendor audit reports, status reports and other information that a periodic audit review would take into consideration.
C. Conducting periodic reviews of the vendor ensures that the agreements within the contract are completed in a satisfactory manner. Without future audit reviews after the contract is signed, service level agreements and the client's requirements for security controls may become less of a focus for the vendor, and the results may slip. Periodic audit reviews allow the client to take a look at the vendor's current state to ensure that the vendor is one with which they want to continue to work.
D. Requiring that performance parameters be stated within the contract is important, but only if periodic reviews are performed to determine that performance parameters are met.

A2-72 Which of the following inputs adds the **MOST** value to the strategic IT initiative decision-making process?

A. The maturity of the project management process
B. The regulatory environment
C. Past audit findings
D. The IT project portfolio analysis

D is the correct answer.

Justification:

A. The maturity of the project management process is more important with respect to managing the day-to-day operations of IT versus performing strategic planning.
B. Regulatory requirements may drive investment in certain technologies and initiatives; however, having to meet regulatory requirements is not typically the main focus of the IT and business strategy.
C. Past audit findings may drive investment in certain technologies and initiatives; however, having to remediate past audit findings is not the main focus of the IT and business strategy.
D. Portfolio analysis provides the best input into the decision-making process relating to planning strategic IT initiatives. An analysis of the IT portfolio provides comparable information of planned initiatives, projects and ongoing IT services, which allows the IT strategy to be aligned with the business strategy.

A2-73 Which of the following does a lack of adequate security controls represent?

A. Threat
B. Asset
C. Impact
D. Vulnerability

D is the correct answer.

Justification:

A. A threat is anything (e.g., object, substance, human) that is capable of acting against an asset in a manner that can result in harm. A threat exists regardless of controls or a lack of controls.
B. An asset is something of either tangible or intangible value that is worth protecting, including people, information, infrastructure, finances and reputation. The asset value is not affected by a lack of controls.
C. Impact represents the outcome or result of a threat exploiting a vulnerability. A lack of controls would lead to a higher impact, but the lack of controls is defined as a vulnerability, not an impact.
D. The lack of adequate security controls represents a vulnerability, exposing sensitive information and data to the risk of malicious damage, attack or unauthorized access by hackers. This can result in a loss of sensitive information and lead to the loss of goodwill for the organization. A succinct definition of risk is provided by the Guidelines for the Management of IT Security published by the International Organization for Standardization (ISO), which defines risk as the "potential that a given threat will exploit the vulnerability of an asset or group of assets to cause loss or damage to the assets." The various elements of the definition are vulnerability, threat, asset and impact. Lack of adequate security functionality in this context is a vulnerability.

A2-74 Which of the following is the **PRIMARY** objective of an IT performance measurement process?

A. Minimize errors
B. Gather performance data
C. Establish performance baselines
D. Optimize performance

D is the correct answer.

Justification:
A. Minimizing errors is an aspect of performance but not the primary objective of performance management.
B. Gathering performance data is necessary to measure IT performance but is not the objective of the process.
C. The performance measurement process compares actual performance with baselines but is not the objective of the process.
D. An IT performance measurement process can be used to optimize performance, measure and manage products/services, assure accountability and make budget decisions.

A2-75 As an outcome of information security governance, strategic alignment provides:

A. Security requirements driven by enterprise requirements.
B. Baseline security following good practices.
C. Institutionalized and commoditized solutions.
D. An understanding of risk exposure.

A is the correct answer.

Justification:
A. Information security governance, when properly implemented, should provide four basic outcomes: strategic alignment, value delivery, risk management and performance measurement. Strategic alignment provides input for security requirements driven by enterprise requirements.
B. Strategic alignment ensures that security aligns with business goals. Providing a standard set of security practices (i.e., baseline security following good practices or institutionalized and commoditized solutions) is a part of value delivery.
C. Value delivery addresses the effectiveness and efficiency of solutions but is not a result of strategic alignment.
D. Risk management is a primary goal of IT governance, but strategic alignment is not focused on understanding risk exposure.

A2-76 Which of the following should be of **GREATEST** concern to an IS auditor when reviewing an information security policy? The policy:

A. Is driven by an IT department's objectives.
B. Is published, but users are not required to read the policy.
C. Does not include information security procedures.
D. Has not been updated in over a year.

A is the correct answer.

Justification:

A. Business objectives drive the information security policy, and the information security policy drives the selection of IT department objectives. A policy driven by IT objectives is at risk of not being aligned with business goals.
B. Policies should be written so that users can understand each policy, and employees should be able to easily access the policies. The fact that users have not read the policy is not the greatest concern because they still may be compliant with the policy.
C. Policies should not contain procedures. Procedures are established to assist with policy implementation and compliance.
D. Policies should be reviewed annually, but they might not necessarily be updated annually unless there are significant changes in the environment such as new laws, rules or regulations.

A2-77 Which of the following IT governance good practices improves strategic alignment?

A. Supplier and partner risk is managed.
B. A knowledge base on customers, products, markets and processes is in place.
C. A structure is provided that facilitates the creation and sharing of business information.
D. Top management mediates between the imperatives of business and technology.

D is the correct answer.

Justification:

A. Supplier and partner risk being managed is a risk management good practice but not a strategic function.
B. A knowledge base on customers, products, markets and processes being in place is an IT value delivery good practice but does not ensure strategic alignment.
C. An infrastructure being provided to facilitate the creation and sharing of business information is an IT value delivery and risk management good practice but is not as effective as top management involvement in business and technology alignment.
D. Top management mediating between the imperatives of business and technology is an IT strategic alignment good practice.

A2-78 Effective IT governance requires organizational structures and processes to ensure that:

A. Risk is maintained at a level acceptable for IT management.
B. The business strategy is derived from an IT strategy.
C. IT governance is separate and distinct from the overall governance.
D. The IT strategy extends the organization's strategies and objectives.

D is the correct answer.

Justification:

A. Risk acceptance levels are set by senior management, not by IT management.
B. The business strategy drives the IT strategy, not the other way around.
C. IT governance is not an isolated discipline; it must become an integral part of the overall enterprise governance.
D. Effective IT governance requires that board and executive management extend governance to IT and provide the leadership, organizational structures and processes that ensure that the organization's IT sustains and extends the organization's strategies and objectives, and that the strategy is aligned with business strategy.

A2-79 Assessing IT risk is **BEST** achieved by:

A. Evaluating threats and vulnerabilities associated with existing IT assets and IT projects
B. Using the organization's past actual loss experience to determine current exposure
C. Reviewing published loss statistics from comparable organizations
D. Reviewing IT control weaknesses identified in audit reports

A is the correct answer.

Justification:

A. To assess IT risk, threats and vulnerabilities need to be evaluated using qualitative or quantitative risk assessment approaches.
B. Basing an assessment on past losses will not adequately reflect new threats or inevitable changes to the firm's IT assets, projects, controls and strategic environment. There are also likely to be problems with the scope and quality of the loss data available to be assessed.
C. Comparable organizations will have differences in their IT assets, control environment and strategic circumstances. Therefore, their loss experience cannot be used to directly assess organizational IT risk.
D. Control weaknesses identified during audits will be relevant in assessing threat exposure and further analysis may be needed to assess threat probability. Depending on the scope of the audit coverage, it is possible that not all of the critical IT assets and projects will have recently been audited, and there may not be a sufficient assessment of strategic IT risk.

A2-80 When segregation of duties concerns exist between IT support staff and end users, what would be a suitable compensating control?

A. Restricting physical access to computing equipment
B. Reviewing transaction and application logs
C. Performing background checks prior to hiring IT staff
D. Locking user sessions after a specified period of inactivity

B is the correct answer.

Justification:

A. IT support staff usually require physical access to computing equipment to perform their job functions. It would not be reasonable to take this away.
B. Reviewing transaction and application logs directly addresses the threat posed by poor segregation of duties. The review is a means of detecting inappropriate behavior and also discourages abuse, because people who may otherwise be tempted to exploit the situation are aware of the likelihood of being caught.
C. Performing background checks is a useful control to ensure IT staff are trustworthy and competent but does not directly address the lack of an optimal segregation of duties.
D. Locking user sessions after a specified period of inactivity acts to prevent unauthorized users from gaining system access, but the issue of a lack of segregation of duties is more the misuse (deliberately or inadvertently) of access privileges that have officially been granted.

A2-81 A top-down approach to the development of operational policies helps to ensure:

A. That they are consistent across the organization.
B. That they are implemented as a part of risk assessment.
C. Compliance with all policies.
D. That they are reviewed periodically.

A is the correct answer.

Justification:

A. Deriving lower-level policies from corporate policies (a top-down approach) aids in ensuring consistency across the organization and consistency with other policies.
B. Policies should be influenced by risk assessment, but the primary reason for a top-down approach is to ensure that the policies are consistent across the organization.
C. A top-down approach, of itself, does not ensure compliance.
D. A top-down approach, of itself, does not ensure that policies are reviewed.

A2-82 An IS auditor reviewing an organization that uses cross-training practices should assess the risk of:

A. dependency on a single person.
B. inadequate succession planning.
C. one person knowing all parts of a system.
D. a disruption of operations.

C is the correct answer.

Justification:

A. Cross-training helps decrease dependence on a single person.
B. Cross-training assists in succession planning.
C. Cross-training is a process of training more than one individual to perform a specific job or procedure. However, before using this approach, it is prudent to assess the risk of any person knowing all parts of a system and the related potential exposures related to abuse of privilege.
D. Cross-training provides for the backup of personnel in the event of an absence and, thereby, provides for the continuity of operations.

A2-83 Which of the following should be of **PRIMARY** concern to an IS auditor reviewing the management of external IT service providers?

A. Minimizing costs for the services provided
B. Prohibiting the provider from subcontracting services
C. Evaluating the process for transferring knowledge to the IT department
D. Determining if the services were provided as contracted

D is the correct answer.

Justification:

A. Minimizing costs, if applicable and achievable (depending on the customer's need), is traditionally not part of an IS auditor's job. This would normally be done by a line management function within the IT department. Furthermore, during an audit, it is too late to minimize the costs for existing provider arrangements.
B. Subcontracting providers could be a concern but would not be the primary concern. This should be addressed in the contract.
C. Transferring knowledge to the internal IT department might be desirable under certain circumstances but should not be the primary concern of an IS auditor when auditing IT service providers and the management thereof.
D. From an IS auditor's perspective, the primary objective of auditing the management of service providers should be to determine if the services that were requested were provided in a way that is acceptable, seamless and in line with contractual agreements.

A2-84 Which of the following **MOST** likely indicates that a customer data warehouse should remain in-house rather than be outsourced to an offshore operation?

A. Time-zone differences can impede communications between IT teams.
B. Telecommunications cost can be much higher in the first year.
C. Privacy laws can prevent cross-border flow of information.
D. Software development may require more detailed specifications.

C is the correct answer.

Justification:

A. Time-zone differences are usually manageable issues for outsourcing solutions.
B. Higher telecommunications costs are a part of the cost-benefit analysis and not usually a reason to retain data in-house.
C. Privacy laws prohibiting the cross-border flow of personally identifiable information make it impossible to locate a data warehouse containing customer information in another country.
D. Software development typically requires more detailed specifications when dealing with offshore operations, but that is not a factor that should prohibit the outsourcing solution.

A2-85 When reviewing an organization's approved software product list, which of the following is the **MOST** important thing to verify?

A. The risk associated with the use of the products is periodically assessed.
B. The latest version of software is listed for each product.
C. Due to licensing issues, the list does not contain open source software.
D. After-hours support is offered.

A is the correct answer.

Justification:

A. Because the business conditions surrounding vendors may change, it is important for an organization to conduct periodic risk assessments of the vendor software list. This may be best incorporated into the IT risk management process.
B. The organization may not be using the latest version of a product.
C. The list may contain open source software depending on the business requirements and associated risk.
D. Support may be provided internally or externally, and technical support should be arranged depending on the criticality of the software.

A2-86 When reviewing the development of information security policies, the **PRIMARY** focus of an IS auditor should be on assuring that these policies:

A. are aligned with globally accepted industry good practices.
B. are approved by the board of directors and senior management.
C. strike a balance between business and security requirements.
D. provide direction for implementing security procedures.

C is the correct answer.

Justification:

A. An organization is not required to base its IT policies on industry good practices. Policies must be based on the culture and business requirements of the organization.
B. It is essential that policies be approved; however, that is not the primary focus during the development of the policies.
C. Because information security policies must be aligned with an organization's business and security objectives, this is the primary focus of the IS auditor when reviewing the development of information security policies.
D. Policies cannot provide direction if they are not aligned with business requirements.

A2-87 On which of the following factors should an IS auditor **PRIMARILY** focus when determining the appropriate level of protection for an information asset?

A. Results of a risk assessment
B. Relative value to the business
C. Results of a vulnerability assessment
D. Cost of security controls

A is the correct answer.

Justification:

A. The appropriate level of protection for an asset is determined based on the risk associated with the asset. The results of the risk assessment are, therefore, the primary information that the IS auditor should review.
B. The relative value of an asset to the business is one element considered in the risk assessment; this alone does not determine the level of protection required.
C. The results of a vulnerability assessment would be useful when creating the risk assessment; however, this would not be the primary focus.
D. The cost of security controls is not a primary factor to consider because the expenditures on these controls are determined by the value of the information assets being protected.

A2-88 From an IT governance perspective, what is the **PRIMARY** responsibility of the board of directors? To ensure that the IT strategy:

A. Is cost-effective.
B. Is future thinking and innovative.
C. Is aligned with the business strategy.
D. Has the appropriate priority level assigned.

C is the correct answer.

Justification:

A. The IT strategy should be cost-effective, but it must align with the business strategy for the strategy to be effective.
B. The IT strategy should be forward thinking and innovative, but it must align with the business strategy to be effective.
C. The board of directors is responsible for ensuring that the IT strategy is aligned with the business strategy.
D. The IT strategy should be appropriately prioritized; however, it must align with the business strategy first and then it will be prioritized.

A2-89 Which of the following is the **MOST** important element for the successful implementation of IT governance?

A. Implementing an IT scorecard
B. Identifying organizational strategies
C. Performing a risk assessment
D. Creating a formal security policy

B is the correct answer.

Justification:

A. A scorecard is an excellent tool to implement a program based on good governance, but the most important factor in implementing governance is alignment with organizational strategies.

B. The key objective of an IT governance program is to support the business; therefore, the identification of organizational strategies is necessary to ensure alignment between IT and corporate governance. Without identification of organizational strategies, the remaining choices—even if implemented—would be ineffective.

C. A risk assessment is important to ensure that the security program is based on areas of highest risk, but risk assessment must be based on organizational strategies.

D. A policy is a key part of security program implementation, but even the policy must be based on organizational strategies.

A2-90 To aid management in achieving IT and business alignment, an IS auditor should recommend the use of:

A. control self-assessments.
B. a business impact analysis.
C. an IT balanced scorecard.
D. business process reengineering.

C is the correct answer.

Justification:

A. Control self-assessments are used to improve monitoring of security controls but are not used to align IT with organizational objectives.

B. A business impact analysis is used to calculate the impact on the business in the event of an incident that affects business operations, but it is not used to align IT with organizational objectives.

C. An IT balanced scorecard provides the bridge between IT objectives and business objectives by supplementing the traditional financial evaluation with measures to evaluate customer satisfaction, internal processes and the ability to innovate.

D. Business process reengineering is an excellent tool to review and improve business processes but is not focused on aligning IT with organizational objectives.

A2-91 Which of the following is the **BEST** reference for an IS auditor to determine a vendor's ability to meet service level agreement requirements for a critical IT security service?

A. Compliance with the master contract
B. Agreed-on key performance indicators
C. Results of business continuity tests
D. Results of independent audit reports

B is the correct answer.

Justification:

A. The master contract typically includes terms, conditions and costs but does not typically include service levels.
B. Key performance indicators are metrics that allow for a means to measure performance. Service level agreements (SLAs) are statements related to expected service levels. For example, an Internet service provider (ISP) may guarantee that their service will be available 99.99 percent of the time.
C. If applicable to the service, results of business continuity tests are typically included as part of the due diligence review.
D. Independent audits report on the financial condition of an organization or the control environment. Reviewing audit reports is typically part of the due diligence review. Even audits must be performed against a set of standards or metrics to validate compliance.

A2-92 To address the risk of operations staff's failure to perform the daily backup, management requires that the systems administrator sign off on the daily backup. This is an example of risk:

A. Avoidance.
B. Transfer.
C. Mitigation.
D. Acceptance.

C is the correct answer.

Justification:

A. Risk avoidance is a strategy that provides for not implementing certain activities or processes that would incur risk.
B. Risk transfer is the strategy that provides for sharing risk with partners or purchasing insurance coverage.
C. Risk mitigation is the strategy that provides for the definition and implementation of controls to address the risk described. By requiring the system's administrator to sign off on the completion of the backups, this is an administrative control that can be validated for compliance.
D. Risk acceptance is a strategy that provides for formal acknowledgment of the existence of a risk but not taking any action to reduce the risk, and the monitoring of that risk.

A2-93 A poor choice of passwords and unencrypted data transmissions over unprotected communications lines are examples of:

A. vulnerabilities.
B. threats.
C. probabilities.
D. impacts.

A is the correct answer.

Justification:

A. Vulnerabilities represent weaknesses of information resources that may be exploited by a threat. Because these are weaknesses that can be addressed by the security specialist, they are examples of vulnerabilities.
B. Threats are circumstances or events with the potential to cause harm to information resources. Threats are usually outside the control of the security specialist.
C. Probabilities represent the likelihood of the occurrence of a threat.
D. Impacts represent the outcome or result of a threat exploiting a vulnerability.

A2-94 An IS auditor is assigned to review IT structures and activities recently outsourced to various providers. Which of the following should the IS auditor determine **FIRST**?

A. An audit clause is present in all contracts.
B. The service level agreement of each contract is substantiated by appropriate key performance indicators.
C. The contractual warranties of the providers support the business needs of the organization.
D. At contract termination, support is guaranteed by each outsourcer for new outsourcers.

C is the correct answer.

Justification:

A. All other choices are important, but the first step is to ensure that the contracts support the business—only then can an audit process be valuable.
B. All service level agreements should be measurable and reinforced through key performance indicators—but the first step is to ensure that the SLAs are aligned with business requirements.
C. The primary requirement is for the services provided by the outsource supplier to meet the needs of the business.
D. Having appropriate controls in place for contract termination are important, but first the IS auditor must be focused on the requirement of the supplier to meet business needs.

A2-95 To gain an understanding of the effectiveness of an organization's planning and management of investments in IT assets, an IS auditor should review the:

A. enterprise data model.
B. IT balanced scorecard.
C. IT organizational structure.
D. historical financial statements.

B is the correct answer.

Justification:

A. An enterprise data model is a document defining the data structure of an organization and how data interrelate. It is useful, but it does not provide information on investments in IT assets.

B. The IT balanced scorecard is a tool that provides the bridge between IT objectives and business objectives by supplementing the traditional financial evaluation with measures to evaluate customer satisfaction, internal processes and the ability to innovate. In this way, the auditor can measure the success of the IT investment and strategy.

C. The IT organizational structure provides an overview of the functional and reporting relationships in an IT entity but does not ensure effectiveness of IT investment.

D. Historical financial statements do not provide information about planning and lack sufficient detail to enable one to fully understand management's activities regarding IT assets. Past costs do not necessarily reflect value, and assets such as data are not represented on the books of accounts.

A2-96 Regarding the outsourcing of IT services, which of the following conditions should be of **GREATEST** concern to an IS auditor?

A. Core activities that provide a differentiated advantage to the organization have been outsourced.
B. Periodic renegotiation is not specified in the outsourcing contract.
C. The outsourcing contract fails to cover every action required by the business.
D. Similar activities are outsourced to more than one vendor.

A is the correct answer.

Justification:

A. An organization's core activities generally should not be outsourced because they are what the organization does best; an IS auditor observing that condition should be concerned.

B. An IS auditor should not be concerned about periodic renegotiation in the outsourcing contract because that is dependent on the term of the contract.

C. Outsourcing contracts cannot be expected to cover every action and detail expected of the parties involved but should cover business requirements.

D. Multisourcing is an acceptable way to reduce risk associated with a single point of failure.

A2-97 For a health care organization, which one of the following reasons **MOST** likely indicates that the patient benefit data warehouse should remain in-house rather than be outsourced to an offshore operation?

A. There are regulations regarding data privacy.
B. Member service representative training cost will be much higher.
C. It is harder to monitor remote databases.
D. Time zone differences could impede customer service.

A is the correct answer.

Justification:
A. Regulations prohibiting the cross-border flow of personally identifiable information may make it impossible to locate a data warehouse containing customer/member information in another country.
B. Training cost is common and manageable regardless of where the data warehouse resides.
C. Remote database monitoring is manageable regardless of where the data warehouse resides.
D. Time zone difference issues are manageable through contract provisions regardless of where the data warehouse resides.

A2-98 The **PRIMARY** control purpose of required vacations or job rotations is to:

A. allow cross-training for development.
B. help preserve employee morale.
C. detect improper or illegal employee acts.
D. provide a competitive employee benefit.

C is the correct answer.

Justification:
A. Although cross-training is a good practice for business continuity, it is not achieved through mandatory vacations.
B. It is a good practice to maintain good employee morale, but this is not a primary reason to have a required vacation policy.
C. The practice of having another individual perform a job function is a control used to detect possible irregularities or fraud.
D. Vacation time is a competitive benefit, but that is not a control.

A2-99 When reviewing the IT strategic planning process, an IS auditor should ensure that the plan:

A. incorporates state of the art technology.
B. addresses the required operational controls.
C. articulates the it mission and vision.
D. specifies project management practices.

C is the correct answer.

Justification:
A. The plan does not need to address state of the art technology; the decision to implement new technology is dependent on the approach to risk and management strategy.
B. The plan does not need to address operational controls because those are too granular for strategic planning.
C. The IT strategic plan must include a clear articulation of the IT mission and vision.
D. The plan should be implemented with proper project management, but the plan does not need to address project management practices.

A2-100 A small organization has only one database administrator (DBA) and one system administrator. The DBA has root access to the UNIX server, which hosts the database application. How should segregation of duties be enforced in this scenario?

A. Hire a second DBA and split the duties between the two individuals.
B. Remove the DBA's root access on all UNIX servers.
C. Ensure that all actions of the DBA are logged and that all logs are backed up to tape.
D. Ensure that database logs are forwarded to a UNIX server where the DBA does not have root access.

D is the correct answer.

Justification:
A. Hiring additional staff is a costly way to ensure segregation of duties.
B. The database administrator (DBA) needs root access to the database servers to install upgrades or patches.
C. The administrator can modify or erase logs prior to the tape backup event.
D. By creating logs that the DBA cannot erase or modify, segregation of duties is enforced.

A2-101 Which of the following user profiles should be of **MOST** concern to an IS auditor when performing an audit of an electronic funds transfer system?

A. Three users with the ability to capture and verify their own messages
B. Five users with the ability to capture and send their own messages
C. Five users with the ability to verify other users and to send their own messages
D. Three users with the ability to capture and verify the messages of other users and to send their own messages

A is the correct answer.

Justification:
A. The ability of one individual to capture and verify their own messages represents an inadequate segregation because messages can be taken as correct and as if they had already been verified. The verification of messages should not be allowed by the person who sent the message.
B. Users may have the ability to send messages but should not be able to verify their own messages.
C. This is an example of separation of duties. A person can send their own message but only verify the messages of other users.
D. The ability to capture and verify the messages of others but only send their own messages is acceptable.

A2-102 Which of the following does an IS auditor **FIRST** reference when performing an IS audit?

A. Implemented procedures
B. Approved policies
C. Internal standards
D. Documented practices

B is the correct answer.

Justification:
A. Procedures are implemented in accordance with policy.
B. Policies are high-level documents that represent the corporate philosophy of an organization. Internal standards, procedures and practices are subordinate to policy.
C. Standards are subordinate to policy.
D. Practices are subordinate to policy.

A2-103 An enterprise selected a vendor to develop and implement a new software system. To ensure that the enterprise's investment in software is protected, which of the following security clauses is **MOST** important to include in the master services agreement?

A. Limitation of liability
B. Service level requirements
C. Software escrow
D. Version control

C is the correct answer.

Justification:

A. A limitation of liability clause protects the financial exposure of the organization but not its software investment.
B. Service level requirements specify financial penalties for not meeting standards, but these do not address issues of vendor insolvency.
C. Software escrow clauses in a contract ensure that the software source code will still be available to the organization in the event of a vendor issue, such as insolvency and copyright issues.
D. Version control is related to the software development life cycle and not the software investment.

A2-104 When implementing an IT governance framework in an organization the **MOST** important objective is:

A. IT alignment with the business
B. Accountability
C. Value realization with IT
D. Enhancing the return on it investments

A is the correct answer.

Justification:

A. The goals of IT governance are to improve IT performance, deliver optimum business value and ensure regulatory compliance. The key practice in support of these goals is the strategic alignment of IT with the business. To achieve alignment, all other choices need to be tied to business practices and strategies.
B. Accountability is important, but the most important objective of IT governance is to ensure that IT investment and oversight is aligned with business requirements.
C. IT must demonstrate value to the organization, but this value is dependent on the ability of IT to align with, and support, business requirements.
D. Enhancing return is a requirement of the IT governance framework, but this requirement is only demonstrated through aligning IT with business requirements.

A2-105 An IS auditor is reviewing an IT security risk management program. Measures of security risk should:

A. address all of the network risk.
B. be tracked over time against the IT strategic plan.
C. consider the entire IT environment.
D. result in the identification of vulnerability tolerances.

C is the correct answer.

Justification:
A. Measures of security risk should not be limited to network risk, but rather focus on those areas with the highest criticality so as to achieve maximum risk reduction at the lowest possible cost.
B. IT strategic plans are not granular enough to provide appropriate measures. Objective metrics must be tracked over time against measurable goals; thus, the management of risk is enhanced by comparing today's results against results from last week, last month and last quarter. Risk measures will profile assets on a network to objectively measure vulnerability risk.
C. When assessing IT security risk, it is important to consider the entire IT environment.
D. Measures of security risk do not identify tolerances.

A2-106 The ultimate purpose of IT governance is to:

A. encourage optimal use of IT.
B. reduce IT costs.
C. decentralize IT resources across the organization.
D. centralize control of IT.

A is the correct answer.

Justification:
A. IT governance is intended to specify the combination of decision rights and accountability that is best for the enterprise. It is different for every enterprise.
B. Reducing IT costs may not be the best IT governance outcome for an enterprise.
C. Decentralizing IT resources across the organization is not always desired, although it may be desired in a decentralized environment.
D. Centralizing control of IT is not always desired. An example of where it might be desired is an enterprise wanting a single point of customer contact.

A2-107 Which of the following is the **MOST** important for an IS auditor to consider when reviewing a service level agreement with an external IT service provider?

A. Payment terms
B. Uptime guarantee
C. Indemnification clause
D. Default resolution

B is the correct answer.

Justification:
A. Payment terms are typically included in the master agreement rather than in the service level agreement (SLA).
B. The most important element of an SLA is the measurable terms of performance, such as uptime agreements.
C. The indemnification clause is typically included in the master agreement rather than in the SLA.
D. The default resolution would only apply in case of a default of the SLA; therefore, it is more important to review the performance conditions of the SLA.

A2-108 The **PRIMARY** objective of implementing corporate governance is to:

A. provide strategic direction.
B. control business operations.
C. align IT with business.
D. implement good practices.

A is the correct answer.

Justification:

A. Corporate governance is a set of management practices to provide strategic direction to the organization as a whole, thereby ensuring that goals are achievable, risk is properly addressed and organizational resources are properly used. Hence, the primary objective of corporate governance is to provide strategic direction.

B. Business operations are directed and controlled based on the strategic direction.

C. Corporate governance applies strategic planning, monitoring and accountability to the entire organization, not just to IT.

D. Governance is applied through the use of good practices, but this is not the objective of corporate governance.

A2-109 Which of the following should be considered **FIRST** when implementing a risk management program?

A. An understanding of the organization's threat, vulnerability and risk profile
B. An understanding of the risk exposures and the potential consequences of compromise
C. A determination of risk management priorities that are based on potential consequences
D. A risk mitigation strategy sufficient to keep risk consequences at an acceptable level

A is the correct answer.

Justification:

A. Implementing risk management, as one of the outcomes of effective information security governance, requires a collective understanding of the organization's threat, vulnerability and risk profile as a first step.

B. An understanding of risk exposure and potential consequences of compromise can be determined only after there is an understanding the organization's threat, vulnerability and risk profile.

C. Risk management priorities that are based on potential consequences can only be developed after the organization's threat, vulnerability and risk profile is determined.

D. Risk mitigation priorities are based on the risk profile, risk acceptance levels and potential mitigating controls. These elements provide a basis for the formulation of strategies for risk mitigation sufficient to keep the consequences from risk at an acceptable level.

A2-110 In the context of effective information security governance, the **PRIMARY** objective of value delivery is to:

A. Optimize security investments in support of business objectives.
B. Implement a standard set of security practices.
C. Institute a standards-based solution.
D. Implement a continuous improvement culture.

A is the correct answer.

Justification:
A. In the context of effective information security governance, value delivery is implemented to ensure optimization of security investments in support of business objectives.
B. The tools and techniques for implementing value delivery include implementation of a standard set of security practices; however, implementation of standards is a means to achieve the objective of supporting value delivery, not the objective itself.
C. Value delivery may be supported through the use of standards-based solutions, but the use of standards-based solutions is not the goal of value delivery.
D. Continuous improvement culture in relation to a security program is a process, not an objective.

A2-111 As a driver of IT governance, transparency of IT's cost, value and risk is primarily achieved through:

A. performance measurement.
B. strategic alignment.
C. value delivery.
D. resource management.

A is the correct answer.

Justification:
A. Performance measurement includes setting and monitoring measurable objectives of that which the IT processes need to deliver (process outcome), and how they deliver it (process capability and performance). Transparency is primarily achieved through performance measurement, because it provides information to the stakeholders on how well the enterprise is performing when compared to objectives.
B. Strategic alignment primarily focuses on ensuring linkage of business and IT plans, not on transparency.
C. Value delivery is about executing the value proposition throughout the delivery cycle. Value delivery ensures that IT investments deliver on promised values but does not ensure transparency of investment.
D. Resource management is about the optimal investment in and proper management of critical IT resources but does not ensure transparency of IT investments.

A2-112 Which of the following should be the **MOST** important consideration when deciding on areas of priority for IT governance implementations?

A. Process maturity
B. Performance indicators
C. Business risk
D. Assurance reports

C is the correct answer.

Justification:

A. The level of process maturity will evolve as the implementation of the IT governance program occurs and may feed into the decision-making process. Those areas that represent real risk to the business should be given priority.
B. The level of process performance will demonstrate the effectiveness of the program but will not be the means to establish priorities for governance. Those areas that represent real risk to the business should be given priority.
C. Priority should be given to those areas that represent a known risk to the enterprise operations.
D. Audit reports will provide assurance of the effectiveness of the implementation of governance but will not determine the priorities for program. Those areas that represent real risk to the business should be given priority.

A2-113 Responsibility for the governance of IT should rest with the:

A. IT strategy committee.
B. Chief information officer.
C. Audit committee.
D. Board of directors.

D is the correct answer.

Justification:

A. The IT strategy committee plays a significant role in the successful implementation of IT governance within an organization, but the ultimate responsibility resides with the board of directors.
B. The chief information officer plays a significant role in the successful implementation of IT governance within an organization, but the ultimate responsibility resides with the board of directors.
C. The audit committee plays a significant role in monitoring and overseeing the successful implementation of IT governance within an organization, but the ultimate responsibility resides with the board of directors.
D. Governance is the set of responsibilities and practices exercised by the board and executive management with the goal of providing strategic direction, ensuring that objectives are achieved, ascertaining that risk is managed appropriately and verifying that the enterprise's resources are used responsibly.

A2-114 Which of the following is normally a responsibility of the chief information security officer?

A. Periodically reviewing and evaluating the security policy
B. Executing user application and software testing and evaluation
C. Granting and revoking user access to IT resources
D. Approving access to data and applications

A is the correct answer.

Justification:
A. The role of the chief information security officer is to ensure that the corporate security policy and controls are adequate to prevent unauthorized access to the enterprise assets, including data, programs and equipment.
B. User application and other software testing and evaluation normally are the responsibility of the staff assigned to development and maintenance.
C. Granting and revoking access to IT resources is usually a function of system, network or database administrators.
D. Approval of access to data and applications is the duty of the data or application owner.

A2-115 When developing a formal enterprise security program, the **MOST** critical success factor is the:

A. Establishment of a review board.
B. Creation of a security unit.
C. Effective support of an executive sponsor.
D. Selection of a security process owner.

C is the correct answer.

Justification:
A. Establishment of a review board is not effective without visible sponsorship of top management.
B. The creation of a security unit is not effective without visible sponsorship of top management.
C. The executive sponsor is in charge of supporting the organization's strategic security program and aids in directing the organization's overall security management activities. Therefore, support by the executive level of management is the most critical success factor.
D. The selection of a security process owner is not effective without visible sponsorship of top management.

A2-116 When reviewing an organization's strategic IT plan, an IS auditor should expect to find:

A. An assessment of the fit of the organization's application portfolio with business objectives.
B. Actions to reduce hardware procurement cost.
C. A listing of approved suppliers of IT contract resources.
D. A description of the technical architecture for the organization's network perimeter security.

A is the correct answer.

Justification:

A. An assessment of how well an organization's application portfolio supports the organization's business objectives is a key component of the overall IT strategic planning process. This assessment drives the demand side of IT planning and should convert into a set of strategic IT intentions. Further assessment can then be made of how well the overall IT organization, encompassing applications, infrastructure, services, management processes, etc. can support the business objectives. The purpose of an IT strategic plan is to set out how IT will be used to achieve or support an organization's business objectives.
B. Operational efficiency initiatives, including cost reduction of purchasing and maintenance activities of systems, belong to tactical planning, not strategic planning.
C. A list of approved suppliers of IT contract resources is a tactical rather than a strategic concern.
D. An IT strategic plan would not normally include detail of a specific technical architecture.

A2-117 When developing a security architecture, which of the following steps should be executed **FIRST**?

A. Developing security procedures
B. Defining a security policy
C. Specifying an access control methodology
D. Defining roles and responsibilities

B is the correct answer.

Justification:

A. Policy is used to provide direction for procedures, standards and baselines. Therefore, developing security procedures should be executed only after defining a security policy.
B. Defining a security policy for information and related technology is the first step toward building a security architecture. A security policy communicates a coherent security standard to users, management and technical staff. Security policies often set the stage in terms of the tools and procedures that are needed for an organization.
C. Specifying an access control methodology is an implementation concern and should be executed only after defining a security policy.
D. Defining roles and responsibilities should be executed only after defining a security policy.

A2-118 Which of the following should an IS auditor recommend to **BEST** enforce alignment of an IT project portfolio with strategic organizational priorities?

A. Define a balanced scorecard for measuring performance.
B. Consider user satisfaction in the key performance indicators.
C. Select projects according to business benefits and risk.
D. Modify the yearly process of defining the project portfolio.

C is the correct answer.

Justification:
A. Measures such as a balanced scorecard are helpful, but do not guarantee that the projects are aligned with business strategy.
B. Key performance indicators are helpful to monitor and measure IT performance, but they do not guarantee that the projects are aligned with business strategy.
C. Prioritization of projects on the basis of their expected benefit(s) to business, and the related risk, is the best measure for achieving alignment of the project portfolio to an organization's strategic priorities.
D. Modifying the yearly process of the project portfolio definition might improve the situation, but only if the portfolio definition process is closely tied to organizational strategies.

A2-119 The **PRIMARY** benefit of implementing a security program as part of a security governance framework is the:

A. Alignment of the IT activities with IS audit recommendations
B. Enforcement of the management of security risk
C. Implementation of the chief information security officer's recommendations
D. Reduction of the cost for IT security

B is the correct answer.

Justification:
A. Recommendations, visions and objectives of the IS auditor are usually addressed within a security program, but they would not be the major benefit.
B. The major benefit of implementing a security program is management's assessment of risk and its mitigation to an appropriate level, and monitoring of the residual risk.
C. Recommendations, visions and objectives of the chief information security officer are usually included within a security program, but they would not be the major benefit.
D. The cost of IT security may or may not be reduced.

A2-120 An organization has a well-established risk management process. Which of the following risk management practices would **MOST** likely expose the organization to the greatest amount of compliance risk?

A. Risk reduction
B. Risk transfer
C. Risk avoidance
D. Risk mitigation

B is the correct answer.

Justification:

A. Risk reduction is a term synonymous with risk mitigation. Risk reduction lowers risk to a level commensurate with the organization's risk appetite. Risk reduction treats the risk, while risk transfer does not always address compliance risk.
B. Risk transfer typically addresses financial risk. For instance, an insurance policy is commonly used to transfer financial risk, while compliance risk continues to exist.
C. Risk avoidance does not expose the organization to compliance risk because the business practice that caused the inherent risk to exist is no longer being pursued.
D. Mitigating risk will still expose the organization to a certain amount of risk. Risk mitigation lowers risk to a level commensurate with the organization's risk appetite. However, risk transference is the best answer because risk mitigation treats the risk, while risk transfer does not necessarily address compliance risk.

A2-121 An employee who has access to highly confidential information resigned. Upon departure, which of the following should be done **FIRST**?

A. Conduct an exit interview with the employee.
B. Ensure succession plans are in place.
C. Revoke the employee's access to all systems.
D. Review the employee's job history.

C is the correct answer.

Justification:

A. It is important to have an exit interview with any employee; however, this would not be the first step to take upon the employee's departure to protect the confidentiality of information.
B. Succession plans are important to prevent disruption of operations. This would address availability and not confidentiality of information.
C. If an employee has dealt with highly classified information, the first step is to revoke their access to all systems, to prevent exfiltration of data and restrict access to the information.
D. Keeping a record of the job history is important; however, its effectiveness may be limited.

A2-122 An organization has outsourced its help desk activities. An IS auditor's **GREATEST** concern when reviewing the contract and associated service level agreement between the organization and vendor should be the provisions for:

A. documentation of staff background checks.
B. independent audit reports or full audit access.
C. reporting the year-to-year incremental cost reductions.
D. reporting staff turnover, development or training.

B is the correct answer.

Justification:
A. Although it is necessary to document the fact that background checks are performed, this is only one of the provisions that should be in place for audits.
B. When the functions of an IT department are outsourced, an IS auditor should ensure that a provision is made for independent audit reports that cover all essential areas, or that the outsourcer has full audit access.
C. Financial measures such as year-to-year incremental cost reductions are desirable to have in a service level agreement (SLA); however, cost reductions are not as important as the availability of independent audit reports or full audit access.
D. An SLA might include human relationship measures such as resource planning, staff turnover, development or training, but this is not as important as the requirements for independent reports or full audit access by the outsourcing organization.

A2-123 An IS auditor identifies that reports on product profitability produced by an organization's finance and marketing departments give different results. Further investigation reveals that the product definition being used by the two departments is different. What should the IS auditor recommend?

A. User acceptance testing occurs for all reports before release into production
B. Organizational data governance practices are put in place
C. Standard software tools are used for report development
D. Management signs off on requirements for new reports

B is the correct answer.

Justification:
A. Recommending that user acceptance testing occur for all reports before release into production does not address the root cause of the problem described.
B. This choice directly addresses the problem. An organization-wide approach is needed to achieve effective management of data assets and reporting standards. This includes enforcing standard definitions of data elements, which is part of a data governance initiative.
C. Recommending standard software tools be used for report development does not address the root cause of the problem described.
D. Recommending that management sign off on requirements for new reports does not address the root cause of the problem described.

A2-124 Which of the following **BEST** supports the prioritization of new IT projects?

A. Internal control self-assessment
B. Information systems audit
C. Investment portfolio analysis
D. Business risk assessment

C is the correct answer.

Justification:
A. Internal control self-assessment (CSA) may highlight noncompliance to the current policy but may not necessarily be the best source for driving the prioritization of IT projects.
B. Like internal CSA, IS audits are mostly a detective control and may provide only part of the picture for the prioritization of IT projects.
C. It is most desirable to conduct an investment portfolio analysis, which will present not only a clear focus on investment strategy but also provide the rationale for terminating nonperforming IT projects.
D. Business risk analysis is part of the investment portfolio analysis but, by itself, is not the best method for prioritizing new IT projects.

A2-125 Which of the following is the **MOST** important IS audit consideration when an organization outsources a customer credit review system to a third-party service provider? The provider:

A. Claims to meet or exceed industry security standards.
B. Agrees to be subject to external security reviews.
C. Has a good market reputation for service and experience.
D. Complies with security policies of the organization.

B is the correct answer.

Justification:
A. Compliance with security standards is important, but there is no way to verify or prove that is the case without an independent review.
B. It is critical that an independent security review of an outsourcing vendor be obtained, because customer credit information will be kept with the vendor.
C. Though long experience in business and good reputation is an important factor to assess service quality, the business cannot outsource to a provider whose security control is weak.
D. Compliance with organizational security policies is important, but there is no way to verify or prove that that is the case without an independent review.

A2-126 After the merger of two organizations, multiple self-developed legacy applications from both organizations are to be replaced by a new common platform. Which of the following is the **GREATEST** risk?

A. Project management and progress reporting is combined in a project management office that is driven by external consultants.
B. The replacement effort consists of several independent projects without integrating the resource allocation in a portfolio management approach.
C. The resources of each of the organizations are inefficiently allocated while they are being familiarized with the other organization's legacy systems.
D. The new platform will force the business areas of both organizations to change their work processes, which will result in extensive training needs.

B is the correct answer.

Justification:
A. In postmerger integration programs, it is common to form project management offices (often staffed with external experts) to ensure standardized and comparable information levels in the planning and reporting structures, and to centralize dependencies of project deliverables or resources.
B. The efforts should be consolidated to ensure alignment with the overall strategy of the postmerger organization. If resource allocation is not centralized, the separate projects are at risk of overestimating the availability of key knowledge resources for the in-house-developed legacy applications.
C. The development of new integrated systems can require some knowledge of the legacy systems to gain an understanding of each business process.
D. In most cases, mergers result in application changes and thus in training needs as organizations and processes change to leverage the intended synergy effects of the merger.

A2-127 During an audit, an IS auditor notices that the IT department of a medium-sized organization has no separate risk management function, and the organization's operational risk documentation only contains a few broadly described types of IT risk. What is the **MOST** appropriate recommendation in this situation?

A. Create an IT risk management department and establish an IT risk framework with the aid of external risk management experts.
B. Use common industry standard aids to divide the existing risk documentation into several individual types of risk which will be easier to handle.
C. No recommendation is necessary because the current approach is appropriate for a medium-sized organization.
D. Establish regular IT risk management meetings to identify and assess risk and create a mitigation plan as input to the organization's risk management.

D is the correct answer.

Justification:
A. A medium-sized organization would normally not have a separate IT risk management department. Moreover, the risk is usually manageable enough so that external help would not be needed.
B. While common risk may be covered by industry standards, they cannot address the specific situation of an organization. Individual types of risk will not be discovered without a detailed assessment from within the organization. Splitting the one risk position into several is not sufficient to manage IT risk.
C. The auditor should recommend a formal IT risk management effort because the failure to demonstrate responsible IT risk management may be a liability for the organization.
D. Establishing regular IT risk management meetings is the best way to identify and assess IT-related risk in a medium-sized organization, to address responsibilities to the respective management and to keep the risk register and mitigation plans up to date.

A2-128 Overall quantitative business risk for a particular threat can be expressed as:

A. A product of the likelihood and magnitude of the impact if a threat successfully exploits a vulnerability.
B. The magnitude of the impact if a threat source successfully exploits the vulnerability.
C. The likelihood of a given threat source exploiting a given vulnerability.
D. The collective judgment of the risk assessment team.

A is the correct answer.

Justification:

A. Overall business risk takes into consideration the likelihood and magnitude of the impact when a threat exploits a vulnerability, and provides the best measure of the risk to an asset.
B. The calculation of risk must consider impact and likelihood of a threat (not a threat source) exploiting a vulnerability.
C. Considering only the likelihood of an exploit and not the impact or damage caused is not sufficient to determine the overall risk.
D. The collective judgment of the risk assessment team is a part of qualitative risk assessment but must be combined with calculations of the impact on the business to determine overall risk.

A2-129 While conducting an IS audit of a service provider for a government program involving confidential information, an IS auditor noted that the service provider delegated a part of the IS work to another subcontractor. Which of the following provides the **MOST** assurance that the requirements for protecting confidentiality of information are met?

A. Monthly committee meetings include the subcontractor's IS manager.
B. Management reviews weekly reports from the subcontractor.
C. Permission is obtained from the government agent regarding the contract.
D. Periodic independent audit of the work delegated to the subcontractor.

D is the correct answer.

Justification:

A. Regular committee meetings are a good monitoring tool for delegated operations; however, independent reviews provide better assurance.
B. Management should not only rely on self-reported information from the subcontractor.
C. Obtaining permission from the government agent is not related to ensuring the confidentiality of information.
D. Periodic independent audits provide reasonable assurance that the requirements for protecting confidentiality of information are not compromised.

A2-130 During an audit, which of the following situations are **MOST** concerning for an organization that significantly outsources IS processing to a private network?

A. The contract does not contain a right-to-audit clause for the third party.
B. The contract was not reviewed by an information security subject matter expert prior to signing.
C. The IS outsourcing guidelines are not approved by the board of directors.
D. There is a lack of well-defined IS performance evaluation procedures.

A is the correct answer.

Justification:

A. Lack of a right-to-audit clause in the contract impacts the IS auditor's ability to perform the IS audit. Hence, the IS auditor is most concerned with such a situation. In the case of outsourcing to a private network, the organization should ensure that the third party has a minimum set of IT security controls in place and that they are operating effectively.
B. Having an information security subject matter expert review a contract is a good practice, but it is not a requirement in all industries.
C. Approval of the IS outsourcing guidelines by the board is a good practice of governance, and lack of approval is an audit issue. However, it does not impact the IS auditor's ability to perform IS audit.
D. Lack of well-defined procedures does not enable objective evaluation of IS performance and is an audit issue. However, it does not result into major risk or repercussions and also does not impact the IS auditor's ability to perform an IS audit.

A2-131 The **MOST** important element for the effective design of an information security policy is the:

A. threat landscape.
B. prior security incidents.
C. emerging technologies.
D. enterprise risk appetite.

D is the correct answer.

Justification:

A. The threat landscape is dynamic. It should be considered when developing policy, but it is not the primary factor as policy is not meant to change as often as the threat landscape.
B. Prior security incidents may provide insight into the risk appetite statement; however, they are more likely to affect security standards and procedures.
C. Emerging technologies are continually evolving. They should be considered when developing policy, but they are not the primary factor as policy is not meant to change as often as technology.
D. The risk appetite is the amount of risk on a broad level that an entity is willing to accept in pursuit of its mission to meet its strategic objectives. The purpose of the information security policy is to manage information risk to an acceptable level, so that the policy is principally aligned with the risk appetite.

A2-132 As result of profitability pressure, senior management of an enterprise decided to keep investments in information security at an inadequate level, which of the following is the **BEST** recommendation of an IS auditor?

A. Use cloud providers for low-risk operations.
B. Revise compliance enforcement processes.
C. Request that senior management accept the risk.
D. Postpone low-priority security procedures.

C is the correct answer.

Justification:
A. The use of cloud providers may or may not provide cost savings or lower risk.
B. Compliance enforcement processes that identify high levels of residual risk are working as intended and should not be revised.
C. Senior management determines resource allocations. Having established that the level of security is inadequate, it is imperative that senior management accept the risk resulting from their decisions.
D. The IS auditor should not recommend postponing any procedures. This is a management decision, and management should first accept the risk.

A2-133 Which of the following insurance types provide for a loss arising from fraudulent acts by employees?

A. Business interruption
B. Fidelity coverage
C. Errors and omissions
D. Extra expense

B is the correct answer.

Justification:
A. Business interruption insurance covers the loss of profit due to the disruption in the operations of an organization.
B. Fidelity insurance covers the loss arising from dishonest or fraudulent acts by employees.
C. Errors and omissions insurance provides legal liability protection in the event that the professional practitioner commits an act that results in financial loss to a client.
D. Extra expense insurance is designed to cover the extra costs of continuing operations following a disaster/disruption within an organization.

A2-134 Errors in audit procedures **PRIMARILY** impact which of the following risks?

A. Detection risk
B. Inherent risk
C. Control risk
D. Business risk

A is the correct answer.

Justification:
A. Detection risk is the probability that the audit procedures may fail to detect existence of a material error or fraud.
B. Inherent risk refers to the risk involved in the nature of business or transaction and is not affected by human error.
C. Control risk is the risk that a material error exists that would not be prevented or detected on a timely basis by the system of internal controls.
D. Business risk is not a component of audit risk.

A2-135 Which of the following is **MOST** important to consider when reviewing the classification levels of information assets?

A. Potential loss
B. Financial cost
C. Potential threats
D. Cost of insurance

A is the correct answer.

Justification:
A. The best basis for asset classification is an understanding of the total losses a business may incur if the asset is compromised. Typically, estimating these losses requires a review of criticality and sensitivity beyond financial cost, such as operational and strategic.
B. The value of an asset can be greater than its monetary cost, such as impact to reputation and brand.
C. The classification of an asset does not change based on potential threats.
D. Insurance would be obtained based on asset classification.

A2-136 Which of the following is of **MOST** interest to an IS auditor reviewing an organization's risk strategy?

A. All risk is mitigated effectively.
B. Residual risk is zero after control implementation.
C. All likely risk is identified and ranked.
D. The organization uses an established risk framework.

C is the correct answer.

Justification:
A. Risk mitigation can only occur after all risk is identified and ranked.
B. It is highly unlikely residual risk would be zero.
C. Risk that is likely to impact the organization should be identified and documented as part of the risk strategy. Without knowing the risk, there is no risk strategy.
D. It is not as important to use an established risk framework as it is to identify and rank all likely risk so that it can be addressed.

A2-137 An enterprise is looking to obtain cloud hosting services from a cloud vendor with a high level of maturity. Which of the following is **MOST** important for the auditor to ensure continued alignment with the enterprise's security requirements?

A. The vendor provides the latest third-party audit report for verification.
B. The vendor provides the latest internal audit report for verification.
C. The vendor agrees to implement controls in alignment with the enterprise.
D. The vendor agrees to provide annual external audit reports in the contract.

D is the correct answer.

Justification:
A. Although the vendor is providing the most recent third-party audit report for review, there is no agreement contractually that would require the vendor to continue to provide annual reports for verification and review.
B. Although the vendor is providing the most recent internal audit report for review, there is no agreement contractually that would require the vendor to continue to provide annual reports for verification and review.
C. Without a clause in the contract, an agreement to implement controls does not provide assurance that controls will continue to be implemented in alignment with the enterprise.
D. The only way to ensure that any potential risk is mitigated today and in the future is to include a clause within the contract that the vendor will provide future external audit reports. Without the audit clause the vendor can choose to forego future audits.

A2-138 An IS auditor is evaluating the IT governance framework of an organization. Which of the following is the **GREATEST** concern?

A. Senior management has limited involvement.
B. Return on investment is not measured.
C. Chargeback of IT cost is not consistent.
D. Risk appetite is not quantified.

A is the correct answer.

Justification:

A. To ensure that the IT governance framework is effectively in place, senior management must be involved and aware of roles and responsibilities. Therefore, it is most essential to ensure the involvement of senior management when evaluating the soundness of IT governance.
B. Ensuring revenue management is a part of the objectives in the IT governance framework. Therefore, it is not effective in verifying the soundness of IT governance.
C. Introduction of a cost allocation system is part of the objectives in an IT governance framework. Therefore, it is not effective in verifying the soundness of IT governance.
D. Estimation of risk appetite is important; however, at the same time, management should ensure that controls are in place. Therefore, checking only on risk appetite does not verify soundness of IT governance.

A2-139 After an organization completed a threat and vulnerability analysis as part of a risk assessment, the final report suggested that an intrusion prevention system (IPS) should be installed at the main Internet gateways and that all business units should be separated via a proxy firewall. Which of the following is the **BEST** method to determine whether the controls should be implemented?

A. A cost-benefit analysis
B. An annual loss expectancy calculation
C. A comparison of the cost of the IPS and firewall and the cost of the business systems
D. A business impact analysis

A is the correct answer.

Justification:

A. In a cost-benefit analysis, the total expected purchase and operational/support costs, and a qualitative value for all actions are weighted against the total expected benefits to choose the best technical, most profitable, least expensive or acceptable risk option.
B. The annual loss expectancy is the expected monetary loss that is estimated for an asset over a one-year period. It is a useful calculation that should be included in determining the necessity of controls but is not sufficient alone.
C. The cost of the hardware assets should be compared to the total value of the information that the asset protects, including the cost of the systems where the data reside and across which data are transmitted.
D. Potential business impact is only one part of the cost-benefit analysis.

A2-140 An IS auditor is reviewing a contract management process to determine the financial viability of a software vendor for a critical business application. An IS auditor should determine whether the vendor being considered:

A. Can deliver on the immediate contract.
B. Is of similar financial standing as the organization.
C. Has significant financial obligations that can impose liability to the organization.
D. Can support the organization in the long term.

D is the correct answer.

Justification:
A. The capability of the organization to support the enterprise should extend beyond the time of execution of the immediate contract. The objective of financial evaluation should not be confined to the immediate contract but should be to provide assurance of sustainability over a longer time frame.
B. Whether the vendor is of similar financial standing as the purchaser is irrelevant to this review.
C. The vendor should not have financial obligations that could impose a liability to the purchaser; the financial obligations are usually from the purchaser to the vendor.
D. The long-term financial viability of a vendor is essential for deriving maximum value for the organization—it is more likely that a financially sound vendor would be in business for a long period of time and thereby more likely to be capable of providing long-term support for the purchased product.

A2-141 Which of the following is the **BEST** way to ensure that organizational policies comply with legal requirements?

A. Inclusion of a blanket legal statement in each policy
B. Periodic review by subject matter experts
C. Annual sign-off by senior management on organizational policies
D. Policy alignment to the most restrictive regulations

B is the correct answer.

Justification:
A. A blanket legal statement in each policy to adhere to all applicable laws and regulations is ineffective because the readers of the policy (internal personnel) will not know which statements are applicable or the specific nature of their requirements. As a result, personnel may lack the knowledge to perform the required activities for legal compliance.
B. Periodic review of policies by personnel with specific knowledge of regulatory and legal requirements best ensures that organizational policies are aligned with legal requirements.
C. Annual sign-off by senior management on an organization's policies helps set the tone at the top but does not ensure that the policies comply with regulatory and legal requirements.
D. Aligning policies to the most restrictive regulations may create an unacceptable financial burden for the organization. This could then lead to securing minimal risk systems to the same degree as those containing sensitive customer data and other information protected by legislation.

A2-142 An IS auditor is reviewing the risk management process. Which of the following is the **MOST** important consideration during this review?

A. Controls are implemented based on cost-benefit analysis.
B. The risk management framework is based on global standards.
C. The approval process for risk response is in place.
D. IT risk is presented in business terms.

D is the correct answer.

Justification:
A. Controls to mitigate risk must be implemented based on cost-benefit analysis; however, the cost-benefit analysis is effective only if risk is presented in business terms.
B. A risk management framework based on global standards helps in ensuring completeness; however, organizations must adapt it to suit specific business requirements.
C. Approvals for risk response come later in the process.
D. For risk management to be effective, it is necessary to align IT risk with business objectives. This can be done by adopting acceptable terminology that is understood by all, and the best way to achieve this is to present IT risk in business terms.

A2-143 An enterprise hosts its data center onsite and has outsourced the management of its key financial applications to a service provider. Which of the following controls **BEST** ensures that the service provider's employees adhere to the security policies?

A. Sign-off is required on the enterprise's security policies for all users.
B. An indemnity clause is included in the contract with the service provider.
C. Mandatory security awareness training is implemented for all users.
D. Security policies should be modified to address compliance by third-party users.

B is the correct answer.

Justification:
A. Having users sign off on policies is a good practice; however, this only puts the onus of compliance on the individual user, not on the organization.
B. Having the service provider sign an indemnity clause will ensure compliance to the enterprise's security policies, because any violations discovered will lead to a financial liability for the service provider. This will also prompt the enterprise to monitor security violations closely.
C. Awareness training is an excellent control but will not ensure that the service provider's employees adhere to policy.
D. Modification of security policy does not ensure compliance by users unless the policies are appropriately communicated to users and enforced, and awareness training is provided.

A2-144 The corporate IT policy for a call center requires that all users be assigned unique user accounts. On discovering that this is not the case for all current users, what is the **MOST** appropriate recommendation?

A. Have the current configuration approved by operations management.
B. Ensure that there is an audit trail for all existing accounts.
C. Implement individual user accounts for all staff.
D. Amend the IT policy to allow shared accounts.

C is the correct answer.

Justification:

A. Having the current configuration approved is a recommendation that is not in compliance with the enterprise's own policy and would violate good practice.
B. Having an audit trail for existing shared accounts would not provide accountability or resolve the problem of noncompliance with policy.
C. Individual user accounts allow for accountability of transactions and should be the most important recommendation, given the current scenario.
D. Shared user IDs do not allow for accountability of transactions and would not reflect good practice.

A2-145 Which of the following reasons **BEST** describes the purpose of a mandatory vacation policy?

A. To ensure that employees are properly cross-trained in multiple functions
B. To improve employee morale
C. To identify potential errors or inconsistencies in business processes
D. To be used as a cost-saving measure

C is the correct answer.

Justification:

A. Ensuring that employees are properly cross-trained in multiple functions improves the skills of employees and provides for succession planning but is not the primary purpose of mandatory vacations.
B. Improving employee morale helps in reducing employee burnout but is not the primary reason for mandatory vacations.
C. Mandatory vacations help uncover potential fraud or inconsistencies. Ensuring that people who have access to sensitive internal controls or processes take a mandatory vacation annually is often a regulatory requirement and, most importantly, a good way to uncover fraud.
D. Mandatory vacations may or may not be a cost-saving measure, depending on the enterprise.

A2-146 The **MOST** important point of consideration for an IS auditor while reviewing an enterprise's project portfolio is that it:

A. Does not exceed the existing IT budget.
B. Is aligned with the investment strategy.
C. Has been approved by the IT steering committee.
D. Is aligned with the business plan.

D is the correct answer.

Justification:

A. It should be identified if the project portfolio exceeds the IT budget, but it is not as critical as ensuring that it is aligned with the business plan.
B. The project portfolio should be aligned with the investment strategy, but it is most important that it is aligned with the business plan.
C. Appropriate approval of the project portfolio should be granted. However, not every enterprise has an IT steering committee, and this is not as critical as ensuring that the projects are aligned with the business plan.
D. Portfolio management takes a holistic view of an enterprise's overall IT strategy, which, in turn, should be aligned with the business strategy. A business plan provides the justification for each of the projects in the project portfolio, and that is the major consideration for an IS auditor.

A2-147 An IS auditor observes that an enterprise has outsourced software development to a third party that is a startup company. To ensure that the enterprise's investment in software is protected, which of the following should be recommended by the IS auditor?

A. Due diligence should be performed on the software vendor.
B. A quarterly audit of the vendor facilities should be performed.
C. There should be a source code escrow agreement in place.
D. A high penalty clause should be included in the contract.

C is the correct answer.

Justification:

A. Although due diligence is a good practice, it does not ensure availability of the source code in the event of vendor failure.
B. Although a quarterly audit of vendor facilities is a good practice, it does not ensure availability of the source code in the event of failure of the start-up vendor.
C. A source code escrow agreement is primarily recommended to help protect the enterprise's investment in software, because the source code will be available through a trusted third party and can be retrieved if the start-up vendor goes out of business.
D. Although a penalty clause is a good practice, it does not provide protection or ensure availability of the source code in the event of vendor bankruptcy.

A2-148 An enterprise's risk appetite is **BEST** established by:

A. The chief legal officer
B. Security management
C. The audit committee
D. The steering committee

D is the correct answer.

Justification:
A. Although chief legal officers can give guidance regarding legal issues on the policy, they cannot determine the risk appetite.
B. The security management team is concerned with managing the security posture but not with determining the posture.
C. The audit committee is not responsible for setting the risk tolerance or appetite of the enterprise.
D. The steering committee is best suited to determine the enterprise's risk appetite because the committee draws its representation from senior management.

A2-149 A financial services enterprise has a small IT department, and individuals perform more than one role. Which of the following practices represents the **GREATEST** risk?

A. The developers promote code into the production environment.
B. The business analyst writes the requirements and performs functional testing.
C. The IT manager also performs systems administration.
D. The database administrator also performs data backups.

A is the correct answer.

Justification:
A. If developers have access to the production environment, there is a risk that untested code can be migrated into the production environment.
B. In situations in which there is no dedicated testing group, the business analyst is often the one to perform testing because the analyst has detailed knowledge of how the system must function as a result of writing the requirements.
C. It is acceptable in a small team for the IT manager to perform system administration, as long as the manager does not also develop code.
D. It may be part of the database administrator's duties to perform data backups.

A2-150 A financial enterprise has had difficulties establishing clear responsibilities between its IT strategy committee and its IT steering committee. Which of the following responsibilities would **MOST** likely be assigned to its IT steering committee?

A. Approving IT project plans and budgets
B. Aligning IT to business objectives
C. Advising on IT compliance risk
D. Promoting IT governance practices

A is the correct answer.

Justification:

A. An IT steering committee typically has a variety of responsibilities, including approving IT project plans and budgets. Issues related to business objectives, risk and governance are responsibilities that are generally assigned to an IT strategy committee, because it provides insight and advice to the board.
B. Aligning IT to business objectives is a task usually assigned to an IT strategy committee. The steering committee would be more involved in approval and monitoring of individual projects and budgets.
C. Issues related to compliance are tasks usually assigned to an IT strategy committee. The steering committee would be more involved in approval and monitoring of individual projects and budgets.
D. IT governance is a task usually assigned to an IT strategy committee. The steering committee would be more involved in approval and monitoring of individual projects and budgets.

A2-151 Which of the following is the **BEST** enabler for strategic alignment between business and IT?

A. A maturity model
B. Goals and metrics
C. Control objectives
D. A responsible, accountable, consulted and informed (RACI) chart

B is the correct answer.

Justification:

A. Maturity models enable assessment of current process capability and could be used for process improvement and measuring the maturity of the alignment process, but they do not directly enable strategic alignment.
B. Goals and metrics ensure that IT goals are set based on business goals, and they are the best enablers of strategic alignment.
C. Control objectives facilitate the implementation of controls in the related processes according to business requirements.
D. RACI charts enable the assignment of responsibility to key functionaries but do not ensure strategic alignment.

A2-152 An IT steering committee should:

A. Include a mix of members from different departments and staff levels.
B. Ensure that information security policies and procedures have been executed properly.
C. Maintain minutes of its meetings and keep the board of directors informed.
D. Be briefed about new trends and products at each meeting by a vendor.

C is the correct answer.

Justification:

A. Only senior management or high-level staff members should be on this committee because of its strategic mission.
B. Ensuring that information security policies and procedures have been executed properly is not a responsibility of this committee, but the responsibility of IT management and the security administrator.
C. It is important to keep detailed IT steering committee minutes to document the decisions and activities of the IT steering committee. The board of directors should be informed about those decisions on a timely basis.
D. A vendor should be invited to meetings only when appropriate.

Page intentionally left blank

DOMAIN 3—INFORMATION SYSTEMS ACQUISITION, DEVELOPMENT AND IMPLEMENTATION (12%)

A3-1 Who should review and approve system deliverables as they are defined and accomplished, to ensure the successful completion and implementation of a new business system application?

A. User management
B. Project steering committee
C. Senior management
D. Quality assurance staff

A is the correct answer.

Justification:

A. **User management assumes ownership of the project and resulting system, allocates qualified representatives to the team and actively participates in system requirements definition, acceptance testing and user training. User management should review and approve system deliverables as they are defined and accomplished, or implemented.**
B. A project steering committee provides overall direction, ensures appropriate representation of the major stakeholders in the project's outcome, reviews project progress regularly and holds emergency meetings when required. A project steering committee is ultimately responsible for all deliverables, project costs and schedules.
C. Senior management demonstrates commitment to the project and approves the necessary resources to complete the project. This commitment from senior management helps ensure involvement by those who are needed to complete the project.
D. Quality assurance staff review results and deliverables within each phase, and at the end of each phase confirm compliance with standards and requirements. The timing of reviews depends on the system development life cycle, the impact of potential deviation methodology used, the structure and magnitude of the system and the impact of potential deviation.

A3-2 Which of the following **BEST** helps to prioritize project activities and determine the time line for a project?

A. A Gantt chart
B. Earned value analysis
C. Program evaluation review technique
D. Function point analysis

C is the correct answer.

Justification:

A. A Gantt chart is a simple project management tool and would help with the prioritization requirement, but it is not as effective as program evaluation review technique (PERT).
B. Earned value analysis is a technique to track project cost versus project deliverables but does not assist in prioritizing tasks.
C. **The PERT method works on the principle of obtaining project time lines based on project events for three likely scenarios—worst, best and normal. The timeline is calculated by a predefined formula and identifies the critical path, which identifies the key activities that must be prioritized.**
D. Function point analysis measures the complexity of input and output and does not help to prioritize project activities.

A3-3 An IS auditor reviewing a series of completed projects finds that the implemented functionality often exceeded requirements and most of the projects ran significantly over budget. Which of these areas of the organization's project management process is the **MOST** likely cause of this issue?

A. Project scope management
B. Project time management
C. Project risk management
D. Project procurement management

A is the correct answer.

Justification:

A. Because the implemented functionality is greater than what was required, the most likely cause of the budget issue is failure to effectively manage project scope. Project scope management is defined as the processes required to ensure that the project includes all of the required work, and only the required work, to complete the project.

B. Project time management is defined as the processes required to ensure timely completion of the project. The issue noted in the question does not mention whether projects were completed on time, so this is not the most likely cause.

C. Project risk management is defined as the processes concerned with identifying, analyzing and responding to project risk. Although the budget overruns mentioned above represent one form of project risk, they appear to be caused by implementing too much functionality, which relates more directly to project scope.

D. Project procurement management is defined as the processes required to acquire goods and services from outside the performing organization. Although purchasing goods and services that are too expensive can cause budget overruns, in this case the key to the question is that implemented functionality is greater than what was required, which is more likely related to project scope.

A3-4 An IS auditor is reviewing the software development process for an organization. Which of the following functions are appropriate for the end users to perform?

A. Program output testing
B. System configuration
C. Program logic specification
D. Performance tuning

A is the correct answer.

Justification:

A. A user can test program output by checking the program input and comparing it with the system output. This task, although usually done by the programmer, can also be done effectively by the user.

B. System configuration is usually too technical to be accomplished by a user and this situation could create security issues. This could introduce a segregation of duties issue.

C. Program logic specification is a very technical task that is normally performed by a programmer. This could introduce a segregation of duties issue.

D. Performance tuning also requires high levels of technical skill and will not be effectively accomplished by a user. This could introduce a segregation of duties issue.

A3-5 An IS auditor is reviewing system development for a health care organization with two application environments—production and test. During an interview, the auditor notes that production data are used in the test environment to test program changes. What is the **MOST** significant potential risk from this situation?

A. The test environment may not have adequate controls to ensure data accuracy.
B. The test environment may produce inaccurate results due to use of production data.
C. Hardware in the test environment may not be identical to the production environment.
D. The test environment may not have adequate access controls implemented to ensure data confidentiality.

D is the correct answer.

Justification:

A. The accuracy of data used in the test environment is not of significant concern as long as these data are representative of the production environment.

B. Using production data in the test environment does not cause test results to be inaccurate. If anything, using production data improves the accuracy of testing processes, because the data most closely mirror the production environment. In spite of that fact, the risk of data disclosure or unauthorized access in the test environment is still significant and, as a result, production data should not be used in the test environment. This is especially important in a health care organization where patient data confidentiality is critical and privacy laws in many countries impose strict penalties on misuse of these data.

C. Hardware in the test environment should mirror the production environment to ensure that testing is reliable. However, this does not relate to the risk from using live data in a test environment. This is not the correct answer because it does not relate to the risk presented in the scenario.

D. In many cases, the test environment is not configured with the same access controls that are enabled in the production environment. For example, programmers may have privileged access to the test environment (for testing), but not to the production environment. If the test environment does not have adequate access control, the production data are subject to risk of unauthorized access and/or data disclosure. This is the most significant risk of the choices listed.

A3-6 The IS auditor is reviewing a recently completed conversion to a new enterprise resource planning system. In the final stage of the conversion process, the organization ran the old and new systems in parallel for 30 days before allowing the new system to run on its own. What is the **MOST** significant advantage to the organization by using this strategy?

A. Significant cost savings over other testing approaches
B. Assurance that new, faster hardware is compatible with the new system
C. Assurance that the new system meets functional requirements
D. Increased resiliency during the parallel processing time

C is the correct answer.

Justification:

A. Parallel operation provides a high level of assurance that the new system functions properly compared to the old system. Parallel operation is generally expensive and does not provide a cost savings over most other testing approaches. In many cases, parallel operation is the most expensive form of system testing due to the need for dual data entry, dual sets of hardware, dual maintenance and dual backups—it is twice the amount of work as running a production system and, therefore, costs more time and money.

B. Hardware compatibility should be determined and tested much earlier in the conversion project and is not an advantage of parallel operation. Compatibility is generally determined based on the application's published specifications and on system testing in a lab environment. Parallel operation is designed to test the application's effectiveness and integrity of application data, not hardware compatibility. In general, hardware compatibility relates more to the operating system level than to a particular application. Although new hardware in a system conversion must be tested under a real production load, this can be done without parallel systems.

C. Parallel operation is designed to provide assurance that a new system meets its functional requirements. This is the safest form of system conversion testing because, if the new system fails, the old system is still available for production use. In addition, this form of testing allows the application developers and administrators to simultaneously run operational tasks (e.g., batch jobs and backups) on both systems, to ensure that the new system is reliable before unplugging the old system.

D. Increased resiliency during parallel processing is a legitimate outcome from this scenario, but the advantage it provides is temporary and minor, so this is not the correct answer.

A3-7 What kind of software application testing is considered the final stage of testing and typically includes users outside of the development team?

A. Alpha testing
B. White box testing
C. Regression testing
D. Beta testing

D is the correct answer.

Justification:

A. Alpha testing is the testing stage just before beta testing. Alpha testing is typically performed by programmers and business analysts, instead of users. Alpha testing is used to identify bugs or glitches that can be fixed before beta testing begins with external users.

B. White box testing is performed much earlier in the software development life cycle than alpha or beta testing. White box testing is used to assess the effectiveness of software program logic, where test data are used to determine procedural accuracy of the programs being tested. In other words, does the program operate the way it is supposed to at a functional level? White box testing does not typically involve external users.

C. Regression testing is the process of re-running a portion of a test scenario to ensure that changes or corrections have not introduced more errors. In other words, the same tests are run after multiple successive program changes to ensure that the "fix" for one problem did not "break" another part of the program. Regression testing is not the last stage of testing and does not typically involve external users.

D. Beta testing is the final stage of testing and typically includes users outside of the development area. Beta testing is a form of user acceptance testing and generally involves a limited number of users who are external to the development effort.

A3-8 During which phase of software application testing should an organization perform the testing of architectural design?

A. Acceptance testing
B. System testing
C. Integration testing
D. Unit testing

C is the correct answer.

Justification:

A. Acceptance testing determines whether the solution meets the requirements of the business and is performed after system staff has completed the initial system test. This testing includes both quality assurance testing and user acceptance testing, although not combined.

B. System testing relates a series of tests by the test team or system maintenance staff to ensure that the modified program interacts correctly with other components. System testing references the functional requirements of the system.

C. Integration testing evaluates the connection of two or more components that pass information from one area to another. The objective is to use unit-tested modules, thus building an integrated structure according to the design.

D. Unit testing references the detailed design of the system and uses a set of cases that focus on the control structure of the procedural design to ensure that the internal operation of the program performs according to specification.

A3-9 Which of the following is an advantage of an integrated test facility?

A. It uses actual master files or dummies, and the IS auditor does not have to review the source of the transaction.
B. Periodic testing does not require separate test processes.
C. It validates application systems and ensures the correct operation of the system.
D. The need to prepare test data is eliminated.

B is the correct answer.

Justification:

A. The integrated test facility (ITF) tests a test transaction as if it were a real transaction and validates that transaction processing is being done correctly. It is not related to reviewing the source of a transaction.
B. An ITF creates a fictitious entity in the database to process test transactions simultaneously with live input. Its advantage is that periodic testing does not require separate test processes. Careful planning is necessary, and test data must be isolated from production data.
C. An ITF does validate the correct operation of a transaction in an application, but it does not ensure that a system is being operated correctly.
D. The ITF is based on the integration of test data into the normal process flow, so test data is still required.

A3-10 An organization is replacing a payroll program that it developed in-house, with the relevant subsystem of a commercial enterprise resource planning (ERP) system. Which of the following would represent the **HIGHEST** potential risk?

A. Undocumented approval of some project changes
B. Faulty migration of historical data from the old system to the new system
C. Incomplete testing of the standard functionality of the ERP subsystem
D. Duplication of existing payroll permissions on the new ERP subsystem

B is the correct answer.

Justification:

A. Undocumented changes (leading to scope creep) are a risk, but the greatest risk is the loss of data integrity when migrating data from the old system to the new system.
B. The most significant risk after a payroll system conversion is loss of data integrity and not being able to pay employees in a timely and accurate manner or have records of past payments. As a result, maintaining data integrity and accuracy during migration is paramount.
C. A lack of testing is always a risk; however, in this case, the new payroll system is a subsystem of an existing commercially available (and therefore probably well-tested) system.
D. Setting up the new system, including access permissions and payroll data, always presents some level of risk; however, the greatest risk is related to the migration of data from the old system to the new system.

A3-11 An enterprise is developing a strategy to upgrade to a newer version of its database software. Which of the following tasks can an IS auditor perform without compromising the objectivity of the IS audit function?

A. Advise on the adoption of application controls to the new database software.
B. Provide future estimates of the licensing expenses to the project team.
C. Recommend to the project manager how to improve the efficiency of the migration.
D. Review the acceptance test case documentation before the tests are carried out.

D is the correct answer.

Justification:

A. Independence can be compromised if the IS auditor advises on the adoption of specific application controls.
B. Independence can be compromised if the IS auditor were to audit the estimate of future expenses used to support a business case for management approval of the project.
C. Advising the project manager on how to increase the efficiency of the migration may compromise the IS auditor's independence.
D. The review of the test cases will facilitate the objective of a successful migration and ensure that proper testing is conducted. An IS auditor can advise as to the completeness of the test cases.

A3-12 During a postimplementation review, which of the following activities should be performed?

A. User acceptance testing
B. Return on investment analysis
C. Activation of audit trails
D. Updates of the state of enterprise architecture diagrams

B is the correct answer.

Justification:

A. User acceptance testing should be performed prior to the implementation (perhaps during the development phase), not after the implementation.
B. Following implementation, a cost-benefit analysis or return on investment should be reperformed to verify that the original business case benefits are delivered.
C. The audit trail should be activated during the implementation of the application.
D. While updating the enterprise architecture diagrams is a good practice, it would not normally be part of a postimplementation review.

A3-13 Which of the following is the **BEST** approach to ensure that sufficient test coverage will be achieved for a project with a strict end date and a fixed time to perform testing?

A. Requirements should be tested in terms of importance and frequency of use.
B. Test coverage should be restricted to functional requirements.
C. Automated tests should be performed through the use of scripting.
D. The number of required test runs should be reduced by retesting only defect fixes.

A is the correct answer.

Justification:

A. The idea is to maximize the usefulness of testing by concentrating on the most important aspects of the system and on the areas where defects represent the greatest risk to user acceptance. A further extension of this approach is to also consider the technical complexity of requirements, because complexity tends to increase the likelihood of defects.

B. The problem with testing only functional requirements is that nonfunctional requirement areas, such as usability and security, which are important to the overall quality of the system, are ignored.

C. Increasing the efficiency of testing by automating test execution is a good idea. However, by itself, this approach does not ensure the appropriate targeting of test coverage and so is not as effective an alternative.

D. Retesting only defect fixes has a considerable risk that it will not detect instances in which defect fixes may have caused the system to regress (i.e., introduced errors in parts of the system that were previously working correctly). For this reason, it is a good practice to undertake formal regression testing after defect fixes have been implemented.

A3-14 By evaluating application development projects against the capability maturity model, an IS auditor should be able to verify that:

A. Reliable products are guaranteed.
B. Programmers' efficiency is improved.
C. Security requirements are designed.
D. Predictable software processes are followed.

D is the correct answer.

Justification:

A. Although the likelihood of success should increase as the software processes mature toward the optimizing level, mature processes do not guarantee a reliable product.

B. The capability maturity model (CMM) does not evaluate technical processes such as programming efficiency.

C. The CMM does not evaluate security requirements or other application controls.

D. By evaluating the organization's development projects against the CMM, an IS auditor determines whether the development organization follows a stable, predictable software development process.

A3-15 An IS auditor is performing a post-implementation review of an organization's system and identifies output errors within an accounting application. The IS auditor determined this was caused by input errors. Which of the following controls should the IS auditor recommend to management?

A. Recalculations
B. Limit checks
C. Run-to-run totals
D. Reconciliations

B is the correct answer.

Justification:

A. A sample of transactions may be recalculated manually to ensure that processing is accomplishing the anticipated task. Recalculations are performed after the output phase.
B. Processing controls should be implemented as close as possible to the point of data entry. Limit checks are one type of input validation check that provides a preventive control to ensure that invalid data cannot be entered because values must fall within a predetermined limit.
C. Run-to-run totals provide the ability to verify data values through the stages of application processing. Run-to-run total verification ensures that data read into the computer were accepted and then applied to the updating process. Run-to-run totals are performed after the output phase.
D. Reconciliation of file totals should be performed on a routine basis. Reconciliations may be performed through the use of a manually maintained account, a file control record or an independent control file. Reconciliations are performed after the output phase.

A3-16 Due to a reorganization, a business application system will be extended to other departments. Which of the following should be of the **GREATEST** concern for an IS auditor?

A. Process owners have not been identified.
B. The billing cost allocation method has not been determined.
C. Multiple application owners exist.
D. A training program does not exist.

A is the correct answer.

Justification:

A. When one application is expanded to multiple departments, it is important to ensure the mapping between the process owner and system functions. The absence of a defined process owner, may cause issues with monitoring or authorization controls.
B. The allocation method of application usage cost is of less importance.
C. The fact that multiple application owners exist is not a concern for an IS auditor as long as process owners have been identified.
D. The fact that a training program does not exist is only be a minor concern for the IS auditor.

A3-17 When auditing the proposed acquisition of a new computer system, an IS auditor should **FIRST** ensure that:

A. A clear business case has been approved by management.
B. Corporate security standards will be met.
C. Users will be involved in the implementation plan.
D. The new system will meet all required user functionality.

A is the correct answer.

Justification:
A. The first concern of an IS auditor is to ensure that the proposal meets the needs of the business. This should be established by a clear business case.
B. Compliance with security standards is essential, but it is too early in the procurement process for this to be an IS auditor's first concern.
C. Having users involved in the implementation process is essential, but it is too early in the procurement process for this to be an IS auditor's first concern.
D. Meeting the needs of the users is essential, and this should be included in the business case presented to management for approval.

A3-18 Which of the following types of risk is **MOST** likely encountered in a software as a service environment?

A. Noncompliance with software license agreements
B. Performance issues due to Internet delivery method
C. Higher cost due to software licensing requirements
D. Higher cost due to the need to update to compatible hardware

B is the correct answer.

Justification:
A. Software as a service (SaaS) is provisioned on a usage basis and the number of users is monitored by the SaaS provider; therefore, there should be no risk of noncompliance with software license agreements.
B. The risk that can be most likely encountered in a SaaS environment is speed and availability issues, because SaaS relies on the Internet for connectivity.
C. The costs for a SaaS solution should be fixed as a part of the services contract and considered in the business case presented to management for approval of the solution.
D. The open design and Internet connectivity allow most SaaS to run on virtually any type of hardware.

A3-19 The most common reason for the failure of information systems to meet the needs of users is that:

A. user needs are constantly changing.
B. the growth of system requirements was forecast inaccurately.
C. the hardware system limits the number of concurrent users.
D. user participation in defining the system's requirements was inadequate.

D is the correct answer.

Justification:
A. Although changing user needs has an effect on the success or failure of many projects, the core problem is usually a lack of getting the initial requirements correct at the beginning of the project.
B. Projects may fail as the needs of the users increase; however, this can be mitigated through better change control procedures.
C. Rarely do hardware limitations affect the usability of the project as long as the requirements were correctly documented at the beginning of the project.
D. Lack of adequate user involvement, especially in the system's requirements phase, will usually result in a system that does not fully or adequately address the needs of the user. Only users can define what their needs are and, therefore, what the system should accomplish.

A3-20 Many IT projects experience problems because the development time and/or resource requirements are underestimated. Which of the following techniques provides the **GREATEST** assistance in developing an estimate of project duration?

A. Function point analysis
B. Program evaluation review technique chart
C. Rapid application development
D. Object-oriented system development

B is the correct answer.

Justification:

A. Function point analysis is a technique for determining the size of a development task based on the number of function points. Function points are factors such as inputs, outputs, inquiries and logical internal files. While this will help determine the size of individual activities, it will not assist in determining project duration because there are many overlapping tasks.
B. A program evaluation review technique (chart will help determine project duration once all the activities and the work involved with those activities are known.
C. Rapid application development is a methodology that enables organizations to develop strategically important systems faster while reducing development costs and maintaining quality.
D. Object-oriented system development is the process of solution specification and modeling but will not assist in calculating project duration.

A3-21 An IS auditor is reviewing IT projects for a large company and wants to determine whether the IT projects undertaken in a given year are those which have been assigned the highest priority by the business and which will generate the greatest business value. Which of the following is **MOST** relevant?

A. A capability maturity model
B. Portfolio management
C. Configuration management
D. Project management body of knowledge

B is the correct answer.

Justification:

A. A capability maturity model (CMM) would not help determine the optimal portfolio of capital projects because it is a means of assessing the relative maturity of the IT processes within an organization: running from Level 0 (Incomplete—Processes are not implemented or fail to achieve their purpose) to Level 5 (Optimizing—Metrics are defined and measured, and continuous improvement techniques are in place).
B. Portfolio management is designed to assist in the definition, prioritization, approval and running of a set of projects within a given organization. These tools offer data capture, workflow and scenario planning functionality, which can help identify the optimum set of projects (from the full set of ideas) to take forward within a given budget.
C. A configuration management database (which stores the configuration details for an organization's IT systems) is an important tool for IT service delivery and, in particular, change management. It may provide information that would influence the prioritization of projects but is not designed for that purpose.
D. The project management body of knowledge is a methodology for the management and delivery of projects. It offers no specific guidance or assistance in optimizing a project portfolio.

A3-22 The reason for establishing a stop or freezing point on the design of a new system is to:

A. prevent further changes to a project in process.
B. indicate the point at which the design is to be completed.
C. require that changes after that point be evaluated for cost-effectiveness.
D. provide the project management team with more control over the project design.

C is the correct answer.

Justification:
A. The stop point is intended to provide greater control over changes but not to prevent them.
B. The stop point is used for project control but not to create an artificial fixed point that requires the design of the project to cease.
C. Projects often tend to expand, especially during the requirements definition phase. This expansion often grows to a point where the originally anticipated cost-benefits are diminished because the cost of the project has increased. When this occurs, it is recommended that the project be stopped or frozen to allow a review of all of the cost-benefits and the payback period.
D. A stop point is used to control requirements, not systems design.

A3-23 Change control for business application systems being developed using prototyping could be complicated by the:

A. iterative nature of prototyping.
B. rapid pace of modifications in requirements and design.
C. emphasis on reports and screens.
D. lack of integrated tools.

B is the correct answer.

Justification:
A. A characteristic of prototyping is its iterative nature, but it does not have an adverse effect on change control.
B. Changes in requirements and design happen so quickly that they are seldom documented or approved.
C. A characteristic of prototyping is its emphasis on reports and screens, but it does not have an adverse effect on change control.
D. Lack of integrated tools is a characteristic of prototyping, but it does not have an adverse effect on change control.

A3-24 An IS auditor performing a review of a major software development project finds that it is on schedule and under budget even though the software developers have worked considerable amounts of unplanned overtime. The IS auditor should:

A. conclude that the project is progressing as planned because dates are being met.
B. question the project manager further to identify whether overtime costs are being tracked accurately.
C. conclude that the programmers are intentionally working slowly to earn extra overtime pay.
D. investigate further to determine whether the project plan may not be accurate.

D is the correct answer.

Justification:
A. Although the project is on time and budget, there may be problems with the project plan because considerable amounts of unplanned overtime have been required.
B. There is a possibility that the project manager has hidden some costs to make the project look better; however, the real problem may be with whether the project plan is realistic, not just the accounting.
C. It is possible that the programmers are trying to take advantage of the time system, but if the overtime has been required to keep the project on track it is more likely that the time lines and expectations of the project are unrealistic.
D. Although the dates on which key projects are completed are important, there may be issues with the project plan if an extraordinary amount of unplanned overtime is required to meet those dates. In most cases, the project plan is based on a certain number of hours, and requiring programmers to work considerable overtime is not a good practice. Although overtime costs may be an indicator that something is wrong with the plan, in many organizations, the programming staff may be salaried, so overtime costs may not be directly recorded.

A3-25 A project development team is considering using production data for its test deck. The team removed sensitive data elements before loading it into the test environment. Which of the following additional concerns should an IS auditor have with this practice?

A. Not all functionality will be tested.
B. Production data are introduced into the test environment.
C. Specialized training is required.
D. The project may run over budget.

A is the correct answer.

Justification:
A. A primary risk of using production data in a test deck is that not all transactions or functionality may be tested if there are no data that meet the requirement.
B. The presence of production data in a test environment is not a concern if the sensitive elements have been scrubbed.
C. Creation of a test deck from production data does not require specialized knowledge, so this is not a concern.
D. The risk of a project running over budget is always a concern, but it is not related to the practice of using production data in a test environment.

A3-26 Which of the following considerations is the **MOST** important while evaluating a business case for the acquisition of a new accounting application?

A. Total cost of ownership of the application
B. The resources required for implementation
C. Return on investment to the company
D. The cost and complexity of security requirements

C is the correct answer.

Justification:

A. Total cost of ownership of the application is important to understand the resource and budget requirements in the short and long term; however, decisions should be based on benefits realization from this investment. Therefore, return on investment (ROI) is the most important consideration.
B. The resources required for implementation of the application are an important consideration; however, decisions should be based on benefits realization from this investment. Therefore, ROI should be carefully considered.
C. The proposed ROI benefits, along with targets or metrics that can be measured, are the most important aspects of a business case. While reviewing the business case, it should be verified that the proposed ROI is achievable, does not make unreasonable assumptions and can be measured for success. (Benefits realization should look beyond project cycles to longer-term cycles that consider the total benefits and total costs throughout the life of the new system.)
D. The cost and complexity of security requirements are important considerations, but they need to be weighed against the proposed benefits of the application. Therefore, ROI is more important.

A3-27 The development of an application has been outsourced to an offshore vendor. Which of the following should be of **GREATEST** concern to an IS auditor?

A. The right to audit clause was not included in the contract.
B. The business case was not established.
C. There was no source code escrow agreement.
D. The contract does not cover change management procedures.

B is the correct answer.

Justification:

A. The lack of the right to audit clause presents a risk to the organization; however, the risk is not as consequential as the lack of a business case.
B. Because the business case was not established, it is likely that the business rationale, risk and risk mitigation strategies for outsourcing the application development were not fully evaluated and the appropriate information was not provided to senior management for formal approval. This situation presents the biggest risk to the organization.
C. If the source code is held by the provider and not provided to the organization, the lack of source code escrow presents a risk to the organization; however, the risk is not as consequential as the lack of a business case.
D. The lack of change management procedures presents a risk to the organization, especially with the possibility of extraordinary charges for any required changes; however, the risk is not as consequential as the lack of a business case.

A3-28 Before implementing controls in a newly developed system, management should **PRIMARILY** ensure that the controls:

A. satisfy a requirement in addressing a risk.
B. do not reduce productivity.
C. are based on a minimized cost analysis.
D. are detective or corrective.

A is the correct answer.

Justification:

A. The purpose of a control is to mitigate a risk; therefore, the primary consideration when selecting a control is that it effectively mitigates an identified risk. When designing controls, it is necessary to consider all of the aspects in choices A through D. In an ideal situation, controls that address all of these aspects would be the best controls. Realistically, it may not be possible to design them all and the cost may be prohibitive; therefore, it is necessary to consider the controls related primarily to the treatment of existing risk in the organization.

B. Controls will often affect productivity and performance; however, this must be balanced against the benefit obtained from the implementation of the control.

C. The most important reason for a control is to mitigate a risk—and the selection of a control is usually based on a cost-benefit analysis, not on selecting just the least expensive control.

D. A good control environment will include preventive, detective and corrective controls.

A3-29 Information for detecting unauthorized input from a user workstation would be **BEST** provided by the:

A. console log printout.
B. transaction journal.
C. automated suspense file listing.
D. user error report.

B is the correct answer.

Justification:

A. A console log printout is not the best because it does not record activity from a specific terminal.

B. The transaction journal records all transaction activity, which then can be compared to the authorized source documents to identify any unauthorized input.

C. An automated suspense file listing lists only transaction activity where an edit error occurred.

D. The user error report lists only input that resulted in an edit error and does not record improper user input.

A3-30 Which of the following has the **MOST** significant impact on the success of an application systems implementation?

A. The prototyping application development methodology
B. Compliance with applicable external requirements
C. The overall organizational environment
D. The software reengineering technique

C is the correct answer.

Justification:

A. The prototyping application development technique reduces the time to deploy systems primarily by using faster development tools that allow a user to see a high-level view of the workings of the proposed system within a short period of time. The use of any one development methodology will have a limited impact on the success of the project.
B. Compliance with applicable external requirements has an impact on the implementation success, but the impact is not as significant as the impact of the overall organizational environments.
C. The overall organizational environment has the most significant impact on the success of applications systems implemented. This includes the alignment between IT and the business, the maturity of the development processes and the use of change control and other project management tools.
D. The software reengineering technique is a process of updating an existing system by extracting and reusing design and program components. This is used to support major changes in the way an organization operates. Its impact on the success of the application systems that are implemented is small compared with the impact of the overall organizational environment.

A3-31 The editing/validation of data entered at a remote site is performed **MOST** effectively at the:

A. central processing site after running the application system.
B. central processing site during the running of the application system.
C. remote processing site after transmission of the data to the central processing site.
D. remote processing site prior to transmission of the data to the central processing site.

D is the correct answer.

Justification:

A. Validating data prior to transmission is the most efficient method and saves the effort of transmitting or processing invalid data. However, due to the risk of errors being introduced during transmission it is also good practice to re-validate the data at the central processing site.
B. Validating data prior to transmission is the most efficient method and saves the effort of transmitting or processing invalid data. However, due to the risk of errors being introduced during transmission it is also good practice to re-validate the data at the central processing site.
C. To validate the data after it has been transmitted is not a valid control.
D. It is important that the data entered from a remote site is edited and validated prior to transmission to the central processing site.

A3-32 The **MAJOR** consideration for an IS auditor reviewing an organization's IT project portfolio is the:

A. IT budget.
B. existing IT environment.
C. business plan.
D. investment plan.

C is the correct answer.

Justification:

A. The IT budget is important to ensure that the resources are being used in the best manner, but this is secondary to the importance of reviewing the business plan.
B. The existing IT environment is important and used to determine gap analysis but is secondary to the importance of reviewing the business plan.
C. One of the most important reasons for which projects get funded is how well a project meets an organization's strategic objectives. Portfolio management takes a holistic view of a company's overall IT strategy. IT strategy should be aligned with the business strategy and, hence, reviewing the business plan should be the major consideration.
D. The investment plan is important to set out project priorities, but secondary to the importance of reviewing the business plan.

A3-33 Regression testing is undertaken **PRIMARILY** to ensure that:

A. system functionality meets customer requirements.
B. a new system can operate in the target environment.
C. applicable development standards have been maintained.
D. applied changes have not introduced new errors.

D is the correct answer.

Justification:

A. Validation testing is used to test the functionality of the system against detailed requirements to ensure that software construction is traceable to customer requirements.
B. Sociability testing is used to see whether the system can operate in the target environment without adverse impacts on the existing systems.
C. Software quality assurance and code reviews are used to determine whether development standards are maintained.
D. Regression testing is used to test for the introduction of new errors in the system after changes have been applied.

A3-34 A proposed transaction processing application will have many data capture sources and outputs in paper and electronic form. To ensure that transactions are not lost during processing, an IS auditor should recommend the inclusion of:

A. validation controls.
B. internal credibility checks.
C. clerical control procedures.
D. automated systems balancing.

D is the correct answer.

Justification:

A. Input and output validation controls are certainly valid controls but will not detect and report lost transactions.
B. Internal credibility checks are valid controls to detect errors in processing but will not detect and report lost transactions.
C. A clerical procedure could be used to summarize and compare inputs and outputs; however, an automated process is less susceptible to error.
D. Automated systems balancing would be the best way to ensure that no transactions are lost as any imbalance between total inputs and total outputs would be reported for investigation and correction.

A3-35 Which of the following should be an IS auditor's **PRIMARY** concern after discovering that the scope of an IS project has changed, and an impact study has not been performed?

A. The time and cost implications caused by the change
B. The risk that regression tests will fail
C. Users not agreeing with the change
D. The project team not having the skills to make the necessary change

A is the correct answer.

Justification:

A. Any scope change might have an impact on duration and cost of the project; that is the reason why an impact study is conducted, and the client is informed of the potential impact on the schedule and cost.
B. A change in scope does not necessarily impact the risk that regression tests will fail.
C. An impact study will not determine whether users will agree with a change in scope.
D. Conducting an impact study could identify a lack of resources such as the project team lacking the skills necessary to make the change; however, this is only part of the impact on the overall time lines and cost to the project due to the change.

A3-36 An IS auditor is reviewing the software development capabilities of an organization that has adopted the agile methodology. The IS auditor would be the **MOST** concerned if:

A. certain project iterations produce proof-of-concept deliverables and unfinished code.
B. application features and development processes are not extensively documented.
C. software development teams continually re-plan each step of their major projects.
D. project managers do not manage project resources, leaving that to project team members.

A is the correct answer.

Justification:

A. The agile software development methodology is an iterative process where each iteration or "sprint" produces functional code. If a development team was producing code for demonstration purposes, this would be an issue because the following iterations of the project build on the code developed in the prior sprint.
B. One focus of agile methodology is to rely more on team knowledge and produce functional code quickly. These characteristics would result in less extensive documentation or documentation embedded in the code itself.
C. After each iteration or "sprint," agile development teams re-plan the project so that unfinished tasks are performed, and resources can be reallocated as needed. The continual re-planning is a key component of agile development methodology.
D. The management of agile software development is different from conventional development approaches in that leaders act as facilitators and allow team members to determine how to manage their own resources to get each sprint completed. Because the team members are performing the work, they are in a good position to understand how much time/effort is required to complete a sprint.

A3-37 Which of the following data validation edits is effective in detecting transposition and transcription errors?

A. Range check
B. Check digit
C. Validity check
D. Duplicate check

B is the correct answer.

Justification:

A. A range check is checking data that matches a predetermined range of allowable values.
B. A check digit is a numeric value that is calculated mathematically and is appended to data to ensure that the original data have not been altered (e.g., an incorrect, but valid, value substituted for the original). This control is effective in detecting transposition and transcription errors.
C. A validity check is programmed checking of the data validity in accordance with predetermined criteria.
D. In a duplicate check, new or fresh transactions are matched to those previously entered to ensure that they are not already in the system.

A3-38 Two months after a major application implementation, management, who assume that the project went well, requests that an IS auditor perform a review of the completed project. The IS auditor's **PRIMARY** focus should be to:

A. determine whether user feedback on the system has been documented.
B. assess whether the planned cost benefits are being measured, analyzed and reported.
C. review controls built into the system to assure that they are operating as designed.
D. review subsequent program change requests.

C is the correct answer.

Justification:

A. The IS auditor should check whether user feedback has been provided, but this is not the most important area for audit.
B. It is important to assess the effectiveness of the project; however, assuring that the production environment is adequately controlled after the implementation is of primary concern.
C. Because management is assuming that the implementation went well, the primary focus of the IS auditor is to test the controls built into the application to assure that they are functioning as designed.
D. Reviewing change requests may be a good idea, but this is more important if the application is perceived to have a problem.

A3-39 Which of the following types of risk could result from inadequate software project baselining?

A. Sign-off delays
B. Software integrity violations
C. Scope creep
D. Inadequate controls

C is the correct answer.

Justification:

A. Sign-off delays may occur due to inadequate software baselining; however, these are most likely caused by scope creep.
B. Software integrity violations can be caused by hardware or software failures, malicious intrusions or user errors. Software baselining does not help prevent software integrity violations.
C. A software baseline is the cutoff point in the design and development of a system. Beyond this point, additional requirements or modifications to the scope must go through formal, strict procedures for approval based on a business cost-benefit analysis. Failure to adequately manage a system through baselining can result in uncontrolled changes in a project's scope and may incur time and budget overruns.
D. Inadequate controls are most likely present in situations in which information security is not duly considered from the beginning of system development; they are not a risk that can be adequately addressed by software baselining.

A3-40 An organization implemented a distributed accounting system, and the IS auditor is conducting a postimplementation review to provide assurance of the data integrity controls. Which of the following choices should the auditor perform **FIRST**?

A. Review user access.
B. Evaluate the change request process.
C. Evaluate the reconciliation controls.
D. Review the data flow diagram.

D is the correct answer.

Justification:

A. The review of user access would be important; however, in terms of data integrity it would be better to review the data flow diagram.
B. The lack of an adequate change control process could impact the integrity of the data; however, the system should be documented first to determine whether the transactions flow to other systems.
C. Evaluating the reconciliation controls would help to ensure data integrity; however, it is more important to understand the data flows of the application to ensure that the reconciliation controls are located in the correct place.
D. The IS auditor should review the application data flow diagram to understand the flow of data within the application and to other systems. This will enable the IS auditor to evaluate the design and effectiveness of the data integrity controls.

A3-41 During the audit of an acquired software package, an IS auditor finds that the software purchase was based on information obtained through the Internet, rather than from responses to a request for proposal. The IS auditor should **FIRST**:

A. test the software for compatibility with existing hardware.
B. perform a gap analysis.
C. review the licensing policy.
D. ensure that the procedure had been approved.

D is the correct answer.

Justification:

A. Because the software package has already been acquired, it is most likely that it is in use and therefore compatible with existing hardware. Further, the first responsibility of the IS auditor is to ensure that the purchasing procedures have been approved.
B. Because there was no request for proposal, there may be no documentation of the expectations of the product and nothing to measure a gap against. The first task for the IS auditor is to ensure that the purchasing procedures were approved.
C. The licensing policy should be reviewed to ensure proper licensing but only after the purchasing procedures are checked.
D. In the case of a deviation from the predefined procedures, an IS auditor should first ensure that the procedure followed for acquiring the software is consistent with the business objectives and has been approved by the appropriate authorities.

A3-42 A failure discovered in which of the following testing stages would have the **GREATEST** impact on the implementation of new application software?

A. System testing
B. Acceptance testing
C. Integration testing
D. Unit testing

B is the correct answer.

Justification:

A. System testing is undertaken by the development team to determine if the combined units of software work together and that the software meets user requirements per specifications. A failure here would be expensive but easier to fix than a failure found later in the testing process.
B. Acceptance testing is the final stage before the software is installed and is available for use. The greatest impact would occur if the software fails at the acceptance testing level because this could result in delays and cost overruns.
C. Integration testing examines the units/modules as one integrated system and unit testing examines the individual units or components of the software. A failure here would be expensive and require re-work of the modules but would not be as expensive as a problem found just prior to implementation.
D. System, integration and unit testing are all performed by the developers at various stages of development; the impact of failure is comparatively less for each than failure at the acceptance testing stage.

A3-43 Which of the following is the **MOST** likely benefit of implementing a standardized infrastructure?

A. Improved cost-effectiveness of IT service delivery and operational support
B. Increased security of the IT service delivery center
C. Reduced level of investment in the IT infrastructure
D. Reduced need for testing future application changes

A is the correct answer.

Justification:

A. A standardized IT infrastructure provides a consistent set of platforms and operating systems across the organization. This standardization reduces the time and effort required to manage a set of disparate platforms and operating systems. In addition, the implementation of enhanced operational support tools (e.g., password management tools, patch management tools and auto provisioning of user access) is simplified. These tools can help the organization reduce the cost of IT service delivery and operational support.
B. A standardized infrastructure results in a more homogeneous environment, which is more prone to attacks.
C. While standardization can reduce support costs, the transition to a standardized kit can be expensive; therefore, the overall level of IT infrastructure investment is not likely to be reduced.
D. A standardized infrastructure may simplify testing of changes, but it does not reduce the need for such testing.

A3-44 Which of the following is the **MOST** important element in the design of a data warehouse?

A. Quality of the metadata
B. Speed of the transactions
C. Volatility of the data
D. Vulnerability of the system

A is the correct answer.

Justification:

A. Quality of the metadata is the most important element in the design of a data warehouse. A data warehouse is a copy of transaction data specifically structured for query and analysis. Metadata describes the data in the warehouse and aims to provide a table of contents to the stored information. Companies that have built warehouses believe that metadata are the most important component of the warehouse.

B. A data warehouse is used for analysis and research, not for production operations, so the speed of transactions is not relevant.

C. Data in a data warehouse is frequently received from many sources and vast amounts of information may be received on an hourly or daily basis. Except to ensure adequate storage capability, this is not a primary concern of the designer.

D. Data warehouses may contain sensitive information, or can be used to research sensitive information, so the security of the data warehouse is important. However, this is not the primary concern of the designer.

A3-45 Ideally, stress testing should be carried out in a:

A. test environment using test data.
B. production environment using live workloads.
C. test environment using live workloads.
D. production environment using test data.

C is the correct answer.

Justification:

A. A test environment should always be used to avoid damaging the production environment, but only testing with test data may not test all aspects of the system adequately.

B. Testing should never take place in a production environment.

C. Stress testing is carried out to ensure that a system can cope with production workloads. Testing with production level workloads is important to ensure that the system will operate effectively when moved into production.

D. It is not advisable to do stress testing in a production environment. Additionally, if only test data are used, there is no certainty that the system was stress tested adequately.

A3-46 Assignment of process ownership is essential in system development projects because it:

A. enables the tracking of the development completion percentage.
B. optimizes the design cost of user acceptance test cases.
C. minimizes the gaps between requirements and functionalities.
D. ensures that system design is based on business needs.

D is the correct answer.

Justification:

A. Process ownership assignment does not have a feature to track the completion percentage of deliverables.
B. Whether the design cost of test cases will be optimized is not determined from the assignment of process ownership. It may help to some extent; however, there are many other factors involved in the design of test cases.
C. For gap minimization, a specific requirements analysis framework should be in place and then applied; however, a gap may be found between the design and the as-built system that could lead to system functionality not meeting requirements. This will be identified during user acceptance testing. Process ownership alone does not have the capability to minimize requirement gaps.
D. The involvement of process owners will ensure that the system will be designed according to the needs of the business processes that depend on system functionality. A sign-off on the design by the process owners is crucial before development begins.

A3-47 The **BEST** time for an IS auditor to assess the control specifications of a new application software package which is being considered for acquisition is during:

A. the internal lab testing phase.
B. testing and prior to user acceptance.
C. the requirements gathering process.
D. the implementation phase.

C is the correct answer.

Justification:

A. During testing, the IS auditor will ensure that the security requirements are met. This is not the time to assess the control specifications.
B. The control specifications will drive the security requirements that are built into the contract and should be assessed before the product is acquired and tested.
C. The best time for the involvement of an IS auditor is at the beginning of the requirements definition of the development or acquisition of applications software. This provides maximum opportunity for review of the vendors and their products. Early engagement of an IS auditor also minimizes the potential of a business commitment to a given solution that might be inadequate and more difficult to overcome as the process continues.
D. During the implementation phase, the IS auditor may check whether the controls have been enabled; however, this is not the time to assess the control requirements.

A3-48 The phases and deliverables of a system development life cycle project should be determined:

A. during the initial planning stages of the project.
B. after early planning has been completed but before work has begun.
C. throughout the work stages, based on risk and exposures.
D. only after all risk and exposures have been identified and the IS auditor has recommended appropriate controls.

A is the correct answer.

Justification:
A. It is extremely important that the project be planned properly, and that the specific phases and deliverables are identified during the early stages of the project. This enables project tracking and resource management.
B. Determining the deliverables and time lines of a project are a part of the early project planning work.
C. The requirements may change over the life of a project, but the initial deliverables should be documented from the beginning of the project.
D. Risk management is a never-ending process, so project planning cannot wait until all risk has been identified.

A3-49 Management observed that the initial phase of a multiphase implementation was behind schedule and over budget. Prior to commencing with the next phase, an IS auditor's **PRIMARY** suggestion for a postimplementation focus should be to:

A. assess whether the planned cost benefits are being measured, analyzed and reported.
B. review control balances and verify that the system is processing data accurately.
C. review the impact of program changes made during the first phase on the remainder of the project.
D. determine whether the system's objectives were achieved.

C is the correct answer.

Justification:
A. While all choices are valid, the postimplementation focus and primary objective should be understanding the impact of the problems in the first phase on the remainder of the project.
B. The review should assess whether the control is working correctly but should focus on the problems that led to project overruns in budget and time.
C. Because management is aware that the project had problems, reviewing the subsequent impact will provide insight into the types and potential causes of the project issues. This will help to identify whether IT has adequately planned for those issues in subsequent projects.
D. Ensuring that the system works is a primary objective for the IS auditor, but in this case because the project planning was a failure, the IS auditor should focus on the reasons for, and impact of, the failure.

A3-50 When implementing an application software package, which of the following presents the **GREATEST** risk?

A. Uncontrolled multiple software versions
B. Source programs that are not synchronized with object code
C. Incorrectly set parameters
D. Programming errors

C is the correct answer.

Justification:

A. Having multiple versions is a problem, but as long as the correct version is implemented, the most serious risk during implementation is to have the parameters for the program set incorrectly.
B. Lack of synchronization between source and object code will be a serious risk for later maintenance of compiled programs, but this will not affect other types of programs and is not the most serious risk at the time of implementation.
C. Parameters that are not set correctly would be the greatest concern when implementing an application software package. Incorrectly set parameters are an immediate problem that could lead to system breach, failure or noncompliance.
D. Programming errors should be found during testing, not at the time of implementation.

A3-51 Which of the following is an advantage of prototyping?

A. The finished system normally has strong internal controls.
B. Prototype systems can provide significant time and cost savings.
C. Change control is often less complicated with prototype systems.
D. Prototyping ensures that functions or extras are not added to the intended system.

B is the correct answer.

Justification:

A. Prototyping often has poor internal controls because the focus is primarily on functionality, not on security.
B. Prototype systems can provide significant time and cost savings through better user interaction and the ability to rapidly adapt to changing requirements; however, they also have several disadvantages, including loss of overall security focus, project oversight and implementation of a prototype that is not yet ready for production.
C. Change control becomes much more complicated with prototyping.
D. Prototyping often leads to functions or extras being added to the system that were not originally intended.

A3-52 The **PRIMARY** objective of performing a postincident review is that it presents an opportunity to:

A. improve internal control procedures.
B. harden the network to industry good practices.
C. highlight the importance of incident response management to management.
D. improve employee awareness of the incident response process.

A is the correct answer.

Justification:

A. A postincident review examines both the cause and response to an incident. The lessons learned from the review can be used to improve internal controls. Understanding the purpose and structure of postincident reviews and follow-up procedures enables the information security manager to continuously improve the security program. Improving the incident response plan based on the incident review is an internal (corrective) control.

B. A postincident review may result in improvements to controls, but its primary purpose is not to harden a network.

C. The purpose of postincident review is to ensure that the opportunity is presented to learn lessons from the incident. It is not intended as a forum to educate management.

D. An incident may be used to emphasize the importance of incident response, but that is not the intention of the postincident review.

A3-53 An advantage of using sanitized live transactions in test data is that:

A. all transaction types will be included.
B. every error condition is likely to be tested.
C. no special routines are required to assess the results.
D. test transactions are representative of live processing.

D is the correct answer.

Justification:

A. Sanitized production data may not contain all transaction types. The test data may need to be modified to ensure that all data types are represented.

B. Not all error types are sure to be tested because most production data will only contain certain types of errors.

C. The results can be tested using normal routines, but that is not a significant advantage of using sanitized live data.

D. Test data will be representative of live processing; however, it is important that all sensitive information in the live transaction file is sanitized to prevent improper data disclosure.

A3-54 An IS auditor's **PRIMARY** concern when application developers wish to use a copy of yesterday's production transaction file for volume tests is that:

A. users may prefer to use contrived data for testing.
B. unauthorized access to sensitive data may result.
C. error handling and credibility checks may not be fully proven.
D. the full functionality of the new process may not necessarily be tested.

B is the correct answer.

Justification:
A. Production data are easier for users to use for comparison purposes.
B. Unless the data are sanitized, there is a risk of disclosing sensitive data.
C. There is a risk that former production data may not test all error routines; however, this is not as serious as the risk of release of sensitive data.
D. Using a copy of production data may not test all functionality, but this is not as serious as the risk of disclosure of sensitive data.

A3-55 Which of the following is the **PRIMARY** purpose for conducting parallel testing?

A. To determine whether the system is cost-effective
B. To enable comprehensive unit and system testing
C. To highlight errors in the program interfaces with files
D. To ensure the new system meets user requirements

D is the correct answer.

Justification:
A. Parallel testing may show that the old system is, in fact, more cost-effective than the new system, but this is not the primary reason for parallel testing.
B. Unit and system testing are completed before parallel testing.
C. Program interfaces with files are tested for errors during system testing.
D. The purpose of parallel testing is to ensure that the implementation of a new system will meet user requirements by comparing the results of the old system with the new system to ensure correct processing.

A3-56 The knowledge base of an expert system that uses questionnaires to lead the user through a series of choices before a conclusion is reached is known as:

A. rules.
B. decision trees.
C. semantic nets.
D. dataflow diagrams.

B is the correct answer.

Justification:
A. Rules refer to the expression of declarative knowledge through the use of if-then relationships.
B. Decision trees use questionnaires to lead a user through a series of choices until a conclusion is reached.
C. Semantic nets consist of a graph in which nodes represent physical or conceptual objects and the arcs describe the relationship between the nodes.
D. A dataflow diagram is used to map the progress of data through a system and examine logic, error handling and data management.

A3-57 An advantage in using a bottom-up versus a top-down approach to software testing is that:

A. interface errors are detected earlier.
B. confidence in the system is achieved earlier.
C. errors in critical modules are detected earlier.
D. major functions and processing are tested earlier.

C is the correct answer.

Justification:

A. Interface errors will not be found until later in the testing process—as a result of integration or system testing.
B. Confidence in the system cannot be obtained until the testing is completed.
C. The bottom-up approach to software testing begins with the testing of atomic units, such as programs and modules, and works upward until a complete system testing has taken place. The advantages of using a bottom-up approach to software testing are the fact that errors in critical modules are found earlier.
D. Bottom-up testing tests individual components and major functions and processing will not be adequately tested until systems and integration testing is completed.

A3-58 During which of the following phases in system development would user acceptance test plans normally be prepared?

A. Feasibility study
B. Requirements definition
C. Implementation planning
D. Post-implementation review

B is the correct answer.

Justification:

A. The feasibility study is too early for such detailed user involvement.
B. During requirements definition, the project team will be working with the users to define their precise objectives and functional needs. At this time, the users should be working with the team to consider and document how the system functionality can be tested to ensure that it meets their stated needs. An IS auditor should know at what point user testing should be planned to ensure that it is most effective and efficient.
C. The implementation planning phase is when the tests are conducted. It is too late in the process to develop the test plan.
D. User acceptance testing should be completed prior to implementation.

A3-59 The use of object-oriented design and development techniques would **MOST** likely:

A. facilitate the ability to reuse modules.
B. improve system performance.
C. enhance control effectiveness.
D. speed up the system development life cycle.

A is the correct answer.

Justification:
A. One of the major benefits of object-oriented design and development is the ability to reuse modules.
B. Object-oriented design is not intended as a method of improving system performance.
C. Control effectiveness is not an objective of object-oriented design and control effectiveness may, in fact, be reduced through this approach.
D. The use of object-oriented design may speed up the system development life cycle (SDLC) for future projects through the reuse of modules, but it will not speed up development of the initial project.

A3-60 Which of the following should be included in a feasibility study for a project to implement an electronic data interchange process?

A. The encryption algorithm format
B. The detailed internal control procedures
C. The necessary communication protocols
D. The proposed trusted third-party agreement

C is the correct answer.

Justification:
A. Encryption algorithms are too detailed for this phase. They would only be outlined, and any cost or performance implications shown.
B. Internal control procedures are too detailed for this phase. They would only be outlined, and any cost or performance implications shown.
C. The communications protocols must be included because there may be significant cost implications if new hardware and software are involved, and risk implications if the technology is new to the organization.
D. Third-party agreements are too detailed for this phase. They would only be outlined, and any cost or performance implications shown.

A3-61 When a new system is to be implemented within a short time frame, it is **MOST** important to:

A. finish writing user manuals.
B. perform user acceptance testing.
C. add last-minute enhancements to functionalities.
D. ensure that the code has been documented and reviewed.

B is the correct answer.

Justification:
A. The completion of the user manuals is less important than the need to test the system adequately.
B. It would be most important to complete the user acceptance testing to ensure that the system to be implemented is working correctly.
C. If time is tight, the last thing one would want to do is add another enhancement because it would be necessary to freeze the code and complete the testing, then make any other changes as future enhancements.
D. It would be appropriate to have the code documented and reviewed, but unless the acceptance testing is completed, there is no guarantee that the system will work correctly and meet user requirements.

A3-62 Once an organization has finished the business process reengineering (BPR) of all its critical operations, an IS auditor would **MOST** likely focus on a review of:

A. pre-BPR process flowcharts.
B. post-BPR process flowcharts.
C. BPR project plans.
D. continuous improvement and monitoring plans.

B is the correct answer.

Justification:
A. An IS auditor must review the process as it is today, not as it was in the past.
B. An IS auditor's task is to identify and ensure that key controls have been incorporated into the reengineered process.
C. Business process reengineering (BPR) project plans are a step within a BPR project.
D. Continuous improvement and monitoring plans are steps within a BPR project.

A3-63 An IS auditor finds that a system under development has 12 linked modules and each item of data can carry up to 10 definable attribute fields. The system handles several million transactions a year. Which of these techniques could an IS auditor use to estimate the size of the development effort?

A. Program evaluation review technique
B. Function point analysis
C. Counting source lines of code
D. White box testing

B is the correct answer.

Justification:
A. Program evaluation review technique is a project management technique used in the planning and control of system projects.
B. Function point analysis is a technique used to determine the size of a development task based on the number of function points. Function points are factors such as inputs, outputs, inquiries and logical internal sites.
C. The number of source lines of code gives a direct measure of program size, but it does not allow for the complexity that may be caused by having multiple, linked modules and a variety of inputs and outputs.
D. White box testing involves a detailed review of the behavior of program code. It is a quality assurance technique suited to simpler applications during the design and building stage of development.

A3-64 A company has contracted with an external consulting firm to implement a commercial financial system to replace its existing system developed in-house. In reviewing the proposed development approach, which of the following would be of **GREATEST** concern?

A. Acceptance testing is to be managed by users.
B. A quality plan is not part of the contracted deliverables.
C. Not all business functions will be available on initial implementation.
D. Prototyping is being used to confirm that the system meets business requirements.

B is the correct answer.

Justification:
A. Acceptance is normally managed by the user area because users must be satisfied that the new system will meet their requirements.
B. A quality plan is an essential element of all projects. It is critical that the contracted supplier be required to produce such a plan. The quality plan for the proposed development contract should be comprehensive and encompass all phases of the development and include which business functions will be included and when.
C. If the system is large, a phased-in approach to implementing the application is a reasonable approach.
D. Prototyping is a valid method of ensuring that the system will meet business requirements.

A3-65 When preparing a business case to support the need of an electronic data warehouse solution, which of the following choices is the **MOST** important to assist management in the decision-making process?

A. Discuss a single solution.
B. Consider security controls.
C. Demonstrate feasibility.
D. Consult the audit department.

C is the correct answer.

Justification:
A. A business case should discuss all possible solutions to a given problem, which would enable management to select the best option. This may include the option not to undertake the project.
B. It may be important to include security considerations in the business case if security is important to the solution and will address the problem; however, the feasibility study is more important and is necessary regardless of the type of problem.
C. The business case should demonstrate feasibility for any potential project. By including a feasibility study in the business case along with a cost-benefit analysis, management can make an informed decision.
D. While the person preparing the business case may consult with the organization's audit department, this would be situational and is not necessary to include in the business case.

A3-66 Functionality is a characteristic associated with evaluating the quality of software products throughout their life cycle, and is **BEST** described as the set of attributes that bear on the:

A. existence of a set of functions and their specified properties.
B. ability of the software to be transferred from one environment to another.
C. capability of software to maintain its level of performance under stated conditions.
D. relationship between the performance of the software and the amount of resources used.

A is the correct answer.

Justification:
A. Functionality is the set of attributes that bears on the existence of a set of functions and their specified properties. The functionality of a system represents the tasks, operations and purpose of the system in achieving its objective (i.e., supporting a business requirement).
B. The ability of the software to be transferred from one environment to another refers to portability.
C. The capability of software to maintain its level of performance under stated conditions refers to reliability.
D. The relationship between the performance of the software and the amount of resources used refers to efficiency.

A3-67 During the development of an application, quality assurance testing and user acceptance testing were combined. The **MAJOR** concern for an IS auditor reviewing the project is that there will be:

A. increased maintenance.
B. improper documentation of testing.
C. improper acceptance of a program.
D. delays in problem resolution.

C is the correct answer.

Justification:
A. The method of testing used will not affect the maintenance of the system.
B. Quality assurance and user acceptance testing are often led by business representatives according to a defined test plan. The combination of these two tests will not affect documentation.
C. The major risk of combining quality assurance testing and user acceptance testing is that the users may apply pressure to accept a program that meets their needs even though it does not meet quality assurance standards.
D. The method of testing should not affect the time lines for problem resolution.

A3-68 The **GREATEST** advantage of rapid application development over the traditional system development life cycle is that it:

A. facilitates user involvement.
B. allows early testing of technical features.
C. facilitates conversion to the new system.
D. shortens the development time frame.

D is the correct answer.

Justification:
A. Rapid application development (RAD) emphasizes greater user involvement to ensure that the system meets user requirements; however, its primary objective is to speed up development.
B. RAD does allow early testing, but this is also true for the traditional system development life cycle models.
C. RAD does not facilitate conversion to a new system.
D. The greatest advantage and core objective of RAD is a shorter time frame for the development of a system.

A3-69 An IS auditor reviewing a proposed application software acquisition should ensure that the:

A. operating system (OS) being used is compatible with the existing hardware platform.
B. planned OS updates have been scheduled to minimize negative impacts on company needs.
C. OS has the latest versions and updates.
D. product is compatible with the current or planned OS.

D is the correct answer.

Justification:

A. If the operating system (OS) is currently being used, it is compatible with the existing hardware platform; if it were incompatible, it would not operate properly.
B. The planned OS updates should be scheduled to minimize negative impacts on the organization, but this is not an issue when considering the acquisition of new software.
C. The installed OS should be equipped with the most recent versions and updates (with sufficient history and stability). Because this is installed, it is not a consideration at the time of considering acquisition of a new application.
D. In reviewing the proposed application, the auditor should ensure that the products to be purchased are compatible with the current or planned OS.

A3-70 Which of the following is of **GREATEST** concern to an IS auditor when performing an audit of a client relationship management system migration project?

A. The technical migration is planned for a Friday preceding a long weekend, and the time window is too short for completing all tasks.
B. Employees pilot-testing the system are concerned that the data representation in the new system is completely different from the old system.
C. A single implementation is planned, immediately decommissioning the legacy system.
D. Five weeks prior to the target date, there are still numerous defects in the printing functionality of the new system's software.

C is the correct answer.

Justification:

A. A weekend can be used as a time buffer so that the new system will have a better chance of being up and running after the weekend.
B. A different data representation does not mean different data presentation at the front end. Even when this is the case, this issue can be solved by adequate training and user support.
C. Major system migrations should include a phase of parallel operation or a phased cut-over to reduce implementation risk. Decommissioning or disposing of the old hardware would complicate any fallback strategy, should the new system not operate correctly.
D. The printing functionality is commonly one of the last functions to be tested in a new system because it is usually the last step performed in any business event. Thus, meaningful testing and the respective error fixing are only possible after all other parts of the software have been successfully tested.

A3-71 Which of the following types of testing would determine whether a new or modified system can operate in its target environment without adversely impacting other existing systems?

A. Parallel testing
B. Pilot testing
C. Interface/integration testing
D. Sociability testing

D is the correct answer.

Justification:

A. Parallel testing is the process of feeding data into two systems—the modified system and an alternate system—and comparing the results. In this approach, the old and new systems operate concurrently for a period of time and perform the same processing functions. This allows a new system to be tested without affecting existing systems.
B. Pilot testing takes place first at one location and is then extended to other locations. The purpose is to see if the new system operates satisfactorily in one place before implementing it at other locations. In most cases the cutover to the new system will disable existing systems.
C. Interface/integration testing is a hardware or software test that evaluates the connection of two or more components that pass information from one area to another. The objective is to take unit-tested modules and build an integrated structure. This will not test in a true production environment.
D. The purpose of sociability testing is to confirm that a new or modified system can operate in its target environment without adversely impacting existing systems. This should cover the platform that will perform primary application processing and interfaces with other systems, as well as changes to the desktop in a client-server or web development.

A3-72 At the end of the testing phase of software development, an IS auditor observes that an intermittent software error has not been corrected. No action has been taken to resolve the error. The IS auditor should:

A. report the error as a finding and leave further exploration to the auditee's discretion.
B. attempt to resolve the error.
C. recommend that problem resolution be escalated.
D. ignore the error because it is not possible to get objective evidence for the software error.

C is the correct answer.

Justification:

A. Recording it as a minor error and leaving it to the auditee's discretion would be inappropriate. Action should be taken before the application goes into production.
B. The IS auditor is not authorized to resolve the error.
C. When an IS auditor observes such conditions, it is best to fully apprise the auditee and suggest that further problem resolutions be attempted including escalation if necessary.
D. Neglecting the error would indicate that the IS auditor has not taken steps to further probe the issue to its logical end.

A3-73 Which of the following is the **GREATEST** risk to the effectiveness of application system controls?

A. Removal of manual processing steps
B. Inadequate procedure manuals
C. Collusion between employees
D. Unresolved regulatory compliance issues

C is the correct answer.

Justification:

A. Automation should remove manual processing steps wherever possible. The only risk would be the removal of manual security controls without replacement with automated controls.
B. The lack of documentation is a problem on many systems but not a serious risk in most cases.
C. Collusion is an active attack where users collaborate to bypass controls such as separation of duties. Such breaches may be difficult to identify because even well-thought-out application controls may be circumvented.
D. Unregulated compliance issues are a risk but do not measure the effectiveness of the controls.

A3-74 An organization is implementing a new system to replace a legacy system. Which of the following conversion practices creates the **GREATEST** risk?

A. Pilot
B. Parallel
C. Direct cutover
D. Phased

C is the correct answer.

Justification:

A. All other alternatives are done gradually and, thus, provide greater recoverability and are less risky. A pilot implementation is the implementation of the system at a single location or region and then a rollout of the system to the rest of the organization after the application and implementation plan have been proven to work correctly at the pilot location.
B. A parallel test requires running both the old and new system in parallel for a time period. This would highlight any problems or inconsistencies between the old and new systems.
C. Direct cutover implies switching to the new system immediately, usually without the ability to revert to the old system in the event of problems. This is the riskiest approach and may cause a significant impact on the organization.
D. A phased approach is used to implement the system in phases or sections—this minimizes the overall risk by only affecting one area at a time.

A3-75 During the requirements definition stage of a proposed enterprise resource planning system, the project sponsor requests that the procurement and accounts payable modules be linked. Which of the following test methods would be the **BEST** to perform?

A. Unit testing
B. Integration testing
C. Sociability testing
D. Quality assurance testing

B is the correct answer.

Justification:

A. Unit testing is a technique that is used to test program logic within a particular program or module and does not specifically address the linkage between software modules. Integration testing is the best answer.

B. Integration testing is a hardware or software test that evaluates the connection of two or more components that pass information from one area to another. The objective is to take unit-tested modules and build an integrated structure dictated by design.

C. Sociability testing confirms that the new or modified system can operate in its target environment without adversely impacting existing systems and does not specifically address the linkage between software modules. Integration testing is the best answer.

D. Quality assurance testing is primarily used to ensure that the logic of the application is correct and does not specifically address the linkage between software modules. Integration testing is the best answer.

A3-76 During a post-implementation review of an enterprise resource management system, an IS auditor would **MOST** likely:

A. review access control configuration.
B. evaluate interface testing.
C. review detailed design documentation.
D. evaluate system testing.

A is the correct answer.

Justification:

A. Reviewing access control configuration would be the first task performed to determine whether security has been appropriately mapped in the system.

B. Because a post-implementation review is done after user acceptance testing and actual implementation, one would not engage in interface testing or detailed design documentation. Evaluating interface testing would be part of the implementation process.

C. The issue of reviewing detailed design documentation is not generally relevant to an enterprise resource management system because these are usually vendor packages with user manuals. Further, because the system has been implemented, the IS auditor would only check the detailed design if there appeared to be a gap between design and functionality.

D. System testing should be performed before final user signoff. The IS auditor should not need to review the system tests post-implementation.

A3-77 An organization recently deployed a customer relationship management application that was developed in-house. Which of the following is the **BEST** option to ensure that the application operates as designed?

A. User acceptance testing
B. Project risk assessment
C. Post-implementation review
D. Management approval of the system

C is the correct answer.

Justification:

A. User acceptance testing (UAT) verifies that the system functionality has been deemed acceptable by the end users of the system; however, a review of UAT will not validate whether the system is performing as designed because UAT would be performed on a subset of system functionality. The UAT review is a part of the post-implementation review.
B. While a risk assessment would highlight the risk of the system, it would not include an analysis to verify that the system is operating as designed.
C. The purpose of a post-implementation review is to evaluate how successfully the project results match original goals, objectives and deliverables. The post-implementation review also evaluates how effective the project management practices were in keeping the project on track.
D. Management approval of the system could be based on reduced functionality and does not verify that the system is operating as designed. Management approval is a part of post-implementation review.

A3-78 In an online transaction processing system, data integrity is maintained by ensuring that a transaction is either completed in its entirety or not at all. This principle of data integrity is known as:

A. isolation.
B. consistency.
C. atomicity.
D. durability.

C is the correct answer.

Justification:

A. Isolation ensures that each transaction is isolated from other transactions; hence, each transaction can only access data if it is not being simultaneously accessed or modified by another process.
B. Consistency ensures that all integrity conditions in the database are maintained with each transaction.
C. The principle of atomicity requires that a transaction be completed in its entirety or not at all. If an error or interruption occurs, all changes made up to that point are backed out.
D. Durability ensures that, when a transaction has been reported back to a user as complete, the resultant changes to the database will survive subsequent hardware or software failures.

A3-79 A company undertakes a business process reengineering project in support of a new and direct marketing approach to its customers. Which of the following would be an IS auditor's main concern about the new process?

A. Whether key controls are in place to protect assets and information resources
B. Whether the system addresses corporate customer requirements
C. Whether the system can meet the performance goals
D. Whether the new system will support separation of duties

A is the correct answer.

Justification:
A. The audit team must advocate the inclusion of the key controls and verify that the controls are in place before implementing the new process.
B. The system must meet the requirements of all customers not just corporate customers. This is not the IS auditor's main concern.
C. The system must meet performance requirements, but this is of secondary concern to the need to ensure that key controls are in place.
D. Separation of duties is a key control—but only one of the controls that should be in place to protect the assets of the organization.

A3-80 A company has implemented a new client-server enterprise resource planning (ERP) system. Local branches transmit customer orders to a central manufacturing facility. Which of the following would **BEST** ensure that the orders are processed accurately, and the corresponding products are produced?

A. Verifying production of customer orders
B. Logging all customer orders in the ERP system
C. Using hash totals in the order transmitting process
D. Approving (production supervisor) orders prior to production

A is the correct answer.

Justification:
A. Verification of the products produced will ensure that the produced products match the orders in the order system.
B. Logging can be used to detect inaccuracies but does not, in itself, guarantee accurate processing.
C. Hash totals will ensure accurate order transmission, but not accurate processing centrally.
D. Production supervisory approval is a time consuming, manual process that does not guarantee proper control.

A3-81 When two or more systems are integrated, the IS auditor must review input/output controls in the:

A. Systems receiving the output of other systems.
B. Systems sending output to other systems.
C. Systems sending and receiving data.
D. Interfaces between the two systems.

C is the correct answer.

Justification:
A. A responsible control is to protect downstream systems from contamination from an upstream system. This requires a system that sends data to review its output and the receiving system to review its input.
B. Systems sending data to other systems should ensure that the data they send are correct, but that would not protect the receiving system from transmission errors.
C. Both of the systems must be reviewed for input/output controls because the output for one system is the input for the other.
D. The interfaces must be set up correctly and provide error controls, but good practice is to review the data before sending and after receipt.

A3-82 An IS auditor recommends that an initial validation control be programmed into a credit card transaction capture application. The initial validation process would **MOST** likely:

A. check to ensure that the type of transaction is valid for the card type.
B. verify the format of the number entered, then locate it on the database.
C. ensure that the transaction entered is within the cardholder's credit limit.
D. confirm that the card is not shown as lost or stolen on the master file.

B is the correct answer.

Justification:
A. The initial validation would not be used to check the transaction type—just the validity of the card number.
B. The initial validation should confirm whether the card is valid. This validity is established through the card number and personal identification number entered by the user.
C. The initial validation is to prove the card number entered is valid—only then can the transaction amount be checked for approval from the bank.
D. The verification that the card has not been reported as lost or stolen is only done after the card number has been validated as correctly entered.

A3-83 A small company cannot segregate duties between its development processes and its change control function. What is the **BEST** way to ensure that the tested code that is moved into production is the same?

A. Release management software
B. Manual code comparison
C. Regression testing in preproduction
D. Management approval of changes

A is the correct answer.

Justification:

A. Automated release management software can prevent unauthorized changes by moving code into production without any manual intervention.

B. Manual code comparison can detect whether the wrong code has been moved into production; however, code comparison does not prevent the code from being migrated and is not as good a control as using release management software. In addition, manual code comparison is not always efficient and requires highly skilled personnel.

C. Regression testing ensures that changes do not break the current system functionality or unwittingly overwrite previous changes. Regression testing does not prevent untested code from moving into production.

D. Although management should approve every change to production, approvals do not prevent untested code from being migrated into the production environment.

A3-84 Which of the following will **BEST** ensure the successful offshore development of business applications?

A. Stringent contract management practices
B. Detailed and correctly applied specifications
C. Awareness of cultural and political differences
D. Post-implementation review

B is the correct answer.

Justification:

A. Contract management practices, although important, will not ensure successful development if the specifications are incorrect.

B. When dealing with offshore operations, it is essential that detailed specifications be created. Language differences and a lack of interaction between developers and physically remote end users could create gaps in communication in which assumptions and modifications may not be adequately communicated. Inaccurate specifications cannot easily be corrected.

C. Cultural and political differences, although important, should not affect the delivery of a good product.

D. Post-implementation review, although important, is too late in the process to ensure successful project delivery and is not as pivotal to the success of the project.

A3-85 When planning to add personnel to tasks imposing time constraints on the duration of a project, which of the following should be revalidated **FIRST**?

A. The project budget
B. The critical path for the project
C. The length of the remaining tasks
D. The personnel assigned to other tasks

B is the correct answer.

Justification:

A. Given that there may be slack time available on some of the other tasks not on the critical path, the resource allocation should be based on the project segments that affect delivery dates.
B. Adding resources may change the route of the critical path, the critical path must be reevaluated to ensure that additional resources will, in fact, shorten the project duration.
C. Given that there may be slack time available on some of the other tasks not on the critical path, a factor such as the length of other tasks may or may not be affected.
D. Depending on the skill level of the resources required or available, the addition of resources may not, in fact, shorten the time line. Therefore, the first step is to examine what resources are required to address the times on the critical path.

A3-86 When reviewing a project where quality is a major concern, an IS auditor should use the project management triangle to explain that:

A. Increases in quality can be achieved, if resource allocation is decreased.
B. Increases in quality are only achieved if resource allocation is increased.
C. Decreases in delivery time can be achieved, if resource allocation is decreased.
D. Decreases in delivery time can only be achieved if quality is decreased.

A is the correct answer.

Justification:

A. The three primary dimensions of a project are determined by the deliverables, the allocated resources and the delivery time. The area of the project management triangle, comprised of these three dimensions, is fixed. Depending on the degree of freedom, changes in one dimension might be compensated by changing either one or both remaining dimensions. Thus, if resource allocation is decreased, an increase in quality can be achieved if a delay in the delivery time of the project will be accepted. The area of the triangle always remains constant.
B. Increases in quality can be achieved if resource allocation is increased or through increases in delivery time, not only through increases in resource allocation.
C. A decrease in both delivery time and resource allocation would mean that quality would have to decrease.
D. A decrease in delivery time may also be addressed through an increase in resource allocation, even if the quality remains constant.

A3-87 Which of the following is a characteristic of timebox management?

A. Not suitable for prototyping or rapid application development
B. Eliminates the need for a quality process
C. Prevents cost overruns and delivery delays
D. Separates system and user acceptance testing

C is the correct answer.

Justification:
A. Timebox management is very suitable for prototyping and rapid application development.
B. Timebox management does not eliminate the need for a quality process.
C. Timebox management, by its nature, sets specific time and cost boundaries. It is effective in controlling costs and delivery time lines by ensuring that each segment of the project is divided into small controllable time frames.
D. Timebox management integrates system and user acceptance testing.

A3-88 The waterfall life cycle model of software development is **MOST** appropriately used when:

A. requirements are well understood and are expected to remain stable, as is the business environment in which the system will operate.
B. requirements are well understood and the project is subject to time pressures.
C. the project intends to apply an object-oriented design and programming approach.
D. the project will involve the use of new technology.

A is the correct answer.

Justification:
A. Historically, the waterfall model has been best suited to stable conditions and well-defined requirements.
B. When the degree of uncertainty of the system to be delivered and the conditions in which it will be used rises, the waterfall model has not been successful. In these circumstances, the various forms of iterative development life cycle gives the advantage of breaking down the scope of the overall system to be delivered, making the requirements gathering and design activities more manageable. The ability to deliver working software earlier also acts to alleviate uncertainty and may allow an earlier realization of benefits.
C. The choice of a design and programming approach is not, itself, a determining factor of the type of software development life cycle that is appropriate.
D. The use of new technology in a project introduces a significant element of risk. An iterative form of development, particularly one of the agile or exploratory methods that focuses on early development of actual working software, is likely to be the better option to manage this uncertainty.

A3-89 Which of the following is **MOST** critical when creating data for testing the logic in a new or modified application system?

A. A sufficient quantity of data for each test case
B. Data representing conditions that are expected in actual processing
C. Completing the test on schedule
D. A random sample of actual data

B is the correct answer.

Justification:
A. The quantity of data for each test case is not as important as having test cases that will address all types of operating conditions.
B. Selecting the right kind of data is key in testing a computer system. The data should not only include valid and invalid data but should be representative of actual processing; quality is more important than quantity.
C. It is more important to have adequate test data than to complete the testing on schedule.
D. It is unlikely that a random sample of actual data would cover all test conditions and provide a reasonable representation of actual data.

A3-90 Which of the following should an IS auditor review to gain an understanding of the effectiveness of controls over the management of multiple projects?

A. Project database
B. Policy documents
C. Project portfolio database
D. Program organization

C is the correct answer.

Justification:
A. A project database may contain the information about control effectiveness for one specific project and updates to various parameters pertaining to the current status of that single project.
B. Policy documents on project management set direction for the design, development, implementation and monitoring of the project.
C. A project portfolio database is the basis for project portfolio management. It includes project data such as owner, schedules, objectives, project type, status and cost. Project portfolio management requires specific project portfolio reports.
D. Program organization is the team required (steering committee, quality assurance, systems personnel, analyst, programmer, hardware support, etc.) to meet the delivery objectives of the projects.

A3-91 Documentation of a business case used in an IT development project should be retained until:

A. the end of the system's life cycle.
B. the project is approved.
C. user acceptance of the system.
D. the system is in production.

A is the correct answer.

Justification:

A. **A business case can and should be used throughout the life cycle of the product. It serves as an anchor for new (management) personnel, helps to maintain focus and provides valuable information on estimates versus actuals. Questions such as "Why do we do that?", "What was the original intent?" and "How did we perform against the plan?" can be answered, and lessons for developing future business cases can be learned.**
B. The business case should be retained even after project approval to provide ability to review and validate the business case once the project is implemented.
C. The business case will be retained throughout the system development life cycle for later reference and validation.
D. Once the system is in production, the business case can be validated to ensure that the promised costs and benefits were correct.

A3-92 During the review of a web-based software development project, an IS auditor realizes that coding standards are not enforced, and code reviews are rarely carried out. This will **MOST** likely increase the likelihood of a successful:

A. Buffer overflow.
B. Brute force attack.
C. Distributed denial-of-service attack.
D. War dialing attack.

A is the correct answer.

Justification:

A. **Poorly written code, especially in web-based applications, is often exploited by hackers using buffer overflow techniques.**
B. A brute force attack is used to crack passwords, but this is not related to coding standards.
C. A distributed denial-of-service attack floods its target with numerous packets, to prevent it from responding to legitimate requests. This is not related to coding standards.
D. War dialing uses modem-scanning tools to hack private branch exchanges or other telecommunications services.

A3-93 Which testing approach is **MOST** appropriate to ensure that internal application interface errors are identified as soon as possible?

A. Bottom-up testing
B. Sociability testing
C. Top-down testing
D. System testing

C is the correct answer.

Justification:
A. A bottom-up approach to testing begins with atomic units, such as programs and modules, and works upward until a complete system test has taken place.
B. Sociability testing takes place at a later stage in the development process.
C. The top-down approach to testing ensures that interface errors are detected early and that testing of major functions is conducted early.
D. System tests take place at a later stage in the development process.

A3-94 When reviewing input controls, an IS auditor observes that, in accordance with corporate policy, procedures allow supervisory override of data validation edits. The IS auditor should:

A. not be concerned because there may be other compensating controls to mitigate the risk.
B. ensure that overrides are automatically logged and subject to review.
C. verify whether all such overrides are referred to senior management for approval.
D. recommend that overrides not be permitted.

B is the correct answer.

Justification:
A. An IS auditor should not assume that compensating controls exist.
B. If input procedures allow overrides of data validation and editing, automatic logging should occur. A management individual who did not initiate the override should review this log.
C. The log may be reviewed by another manager but does not require senior management approval.
D. As long as the overrides are policy-compliant, there is no need for senior management approval or a blanket prohibition.

A3-95 To minimize the cost of a software project, quality management techniques should be applied:

A. as close to their writing (i.e., point of origination) as possible.
B. primarily at project start to ensure that the project is established in accordance with organizational governance standards.
C. continuously throughout the project with an emphasis on finding and fixing defects primarily through testing to maximize the defect detection rate.
D. mainly at project close-down to capture lessons learned that can be applied to future projects.

C is the correct answer.

Justification:

A. Quality assurance (QA) should start as early as possible but continue through the entire development process.
B. Only performing QA during the start of the project will not detect problems that appear later in the development cycle.
C. Although it is important to properly establish a software development project, quality management should be effectively practiced throughout the project. The major source of unexpected costs on most software projects is rework. The general rule is that the earlier in the development life cycle that a defect occurs, and the longer it takes to find and fix that defect, the more effort will be needed to correct it. A well-written quality management plan is a good start, but it must also be actively applied. Simply relying on testing to identify defects is a relatively costly and less effective way of achieving software quality. For example, an error in requirements discovered in the testing phase can result in scrapping significant amounts of work.
D. Capturing lessons learned will be too late for the current project. Additionally, applying quality management techniques throughout a project is likely to yield its own insights into the causes of quality problems and assist in staff development.

A3-96 When identifying an earlier project completion time, which is to be obtained by paying a premium for early completion, the activities that should be selected are those:

A. whose sum of activity time is the shortest.
B. that have zero slack time.
C. that give the longest possible completion time.
D. whose sum of slack time is the shortest.

B is the correct answer.

Justification:

A. Attention should focus on the tasks within the critical path that have no slack time.
B. A critical path's activity time is longer than that for any other path through the network. This path is important because if everything goes as scheduled, its length gives the shortest possible completion time for the overall project. Activities on the critical path become candidates for crashing (i.e., for reduction in their time by payment of a premium for early completion). Activities on the critical path have zero slack time and conversely, activities with zero slack time are on a critical path. By successively relaxing activities on a critical path, a curve showing total project costs versus time can be obtained.
C. The critical path is the longest time length of the activities but is not based on the longest time of any individual activity.
D. A task on the critical path has no slack time.

A3-97 An IS auditor is assigned to audit a software development project, which is more than 80 percent complete, but has already overrun time by 10 percent and costs by 25 percent. Which of the following actions should the IS auditor take?

A. Report that the organization does not have effective project management
B. Recommend the project manager be changed
C. Review the IT governance structure
D. Review the business case and project management

D is the correct answer.

Justification:

A. The organization may have effective project management practices and still be behind schedule or over budget.
B. There is no indication that the project manager should be changed without looking into the reasons for the overrun.
C. The organization may have sound IT governance and still be behind schedule or over budget.
D. Before making any recommendations, an IS auditor needs to understand the project and the factors that have contributed to bringing the project over budget and over schedule.

A3-98 Which of the following should an IS auditor review to understand project progress in terms of time, budget and deliverables for early detection of possible overruns and for projecting estimates at completion?

A. Function point analysis
B. Earned value analysis
C. Cost budget
D. Program evaluation and review technique

B is the correct answer.

Justification:

A. Function point analysis is an indirect measure of software size and complexity and, therefore, does not address the elements of time and budget.
B. Earned value analysis (EVA) is an industry standard method for measuring a project's progress at any given point in time, forecasting its completion date and final cost, and analyzing variances in the schedule and budget as the project proceeds. It compares the planned amount of work with what has actually been completed to determine if the cost, schedule and work accomplished are progressing in accordance with the plan. EVA works most effectively if a well-formed work breakdown structure exists.
C. Cost budgets do not address time.
D. Program evaluation and review technique aids time and deliverables management but lacks projections for estimates at completion and overall financial management.

A3-99 Which of the following system and data conversion strategies provides the **GREATEST** redundancy?

A. Direct cutover
B. Pilot study
C. Phased approach
D. Parallel run

D is the correct answer.

Justification:

A. Direct cutover is actually quite risky because it does not provide for a "shake down period" nor does it provide an easy fallback option.
B. A pilot study approach is performed incrementally, making rollback procedures difficult to execute.
C. A phased approach is performed incrementally, making rollback procedures difficult to execute.
D. Parallel runs are the safest—though the most expensive—approach because both the old and new systems are run, thus incurring what might appear to be double costs.

A3-100 Which of the following should be developed during the requirements definition phase of a software development project to address aspects of software testing?

A. Test data covering critical applications
B. Detailed test plans
C. Quality assurance test specifications
D. User acceptance test specifications

D is the correct answer.

Justification:

A. Test data will usually be created during the system testing phase.
B. Detailed test plans are created during system testing.
C. Quality assurance test specifications are set out later in the development process.
D. A key objective in any software development project is to ensure that the developed software will meet the business objectives and the requirements of the user. The users should be involved in the requirements definition phase of a development project and user acceptance test specification should be developed during this phase.

A3-101 At the completion of a system development project, a post-project review should include which of the following?

A. Assessing risk that may lead to downtime after the production release
B. Identifying lessons learned that may be applicable to future projects
C. Verifying that the controls in the delivered system are working
D. Ensuring that test data are deleted

B is the correct answer.

Justification:

A. An assessment of potential downtime should be made with the operations group and other specialists before implementing a system.
B. A project team has something to learn from each and every project. As risk assessment is a key issue for project management, it is important for the organization to accumulate lessons learned and integrate them into future projects.
C. Verifying that controls are working should be covered during the acceptance test phase and possibly, again in the post-implementation review. The post-project review will focus on project-related issues.
D. Test data should be retained for future regression testing.

A3-102 An IS auditor has been asked to participate in project initiation meetings for a critical project. The IS auditor's **MAIN** concern should be that the:

A. complexity and risk associated with the project have been analyzed.
B. resources needed throughout the project have been determined.
C. technical deliverables have been identified.
D. a contract for external parties involved in the project has been completed.

A is the correct answer.

Justification:
A. Understanding complexity and risk, and actively managing these throughout a project are critical to a successful outcome.
B. The resources needed will be dependent on the complexity of the project.
C. It is too early to identify the technical deliverables.
D. Not all projects will require contracts with external parties.

A3-103 From a risk management point of view, the **BEST** approach when implementing a large and complex IT infrastructure is:

A. a major deployment after proof of concept.
B. prototyping and a one-phase deployment.
C. a deployment plan based on sequenced phases.
D. to simulate the new infrastructure before deployment.

C is the correct answer.

Justification:
A. A major deployment would pose a higher risk of implementation failure.
B. Prototyping may reduce development failure, but a large environment will usually require a phased approach.
C. When developing a large and complex IT infrastructure, a good practice is to use a phased approach to fit the entire system together. This will provide greater assurance of quality results.
D. It is not usually feasible to simulate a large and complex IT infrastructure prior to deployment.

A3-104 When reviewing an active project, an IS auditor observed that the business case was no longer valid because of a reduction in anticipated benefits and increased costs. The IS auditor should recommend that the:

A. project be discontinued.
B. business case be updated and possible corrective actions be identified.
C. project be returned to the project sponsor for re-approval.
D. project be completed and the business case be updated later.

B is the correct answer.

Justification:
A. An IS auditor should not recommend discontinuing or completing the project before reviewing an updated business case.
B. The IS auditor should recommend that the business case be kept current throughout the project because it is a key input to decisions made throughout the life of any project.
C. The project cannot be returned to the sponsor until the business case has been updated.
D. An IS auditor should not recommend completing the project before reviewing an updated business case and ensuring approval from the project sponsor.

A3-105 Which of the following is an advantage of the top-down approach to software testing?

A. Interface errors are identified early
B. Testing can be started before all programs are complete
C. It is more effective than other testing approaches
D. Errors in critical modules are detected sooner

A is the correct answer.

Justification:

A. The advantage of the top-down approach is that tests of major functions are conducted early, thus enabling the detection of interface errors sooner.
B. That testing can be started before all programs are complete is an advantage of the bottom-up approach to system testing.
C. The most effective testing approach is dependent on the environment being tested.
D. Detecting errors in critical modules sooner is an advantage of the bottom-up approach to system testing.

A3-106 During the system testing phase of an application development project the IS auditor should review the:

A. conceptual design specifications.
B. vendor contract.
C. error reports.
D. program change requests.

C is the correct answer.

Justification:

A. A conceptual design specification is a document prepared during the requirements definition phase. The system testing will be based on a test plan.
B. A vendor contract is prepared during a software acquisition process and may be reviewed to ensure that all the deliverables in the contract have been delivered, but the most important area of review is the error reports.
C. Testing is crucial in determining that user requirements have been validated. The IS auditor should be involved in this phase and review error reports for their precision in recognizing erroneous data and review the procedures for resolving errors.
D. Program change requests would be reviewed normally as a part of the post-implementation phase.

A3-107 Which of the following would be the **MOST** cost-effective recommendation for reducing the number of defects encountered during software development projects?

A. Increase the time allocated for system testing
B. Implement formal software inspections
C. Increase the development staff
D. Require the sign-off of all project deliverables

B is the correct answer.

Justification:

A. Allowing more time for testing may discover more defects; however, little is revealed as to why the quality problems are occurring, and the cost of the extra testing and the cost of rectifying the defects found will be greater than if they had been discovered earlier in the development process.
B. Inspections of code and design are a proven software quality technique. An advantage of this approach is that defects are identified before they propagate through the development life cycle. This reduces the cost of correction because less rework is involved.
C. The ability of the development staff can have a bearing on the quality of what is produced; however, replacing staff can be expensive and disruptive, and the presence of a competent staff cannot guarantee quality in the absence of effective quality management processes.
D. Sign-off of deliverables may help detect defects if signatories are diligent about reviewing deliverable content; however, this is difficult to enforce and may occur too late in the process to be cost-effective. Deliverable reviews normally do not go down to the same level of detail as software inspections.

A3-108 An IS auditor invited to a project development meeting notes that no project risk has been documented. When the IS auditor raises this issue, the project manager responds that it is too early to identify risk and that, if risk starts impacting the project, a risk manager will be hired. The appropriate response of the IS auditor would be to:

A. stress the importance of spending time at this point in the project to consider and document risk and to develop contingency plans.
B. accept the project manager's position because the project manager is accountable for the outcome of the project.
C. offer to work with the risk manager when one is appointed.
D. inform the project manager that the IS auditor will conduct a review of the risk at the completion of the requirements definition phase of the project.

A is the correct answer.

Justification:

A. The majority of project risk can be identified before a project begins, allowing mitigation/avoidance plans to be put in place to deal with this risk. A project should have a clear link back to corporate strategy, enterprise risk management, and tactical plans to support this strategy. The process of setting corporate strategy, setting objectives and developing tactical plans should include the consideration of risk.
B. The project manager cannot accept responsibility for risk acceptance. The risk must be addressed continuously—starting as early in the process as possible.
C. Appointing a risk manager is a good practice but waiting until the project has been impacted by risk is misguided. Risk management needs to be forward looking; allowing risk to evolve into issues that adversely impact the project represents a failure of risk management. With or without a risk manager, persons within and outside of the project team need to be consulted and encouraged to comment when they believe new risk has emerged or risk priorities have changed. The IS auditor has an obligation to the project sponsor and the organization to advise on appropriate project management practices. Waiting for the possible appointment of a risk manager represents an unnecessary and dangerous delay to implement risk management.
D. IS auditors cannot provide risk review without impairing their independence.

A3-109 The **MAIN** purpose of a transaction audit trail is to:

A. reduce the use of storage media.
B. determine accountability and responsibility for processed transactions.
C. help an IS auditor trace transactions.
D. provide useful information for capacity planning.

B is the correct answer.

Justification:
A. Enabling audit trails increases the use of disk space.
B. Enabling audit trails aids in establishing the accountability and responsibility for processed transactions by tracing them through the information system.
C. A transaction log file would be used to trace transactions, but the primary purpose of an audit trail is to support accountability, not to support the work of the IS auditor.
D. The objective of capacity planning is the efficient and effective use of IT resources and requires information such as central processing unit utilization, bandwidth and the number of users.

A3-110 An organization is implementing an enterprise resource planning application. Of the following, who is **PRIMARILY** responsible for overseeing the project to ensure that it is progressing in accordance with the project plan and that it will deliver the expected results?

A. Project sponsor
B. System development project team
C. Project steering committee
D. User project team

C is the correct answer.

Justification:
A. A project sponsor is typically the senior manager in charge of the primary business unit that the application will support. The sponsor provides funding for the project and works closely with the project manager to define the critical success factors or metrics for the project. The project sponsor is not responsible for reviewing the progress of the project.
B. A system development project team completes the assigned tasks, works according to the instructions of the project manager and communicates with the user project team. The SDPT is not responsible for overseeing the progress of the project.
C. A project steering committee that provides an overall direction for the enterprise resource planning (ERP) implementation project is responsible for reviewing the project's progress to ensure that it will deliver the expected results.
D. A user project team (UPT) completes the assigned tasks, communicates effectively with the system development team and works according to the advice of the project manager. A UPT is not responsible for reviewing the progress of the project.

A3-111 A legacy payroll application is migrated to a new application. Which of the following stakeholders should be **PRIMARILY** responsible for reviewing and signing-off on the accuracy and completeness of the data before going live?

A. IS auditor
B. Database administrator
C. Project manager
D. Data owner

D is the correct answer.

Justification:

A. An IS auditor should ensure that there is a review and sign-off by the data owner during the data conversion stage of the project.
B. A database administrator's primary responsibility is to maintain the integrity of the database and make the database available to users. A database administrator is not responsible for reviewing migrated data.
C. A project manager provides day-to-day management and leadership of the project but is not responsible for the accuracy and integrity of the data.
D. During the data conversion stage of a project, the data owner is primarily responsible for reviewing and signing-off that the data are migrated completely and accurately and are valid. An IS auditor is not responsible for reviewing and signing-off on the accuracy of the converted data.

A3-112 An organization is migrating from a legacy system to an enterprise resource planning system. While reviewing the data migration activity, the **MOST** important concern for the IS auditor is to determine that there is a:

A. correlation of semantic characteristics of the data migrated between the two systems.
B. correlation of arithmetic characteristics of the data migrated between the two systems.
C. correlation of functional characteristics of the processes between the two systems.
D. relative efficiency of the processes between the two systems.

A is the correct answer.

Justification:

A. Due to the fact that the two systems could have a different data representation, including the database schema, the IS auditor's main concern should be to verify that the interpretation of the data (structure) is the same in the new as it was in the old system.
B. Arithmetic characteristics represent aspects of data structure and internal definition in the database and, therefore, are less important than the semantic characteristics.
C. A review of the correlation of the functional characteristics between the two systems is not relevant to a data migration review.
D. A review of the relative efficiencies of the processes between the two systems is not relevant to a data migration review.

A3-113 Normally, it would be essential to involve which of the following stakeholders in the initiation stage of a project?

A. System owners
B. System users
C. System designers
D. System builders

A is the correct answer.

Justification:

A. **System owners are the information systems (project) sponsors or chief advocates. They normally are responsible for initiating and funding projects to develop, operate and maintain information systems.**
B. System users are the individuals who use or are affected by the information system. Their requirements are crucial in the requirements definition, design and testing stages of a project.
C. System designers translate business requirements and constraints into technical solutions.
D. System builders construct the system based on the specifications from the systems designers. In most cases, the designers and builders are one and the same.

A3-114 A project manager for a project that is scheduled to take 18 months to complete announces that the project is in a healthy financial position because, after six months, only one-sixth of the budget has been spent. The IS auditor should **FIRST** determine:

A. the amount of progress achieved compared to the project schedule.
B. if the project budget can be reduced.
C. if the project could be brought in ahead of schedule.
D. if the budget savings can be applied to increase the project scope.

A is the correct answer.

Justification:

A. **Cost performance of a project cannot be properly assessed in isolation of schedule performance. Cost cannot be assessed simply in terms of elapsed time on a project.**
B. To properly assess the project budget position, it is necessary to know how much progress has actually been made and, given this, what level of expenditure would be expected. It is possible that project expenditure appears to be low because actual progress has been slow. Until the analysis of project against schedule has been completed, it is impossible to know whether there is any reason to reduce budget. If the project has slipped behind schedule, then not only may there be no spare budget, but it is possible that extra expenditure may be needed to retrieve the slippage. The low expenditure could actually be representative of a situation where the project is likely to miss deadlines rather than potentially come in ahead of time.
C. If the project is found to be ahead of budget after adjusting for actual progress, this is not necessarily a good outcome because it points to flaws in the original budgeting process; and, as said previously, until further analysis is undertaken, it cannot be determined whether any spare funds actually exist.
D. If the project is behind schedule, adding scope may be the wrong thing to do.

A3-115 The **MAJOR** advantage of a component-based development approach is the:

A. ability to manage an unrestricted variety of data types.
B. provision for modeling complex relationships.
C. capacity to meet the demands of a changing environment.
D. support of multiple development environments.

D is the correct answer.

Justification:

A. The data types must be defined within each component, and it is not sure that any component will be able to handle multiple data types.
B. Component-based development is no better than many other development methods at modeling complex relationships.
C. Component-based development is one of the methodologies that can be effective at meeting changing requirements, but this is not its primary benefit or purpose.
D. Component-based development that relies on reusable modules can increase the speed of development. Software developers can then focus on business logic.

A3-116 The specific advantage of white box testing is that it:

A. verifies a program can operate successfully with other parts of the system.
B. ensures a program's functional operating effectiveness without regard to the internal program structure.
C. determines procedural accuracy or conditions of a program's specific logic paths.
D. examines a program's functionality by executing it in a tightly controlled or virtual environment with restricted access to the host system.

C is the correct answer.

Justification:

A. Verifying the program can operate successfully with other parts of the system is sociability testing.
B. Testing the program's functionality without knowledge of internal structures is black box testing.
C. White box testing assesses the effectiveness of software program logic. Specifically, test data are used in determining procedural accuracy or conditions of a program's logic paths.
D. Controlled testing of programs in a semi-debugged environment, either heavily controlled step-by-step or via monitoring in virtual machines, is sand box testing.

A3-117 Following good practices, formal plans for implementation of new information systems are developed during the:

A. development phase.
B. design phase.
C. testing phase.
D. deployment phase.

B is the correct answer.

Justification:

A. The implementation plans are updated during the development of the system, but the plans were already addressed during the design phase.
B. The method of implementation may affect the design of the system. Therefore, planning for implementation should begin well in advance of the actual implementation date. A formal implementation plan should be constructed in the design phase and revised as the development progresses.
C. The testing phase focuses on testing the system and is not concerned with implementation planning.
D. The deployment phase implements the system according to the plans set out earlier in the design phase.

A3-118 An IS auditor is reviewing a project that is using an agile software development approach. Which of the following should the IS auditor expect to find?

A. Use of a capability maturity model
B. Regular monitoring of task-level progress against schedule
C. Extensive use of software development tools to maximize team productivity
D. Postiteration reviews that identify lessons learned for future use in the project

D is the correct answer.

Justification:

A. The capability maturity model places heavy emphasis on predefined formal processes and formal project management and software development deliverables, while agile software development projects, by contrast, rely on refinement of process as dictated by the particular needs of the project and team dynamics.
B. Task-level tracking is not used because daily meetings identify challenges and impediments to the project.
C. Agile projects make use of suitable development tools; however, tools are not seen as the primary means of achieving productivity. Team harmony, effective communications and collective ability to solve challenges are of greater importance.
D. A key tenet of the agile approach to software project management is ongoing team learning to refine project management and software development processes as the project progresses. One of the best ways to achieve this is that the team considers and documents what worked well and what could have worked better at the end of each iteration and identifies improvements to be implemented in subsequent iterations. Additionally, less importance is placed on formal paper-based deliverables, with the preference being effective informal communication within the team and with key outside contributors. Agile projects produce releasable software in short iterations, typically ranging from four to eight weeks. This, in itself, instills considerable performance discipline within the team. This, combined with short daily meetings to agree on what the team is doing and the identification of any impediments, renders task-level tracking against a schedule redundant.

A3-119 An organization sells books and music online at its secure web site. Transactions are transferred to the accounting and delivery systems every hour to be processed. Which of the following controls **BEST** ensures that sales processed on the secure web site are transferred to both the delivery and accounting systems?

A. Transaction totals are recorded on a daily basis in the sales systems. Daily sales system totals are aggregated and totaled.
B. Transactions are automatically numerically sequenced. Sequences are checked and gaps in continuity are accounted for.
C. Processing systems check for duplicated transaction numbers. If a transaction number is duplicated (already present), it is rejected.
D. System time is synchronized hourly using a centralized time server. All transactions have a date/time stamp.

B is the correct answer.

Justification:

A. Totaling transactions on the sales system does not address the transfer of data from the online systems to the accounting system, but rather considers only the sales system.
B. Automatic numerical sequencing is the only option that accounts for completeness of transactions because any missing transactions would be identified by a gap.
C. Checking for duplicates is a valid control; however, it does not address whether the sales transactions processed are complete (ensuring that all transactions are recorded).
D. A date/time stamp does not help account for transactions that are missing or incomplete by the accounting and delivery department.

A3-120 Which of the following techniques would **BEST** help an IS auditor gain reasonable assurance that a project can meet its target date?

A. Estimation of the actual end date based on the completion percentages and estimated time to complete, taken from status reports
B. Confirmation of the target date based on interviews with experienced managers and staff involved in the completion of the project deliverables
C. Extrapolation of the overall end date based on completed work packages and current resources
D. Calculation of the expected end date based on current resources and remaining available project budget

C is the correct answer.

Justification:
A. The IS auditor cannot count on the accuracy of data in status reports for reasonable assurance.
B. Interviews are a valuable source of information but will not necessarily identify any project challenges because the people being interviewed are involved in project.
C. Direct observation of results is better than estimations and qualitative information gained from interviews or status reports. Project managers and involved staff tend to underestimate the time needed for completion and the necessary time buffers for dependencies between tasks, while overestimating the completion percentage for tasks underway (i.e., 80:20 rule).
D. The calculation based on remaining budget does not consider the speed at which the project has been progressing.

A3-121 An IS auditor finds that user acceptance testing of a new system is being repeatedly interrupted by defect fixes from the developers. Which of the following would be the **BEST** recommendation for an IS auditor to make?

A. Consider the feasibility of a separate user acceptance environment
B. Schedule user testing to occur at a given time each day
C. Implement a source code version control tool
D. Only retest high-priority defects

A is the correct answer.

Justification:
A. A separate environment or environments is normally necessary for testing to be efficient and effective and to ensure the integrity of production code. It is important that the development and test code bases be separate. When defects are identified they can be fixed in the development environment, without interrupting testing, before being migrated in a controlled manner to the test environment. A separate test environment can also be used as the final staging area from which code is migrated to production. This enforces a separation between development and production code. The logistics of setting up and refreshing customized test data is easier if a separate environment is maintained.
B. If developers and testers are sharing the same environment, they have to work effectively at separate times of the day. It is unlikely that this would provide optimum productivity.
C. Use of a source code control tool is a good practice, but it does not properly mitigate the lack of an appropriate test environment.
D. Even low priority fixes run the risk of introducing unintended results when combined with the rest of the system code. To prevent this, regular regression testing covering all code changes should occur. A separate test environment makes the logistics of regression testing easier to manage.

A3-122 An IS auditor has found time constraints and expanded needs to be the root causes for recent violations of corporate data definition standards in a new business intelligence project. Which of the following is the **MOST** appropriate suggestion for an auditor to make?

A. Achieve standards alignment through an increase of resources devoted to the project
B. Align the data definition standards after completion of the project
C. Delay the project until compliance with standards can be achieved
D. Enforce standard compliance by adopting punitive measures against violators

A is the correct answer.

Justification:

A. Provided that data architecture, technical and operational requirements are sufficiently documented, the alignment to standards could be treated as a specific work package assigned to new project resources.

B. The usage of nonstandard data definitions would lower the efficiency of the new development and increase the risk of errors in critical business decisions. To change data definition standards after project conclusion is risky and is not a viable solution.

C. Delaying the project would be an inappropriate suggestion because of business requirements or the likely damage to entire project profitability.

D. Punishing the violators would be outside the authority of the auditor and inappropriate until the reason for the violations have been determined.

A3-123 What is the **PRIMARY** reason that an IS auditor would verify that the process of post-implementation review of an application was completed after a release?

A. To make sure that users are appropriately trained
B. To verify that the project was within budget
C. To check that the project meets expectations
D. To determine whether proper controls were implemented

C is the correct answer.

Justification:

A. Post-implementation review does not target verifying user training needs.

B. Project costs are monitored during development and are not the primary reason for a post-implementation review.

C. The objective of a post-implementation review is to reveal whether the implementation of a system has achieved planned objectives (i.e., meets business objectives and risk acceptance criteria).

D. While an IS auditor would be interested in ensuring that proper controls were implemented, the most important consideration would be that the project meets expectations.

A3-124 An IS auditor is reviewing an enterprise's system development testing policy. Which of the following statements concerning use of production data for testing would the IS auditor consider to be **MOST** appropriate?

A. Senior IS and business management must approve use before production data can be used for testing
B. Production data can be used if they are copied to a secure test environment
C. Production data can never be used. All test data must be developed and based on documented test cases
D. Production data can be used provided that confidentiality agreements are in place

A is the correct answer.

Justification:

A. **There is risk associated with the use of production data for testing. These include compromising customer or employee confidentiality (which may also involve breaching legislation) and corrupting production of the data. Additionally, there are certain cases in which effective testing requires specifically designed data. There are other cases in which using production data would provide insights that are difficult or impossible to get from manufactured test data. One example is testing of interfaces to legacy systems. Management information systems are a further example where access to "real" data is likely to enhance testing. Some flexibility on the use of production data is likely to be the best option. In addition to obtaining senior management approval, conditions that mitigate the risk associated with using production data can be agreed on, such as masking names and other identifying fields to protect privacy.**

B. Copying production data to a secure environment is a good practice, but this should only be done with the approval of management. Management must accept the risk of using production data for testing.

C. Creating a complete set of test data would be an ideal situation but is not always possible due to the volume of test data that would be required.

D. Production data could only be used with management's permission. Then it can be appropriate to require the use of confidentiality agreements.

A3-125 An enterprise is developing a new procurement system, and things are behind schedule. As a result, it is proposed that the time originally planned for the test phase be shortened. The project manager asks the IS auditor for recommendations to mitigate the risk associated with reduced testing. Which of the following is a suitable risk mitigation strategy?

A. Test and release a pilot with reduced functionality
B. Fix and retest the highest-severity functional defects
C. Eliminate planned testing by the development team, and proceed straight to acceptance testing
D. Implement a test tool to automate defect tracking

A is the correct answer.

Justification:

A. Testing and releasing a pilot with reduced functionality reduces risk in a number of ways. Reduced functionality should result in fewer overall test cases to run and defects to fix and retest, and in less regression testing. A pilot release made available to a select group of users will reduce the risk associated with a full implementation. All of the benefits of releasing the system to the full user population will not be realized, but some benefits should start to flow. Additionally, some useful comments from real users should be obtained to guide what extra functionality and other improvements need to be included in a full release.

B. When testing starts, a significant number of defects is likely to exist. Focusing only on the highest-severity functional defects runs the risk that other important aspects such as usability problems and nonfunctional requirements of performance and security will be ignored. The system may go live, but users may struggle to use the system as intended to realize business benefits.

C. Eliminating testing by development is usually a bad idea. Before system acceptance testing begins, some prior testing should occur to establish that the system is ready to proceed to acceptance evaluation. If prior testing by the development team does not occur, there is a considerable risk that the software will have a significant number of low-level defects, such as transactions that cause the system to hang and unintelligible error messages. This can prove frustrating for users or testers tasked with acceptance testing and, ultimately, could cause the overall test time to increase rather than decrease.

D. The use of a defect tracking tool could help in improving test efficiency, but it does not address the fundamental risk caused by reducing the testing effort on a system in which quality is uncertain. Given the build problems experienced, there is reason to suspect that quality problems could exist.

A3-126 An IS auditor is involved in the reengineering process that aims to optimize IT infrastructure. Which of the following will **BEST** identify the issues to be resolved?

A. Self-assessment
B. Reverse engineering
C. Prototyping
D. Gap analysis

D is the correct answer.

Justification:

A. Self-assessment may be one of the viable options with which to start; however, the results only indicate current conditions, not desired state, and tend to become subjective.

B. Reverse engineering is a technique applied to analyze how a device or program works and is not appropriate here.

C. Prototyping is applied to ensure that user requirements are met prior to being engaged in a full-blown development process.

D. Gap analysis would be the best method to identify issues that need to be addressed in the reengineering process. Gap analysis indicates which parts of current processes conform to good practices (desired state) and which do not.

A3-127 An IS audit group has been involved in the integration of an automated audit tool kit with an existing enterprise resource planning system. Due to ERP performance issues, the audit tool kit is not permitted to go live. What should the IS auditor's **BEST** recommendation be?

A. Review the implementation of selected integrated controls
B. Request additional IS audit resources
C. Request vendor technical support to resolve performance issues
D. Review the results of stress tests during user acceptance testing

D is the correct answer.

Justification:

A. Reviewing the implementation of selected integrated controls validates the technical design and the control objective, but integrated controls over transactional tables consume large resources. They should be reviewed carefully to determine whether they are mandatory or can be implemented and integrated for only specific transactions over the enterprise resource planning application.
B. The inability to implement the automated tool may necessitate additional audit resources because many audits will require more manual effort; however, the first step should be to try to resolve the performance issues.
C. Requesting vendor technical support to resolve performance issues is a good option, but not the first recommendation.
D. The appropriate recommendation is to review the results of stress tests during user acceptance testing that demonstrated the performance issues.

A3-128 What is the **BEST** method to facilitate successful user testing and acceptance of a new enterprise resource planning payroll system that is replacing an existing legacy system?

A. Multiple testing
B. Parallel testing
C. Integration testing
D. Prototype testing

B is the correct answer.

Justification:

A. Multiple testing will not compare results from the old and new systems.
B. Parallel testing is the best method for testing data results and system behavior because it allows the users to compare results from both systems before decommissioning the legacy system. Parallel testing also results in better user adoption of the new system.
C. Integration testing refers to how the system interacts with other systems, and it is not performed by end users.
D. Prototype testing is used during design and development to ensure that user input is received; however, this method is not used for acquired systems or during user acceptance testing.

A3-129 A rapid application development methodology has been selected to implement a new enterprise resource planning system. All of the project activities have been assigned to the contracted consulting company because internal employees are not available. What is the IS auditor's **FIRST** step to compensate for the lack of resources?

A. Review the project plan and approach
B. Ask the vendor to provide additional external staff
C. Recommend that the company hire more people
D. Stop the project until all human resources are available

A is the correct answer.

Justification:

A. **Rapid methodologies require available resources with good expertise and a fast decision-making process because the plan duration is usually short. Reviewing the project plan and approach is the best recommendation to make the appropriate changes to compensate for the missing end users.**
B. Adding external people to the project will not resolve the problem because they will not be able to decide on behalf of the internal employees who are usually end users from the business side.
C. Hiring new people will take time and does not guarantee the readiness of new hires to make appropriate decisions in this project.
D. Stopping the project could be a good option but reviewing the project and considering all of the aspects should be done first.

A3-130 An IS auditor who is auditing the software acquisition process will ensure that the:

A. contract is reviewed and approved by the legal counsel before it is signed.
B. requirements cannot be met with the systems already in place.
C. requirements are found to be critical for the business.
D. user participation is adequate in the process.

A is the correct answer.

Justification:

A. **The process to review and approve the contract is one of the most important steps in the software acquisition process. An IS auditor should verify that legal counsel reviewed and approved the contract before management signs the contract.**
B. Existing systems may meet the requirements, but management may choose to acquire software for other reasons.
C. Not all of the requirements in the contract need to support critical business needs; some requirements may be there for ease-of-use or other purposes.
D. User participation is not necessarily required in the software acquisition process. Instead, users would most likely participate in requirements definition and user acceptance testing.

A3-131 Which of the following controls helps prevent duplication of vouchers during data entry?

A. A range check
B. Transposition and substitution
C. A sequence check
D. A cyclic redundancy check

C is the correct answer.

Justification:
A. A range check works over a range of numbers. Even if the same voucher number reappears, it will satisfy the range and, therefore, not be useful.
B. Transposition and substitution are used in encoding but will not help in establishing unique voucher numbers.
C. A sequence check involves increasing the order of numbering and would validate whether the vouchers are in sequence and, thus, prevent duplicate vouchers.
D. A cyclic redundancy check is used for completeness of data received over the network but is not useful in application code level validations.

A3-132 Which of the following test techniques would the IS auditor use to identify specific program logic that has not been tested?

A. A snapshot
B. Tracing and tagging
C. Logging
D. Mapping

D is the correct answer.

Justification:
A. A snapshot records the flow of designated transactions through logic paths within programs.
B. Tracing and tagging shows the trail of instructions executed during an application.
C. Logging is the activity of recording specific tasks for future review.
D. Mapping identifies specific program logic that has not been tested and analyzes programs during execution to indicate whether program statements have been executed.

A3-133 The **PRIMARY** objective of conducting a post-implementation review for a business process automation project is to:

A. ensure that the project meets the intended business requirements.
B. evaluate the adequacy of controls.
C. confirm compliance with technological standards.
D. confirm compliance with regulatory requirements.

A is the correct answer.

Justification:
A. Ensuring that the project meets the intended business requirements is the primary objective of a post-implementation review.
B. Evaluating the adequacy of controls may be part of the review but is not the primary objective.
C. Confirming compliance with technological standards is normally not part of the post-implementation review because this should be addressed during the design and development phase.
D. Confirming compliance with regulatory requirements is normally not part of the post-implementation review because this should be addressed during the design and development phase.

A3-134 While evaluating the "out of scope" section specified in a project plan, an IS auditor should ascertain whether the section:

A. effectively describes unofficial project objectives.
B. effectively describes project boundaries.
C. clearly states the project's "nice to have" objectives.
D. provides the necessary flexibility to the project team.

B is the correct answer.

Justification:

A. Out-of-scope items are not part of the project. There should be no unofficial project objectives. Reasonable objectives should be considered by the project leadership and either accepted (in scope) or rejected (out of scope).

B. The purpose of the out of scope section is to make clear to readers what items are not considered project objectives so that all project stakeholders understand the project boundaries and what is in scope versus out of scope. This applies to all types of projects, including individual audits.

C. Out-of-scope items are not part of the project, while nice to have items may be included in the project objectives. However, they may be the last priority on the list of all project objectives.

D. Out-of-scope items are not part of the project; the project team's flexibility regarding project objectives should be managed through a robust change request process. This is particularly important to avoid scope creep.

A3-135 An IS auditor assesses the project management process for an internal software development project. In respect to the software functionality, the IS auditor should look for sign-off by:

A. the project manager.
B. systems development management.
C. business unit management.
D. the quality assurance team.

C is the correct answer.

Justification:

A. The project manager provides day-to-day management and leadership of the project and ensures that project activities remain in line with the overall direction. The project manager cannot sign off on project requirements; that would be a violation of separation of duties.

B. Systems development management provides technical support for hardware and software environments.

C. Business unit management assumes ownership of the project and the resulting system. It is responsible for acceptance testing and confirming that the required functions are available in the software.

D. The quality assurance team ensures the quality of the project by measuring adherence to the organization's system development life cycle. They will conduct testing but not sign off on the project requirements.

A3-136 Which of the following is **MOST** relevant to an IS auditor evaluating how the project manager has monitored the progress of the project?

A. Critical path diagrams
B. Program evaluation review technique diagrams
C. Function point analysis
D. Gantt charts

D is the correct answer.

Justification:

A. Critical path diagrams are used to determine the critical path for the project that represents the shortest possible time required for completing the project.
B. Program evaluation review technique diagrams are a critical path method technique in which three estimates (as opposed to one) of time lines required to complete activities are used to determine the critical path.
C. Function point analysis is a technique used to determine the size of a development task, based on the number of function points.
D. Gantt charts help to identify activities that have been completed early or late through comparison to a baseline. Progress of the entire project can be read from the Gantt chart to determine whether the project is behind, ahead of or on schedule.

A3-137 While reviewing an ongoing project, the IS auditor notes that the development team has spent eight hours of activity on the first day against a budget of 24 hours (over three days). The projected time to complete the remainder of the activity is 20 hours. The IS auditor should report that the project:

A. is behind schedule.
B. is ahead of schedule.
C. is on schedule.
D. cannot be evaluated until the activity is completed.

A is the correct answer.

Justification:

A. Earned value analysis (EVA) is based on the premise that if a project task is assigned 24 hours for completion, it can be reasonably completed during that time frame. According to EVA, the project is behind schedule because the value of the eight hours spent on the task should be only four hours, considering that 20 hours of effort remain to be completed.
B. The project is not ahead of schedule because the work remaining exceeds the time allotted.
C. The project is not on schedule because only 16 hours remain to do 20 hours work.
D. The amount of work left has been evaluated at 20 hours and the time left on the project is 16 hours, so the auditor can evaluate the current status of the project.

A3-138 Which of the following **BEST** helps an IS auditor evaluate the quality of programming activities related to future maintenance capabilities?

A. The programming language
B. The development environment
C. A version control system
D. Program coding standards

D is the correct answer.

Justification:

A. The programming language may be a concern if it is not a commonly used language; however, program coding standards are more important.
B. The development environment may be relevant to evaluate the efficiency of the program development process but not future maintenance of the program.
C. A version control system helps manage software code revisions; however, it does not ensure that coding standards are consistently applied.
D. Program coding standards are required for efficient program maintenance and modifications. To enhance the quality of programming activities and future maintenance capabilities, program coding standards should be applied. Program coding standards are essential to writing, reading and understanding code, simply and clearly, without having to refer back to design specifications.

A3-139 During a system development life cycle audit of a human resources and payroll application, the IS auditor notes that the data used for user acceptance testing have been masked. The purpose of masking the data is to ensure the:

A. confidentiality of the data.
B. accuracy of the data.
C. completeness of the data.
D. reliability of the data.

A is the correct answer.

Justification:

A. Masking is used to ensure the confidentiality of data, especially in a user acceptance testing exercise in which the testers have access to data that they would not have access to in normal production environments.
B. Masking does not ensure accuracy of the data. If the underlying data are inaccurate, the masked data also would be inaccurate.
C. Masking does not ensure completeness of the data. If the underlying data are incomplete, the masked data also would be incomplete.
D. Masking does not ensure reliability of the data. If the underlying data are unreliable, the masked data also would be unreliable.

A3-140 Which of the following helps an IS auditor evaluate the quality of new software that is developed and implemented?

A. The reporting of the mean time between failures over time
B. The overall mean time to repair failures
C. The first report of the mean time between failures
D. The overall response time to correct failures

C is the correct answer.

Justification:

A. The mean time between failures that are repetitive includes the inefficiency in fixing the first reported failures and is a reflection on the response team or help desk team in fixing the reported issues.
B. The mean time to repair is a reflection on the response team or help desk team in addressing reported issues.
C. The mean time between failures that are first reported represents flaws in the software that are reported by users in the production environment. This information helps the IS auditor in evaluating the quality of the software that is developed and implemented.
D. The response time reflects the agility of the response team or the help desk team in addressing reported issues.

A3-141 Which of the following carries the **LOWEST** risk when managing failures while transitioning from legacy applications to new applications?

A. Phased changeover
B. Abrupt changeover
C. Rollback procedure
D. Parallel changeover

D is the correct answer.

Justification:

A. Phased changeover involves the changeover from the old system to the new system in a phased manner. Therefore, at no time will the old system and the new system both be fully operational as one integrated system.
B. In abrupt changeover, the new system is changed from the old system on a cutoff date and time, and the old system is discontinued after changeover to the new system takes place. Therefore, the old system is not available as a backup if there are problems when the new system is implemented.
C. Rollback procedures involve restoring all systems to their previous working state; however, parallel changeover is the better strategy.
D. Parallel changeover involves first running the old system, then running both the old and new systems in parallel, and finally fully changing to the new system after gaining confidence in the functionality of the new system.

A3-142 Which of the following **BEST** helps an IS auditor assess and measure the value of a newly implemented system?

A. Review of business requirements
B. System certification
C. Post-implementation review
D. System accreditation

C is the correct answer.

Justification:

A. While reviewing the business requirements is important, only a post-implementation review provides evidence that the project met the business requirements.
B. System certification involves performing a comprehensive assessment against a standard of management, operational and technical controls in an information system to examine the level of compliance in meeting certain requirements such as standards, policies, processes, procedures, work instructions and guidelines.
C. One key objective of a post-implementation review is to evaluate the projected cost-benefits or the return on investment measurements.
D. System accreditation is an official management decision to authorize operation of an information system and to explicitly accept the risk to the organization's operations, assets or individuals based on the implementation of an agreed-on set of requirements and security controls.

A3-143 A large industrial organization is replacing an obsolete legacy system and evaluating whether to buy a custom solution or develop a system in-house. Which of the following will **MOST** likely influence the decision?

A. Technical skills and knowledge within the organization related to sourcing and software development
B. Privacy requirements as applied to the data processed by the application
C. Whether the legacy system being replaced was developed in-house
D. The users not devoting reasonable time to define the functionalities of the solution

A is the correct answer.

Justification:

A. Critical core competencies will most likely be carefully considered before outsourcing the planning phase of the application.
B. Privacy regulations would apply to both solutions.
C. While individuals with knowledge of the legacy system are helpful, they may not have the technical skills to build a new system. Therefore, this is not the primary factor influencing the make versus buy decision.
D. Unclear business requirements (functionalities) will similarly affect either development process but are not the primary factor influencing the make versus buy decision.

A3-144 A company's development team does not follow generally accepted system development life cycle practices. Which of the following is **MOST** likely to cause problems for software development projects?

A. Functional verification of the prototypes is assigned to end users
B. The project is implemented while minor issues are open from user acceptance testing
C. Project responsibilities are not formally defined at the beginning of a project
D. Program documentation is inadequate

C is the correct answer.

Justification:
A. Prototypes are verified by users.
B. User acceptance testing is seldom completely successful. If errors are not critical, they may be corrected after implementation without seriously affecting usage.
C. Errors or lack of attention in the initial phases of a project may cause costly errors and inefficiencies in later phases. Proper planning is required at the beginning of a project.
D. Lack of adequate program documentation, while a concern, is not as big a risk as the lack of assigned responsibilities during the initial stages of the project.

A3-145 An IS auditor has been asked to review the implementation of a customer relationship management system for a large organization. The IS auditor discovered the project incurred significant over-budget expenses and scope creep caused the project to miss key dates. Which of the following should the IS auditor recommend for future projects?

A. Project management training
B. A software baseline
C. A balanced scorecard
D. Automated requirements software

B is the correct answer.

Justification:
A. While project management training is a good practice, it does not necessarily prevent scope creep without the use of a software baseline and a robust requirements change process.
B. Use of a software baseline provides a cutoff point for the design of the system and allows the project to proceed as scheduled without being delayed by scope creep.
C. A balanced scorecard is a coherent set of performance measures organized into four categories that includes traditional financial measures, but adds customer, internal business process, and learning and growth perspectives. It does not prevent scope creep.
D. Use of automated requirements software does not decrease the risk of scope creep.

A3-146 Which of the following is the **BEST** indicator that a newly developed system will be used after it is in production?

A. Regression testing
B. User acceptance testing
C. Sociability testing
D. Parallel testing

B is the correct answer.

Justification:

A. Regression test results do not assist with the user experience and are primarily concerned with new functionality or processes and whether those changes altered or broke previous functionality.
B. User acceptance testing is undertaken to provide confidence that a system or system component operates as intended, to provide a basis for evaluating the implementation of the requirements or to demonstrate the effectiveness or efficiency of the system or component. If the results of the testing are poor, then the system is unlikely to be adopted by the users.
C. Sociability test results indicate how the application works with other components within the environment and is not indicative of the user experience.
D. Parallel testing is performed when the comparison of two applications is needed but will not provide feedback on user satisfaction.

A3-147 The project steering committee is ultimately responsible for:

A. day-to-day management and leadership of the project.
B. allocating the funding for the project.
C. project deliverables, costs and timetables.
D. ensuring that system controls are in place.

C is the correct answer.

Justification:

A. Day-to-day management and leadership of the project is the function of the project manager.
B. Providing the funding for the project is the function of the project sponsor.
C. The project steering committee provides overall direction; ensures appropriate representation of the major stakeholders in the project's outcome; and takes ultimate responsibility for the deliverables, costs and timetables.
D. Ensuring that system controls are in place is the function of the project security officer.

A3-148 Which of the following **BEST** helps ensure that deviations from the project plan are identified?

A. A project management framework
B. A project management approach
C. A project resource plan
D. Project performance criteria

D is the correct answer.

Justification:

A. Establishment of a project management framework identifies the scope and boundaries of managing projects and the consistent method to be applied when initiating a project but does not define the criteria used to measure project success.
B. A project management approach defines guidelines for project management processes and deliverables but does not define the criteria used to measure project success.
C. A project resource plan defines the responsibilities, relationships, authorities and performance criteria of project team members but does not wholly define the criteria used to measure project success.
D. To identify deviations from the project plan, project performance criteria must be established as a baseline. Successful completion of the project plan is indicative of project success.

A3-149 An IS auditor is reviewing a project for the implementation of a mission-critical system and notes that, instead of parallel implementation, the team opted for an immediate cutover to the new system. Which of the following is the **GREATEST** concern?

A. The implementation phase of the project has no back out plan
B. User acceptance testing was not properly documented
C. Software functionality tests were completed, but stress testing was not performed
D. The go-live date is over a holiday weekend when key IT staff are on vacation

A is the correct answer.

Justification:

A. One of the benefits of deploying a new system in parallel with an existing system is that the original system can always be used as a back out plan. In an immediate cutover scenario, not having a back out plan can create significant issues because it can take considerable time and cost to restore operations to the prior state if there is no viable plan to do so.
B. The documentation of user acceptance testing is a much less important concern than not having a viable back out plan.
C. The lack of stress testing is a much less important concern than not having a viable back out plan.
D. If there are support issues, having the go-live date happen over a holiday weekend may create some delays, but project managers should account for this to ensure that the required staff are available as needed. The greater risk is if there is no back out plan.

A3-150 Which of the following software testing methods provides the **BEST** feedback on how software will perform in the live environment?

A. Alpha testing
B. Regression testing
C. Beta testing
D. White box testing

C is the correct answer.

Justification:

A. Alpha testing is often performed only by users within the organization developing the software. Alpha testing generally involves a software version that does not contain all the features of the final product and may be a simulated test.
B. Regression testing is used to determine whether system changes have introduced new errors to existing functionality.
C. Beta testing follows alpha testing and involves real-world exposure with external user involvement. Beta testing is the last stage of testing and involves sending the beta version of the product to independent beta test sites or offering it free to interested users.
D. White box testing is used to assess the effectiveness of program logic.

A3-151 Which of the following is the **BEST** method of controlling scope creep in a system development project?

A. Defining penalties for changes in requirements
B. Establishing a software baseline
C. Adopting a matrix project management structure
D. Identifying the critical path of the project

B is the correct answer.

Justification:

A. While defining penalties for changes in requirements may help to prevent scope creep, software baselining is a better way to accomplish this goal.
B. Software baselining, the cutoff point in the design phase, occurs after a rigorous review of user requirements. Any changes thereafter will undergo strict formal change control and approval procedures. Scope creep refers to uncontrolled change within a project resulting from improperly managed requirements.
C. In a matrix project organization, management authority is shared between the project manager and the department heads. Adopting a matrix project management structure will not address the problem of scope creep.
D. Although the critical path is important, it will change over time and will not control scope creep.

A3-152 The **PRIMARY** purpose of a post-implementation review is to ascertain that:

A. The lessons learned have been documented.
B. Future enhancements can be identified.
C. The project has been delivered on time and budget.
D. Project objectives have been met.

D is the correct answer.

Justification:

A. It is important to ensure that lessons learned during the project are not forgotten; however, it is more important to ascertain whether the project solved the problem it was designed to address.
B. Identifying future enhancements is not the primary objective of a post-implementation review.
C. Although it is important to review whether the project was completed on time and budget, it is more important to determine whether the project met the business needs.
D. A project manager performs a post-implementation review to obtain feedback regarding the project deliverables and business needs and to determine whether the project has successfully met them.

A3-153 Results of a post-implementation review indicate that only 75 percent of the users can log in to the application concurrently. Which of the following could have **BEST** discovered the identified weakness of the application?

A. Load testing
B. Stress testing
C. Recovery testing
D. Volume testing

A is the correct answer.

Justification:

A. Load testing evaluates the performance of the software under normal and peak conditions. Because this application is not supporting normal numbers of concurrent users, the load testing must not have been adequate.
B. Stress testing determines the capacity of the software to cope with an abnormal number of users or simultaneous operations. Because the number of concurrent users in this question is within normal limits, the answer is load testing, not stress testing.
C. Recovery testing evaluates the ability of a system to recover after a failure.
D. Volume testing evaluates the impact of incremental volume of records (not users) on a system.

A3-154 An IS auditor reviewing the IT project management process is reviewing a feasibility study for a critical project to build a new data center. The IS auditor is **MOST** concerned about the fact that:

A. it has not been determined how the project fits into the overall project portfolio.
B. the organizational impact of the project has not been assessed.
C. not all IT stakeholders have been given an opportunity to provide input.
D. the environmental impact of the data center has not been considered.

B is the correct answer.

Justification:

A. While projects must be assigned a priority and managed as a portfolio, this most likely occurs after the feasibility study determines that the project is viable.
B. The feasibility study determines the strategic benefits of the project. Therefore, the result of the feasibility study determines the organizational impact—a comparison report of costs, benefits, risk, etc. The project portfolio is a part of measuring the organizational strategy.
C. A feasibility study is ordinarily conducted by those with the knowledge to make the decision because the involvement of the entire IT organization is not needed.
D. The environmental impact should be part of the feasibility study however the organizational impact is more important.

Page intentionally left blank

DOMAIN 4—INFORMATION SYSTEMS OPERATIONS AND BUSINESS RESILIENCE (23%)

A4-1 An organization is considering using a new IT service provider. From an audit perspective, which of the following would be the **MOST** important item to review?

A. References from other clients for the service provider
B. The physical security of the service provider site
C. The proposed service level agreement with the service provider
D. Background checks of the service provider's employees

C is the correct answer.

Justification:

A. A due diligence activity such as reviewing references from other clients is a good practice, but the service level agreement (SLA) would be most critical because it would define what specific levels of performance would be required and make the provider contractually obligated to deliver what was promised.

B. A due diligence activity such as reviewing physical security controls is a good practice, but the SLA would be most critical because it would define what specific levels of security would be required and make the provider contractually obligated to deliver what was promised.

C. When contracting with a service provider, it is a good practice to enter into an SLA with the provider. An SLA is a guarantee that the provider will deliver the services according to the contract. The IS auditor will want to ensure that performance and security requirements are clearly stated in the SLA.

D. A due diligence activity such as the use of background checks for the service provider's employees is a good practice, but the SLA would be most critical because it would define what specific levels of security and labor practices would be required and make the provider contractually obligated to deliver what was promised.

A4-2 An IS auditor is to assess the suitability of a service level agreement (SLA) between the organization and the supplier of outsourced services. To which of the following observations should the IS auditor pay the **MOST** attention? The SLA does not contain a:

A. transition clauses from the old supplier to a new supplier or back to internal in the case of expiration or termination.
B. late payment clause between the customer and the supplier.
C. contractual commitment for service improvement.
D. dispute resolution procedure between the contracting parties.

A is the correct answer.

Justification:

A. The delivery of IT services for a specific customer always implies a close linkage between the client and the supplier of the service. If there are no contract terms to specify how the transition to a new supplier may be performed, there is the risk that the old supplier may simply "pull the plug" if the contract expires or is terminated or may not make data available to the outsourcing organization or new supplier. This would be the greatest risk to the organization.

B. Contractual issues regarding payment, service improvement and dispute resolution are important but not as critical as ensuring that service disruption, data loss, data retention, or other significant events occur in the event that the organization switches to a new firm providing outsourced services.

C. The service level agreement (SLA) should address performance requirements and metrics to report on the status of services provided; it's nice to have commitment for performance improvement, although it's not mandated.

D. The SLA should address a dispute resolution procedure and specify the jurisdiction in case of a legal dispute, but this is not the most critical part of an SLA.

A4-3 An IS auditor reviewing a new outsourcing contract with a service provider would be **MOST** concerned if which of the following was missing?

A. A clause providing a "right to audit" the service provider
B. A clause defining penalty payments for poor performance
C. Predefined service level report templates
D. A clause regarding supplier limitation of liability

A is the correct answer.

Justification:

A. **The absence of a "right to audit" clause or other form of attestation that the supplier was compliant with a certain standard would potentially prevent the IS auditor from investigating any aspect of supplier performance moving forward, including control deficiencies, poor performance and adherence to legal requirements. This would be a major concern for the IS auditor because it would be difficult for the organization to assess whether the appropriate controls had been put in place.**

B. While a clear definition of penalty payment terms is desirable, not all contracts require the payment of penalties for poor performance, and when performance penalties are required, these penalties are often subject to negotiation on a case-by-case basis. As such, the absence of this information would not be as significant as a lack of right to audit.

C. While the inclusion of service level report templates would be desirable, as long as the requirement for service level reporting is included in the contract, the absence of predefined templates for reporting is not a significant concern.

D. The absence of a limitation of liability clause for the service provider would, theoretically, expose the provider to unlimited liability. This would be to the advantage of the outsourcing company so, while the IS auditor might highlight the absence of such a clause, it would not constitute a major concern.

A4-4 When reviewing the desktop software compliance of an organization, the IS auditor should be **MOST** concerned if the installed software:

A. was installed, but not documented in the IT department records.
B. was being used by users not properly trained in its use.
C. is not listed in the approved software standards document.
D. license will expire in the next 15 days.

C is the correct answer.

Justification:

A. All software, including licenses, should be documented in IT department records, but this is not as serious as the violation of policy in installing unapproved software.

B. Discovering that users have not been formally trained in the use of a software product is common, and while not ideal, most software includes help files and other tips that can assist in learning how to use the software effectively.

C. **The installation of software that is not allowed by policy is a serious violation and could put the organization at security, legal and financial risk. Any software that is allowed should be part of a standard software list. This is the first thing to review because this would also indicate compliance with policies.**

D. A software license that is about to expire is not a risk if there is a process in place to renew it.

A4-5 An IS auditor of a health care organization is reviewing contractual terms and conditions of a third-party cloud provider being considered to host patient health information. Which of the follow contractual terms would be the **GREATEST** risk to the customer organization?

A. Data ownership is retained by the customer organization.
B. The third-party provider reserves the right to access data to perform certain operations.
C. Bulk data withdrawal mechanisms are undefined.
D. The customer organization is responsible for backup, archive and restore.

B is the correct answer.

Justification:

A. The customer organization would want to retain data ownership and, therefore, this would not be a risk.
B. Some service providers reserve the right to access customer information (third-party access) to perform certain transactions and provide certain services. In the case of protected health information, regulations may restrict certain access. Organizations must review the regulatory environment in which the cloud provider operates because it may have requirements or restrictions of its own. Organizations must then determine whether the cloud provider provides appropriate controls to ensure that data are appropriately secure.
C. An organization may eventually wish to discontinue its service with a third-party cloud-based provider. The organization would then want to remove its data from the system and ensure that the service provider clears the system (including any backups) of its data. Some providers do not offer automated or bulk data withdrawal mechanisms, which the organization needs to migrate its data. These aspects should be clarified prior to using a third-party provider.
D. An organization may need to plan its own data recovery processes and procedures if the service provider does not make this available or the organization has doubts about the service provider's processes. This would only be a risk if the customer organization was unable to perform these activities itself.

A4-6 Which of the following recovery strategies is **MOST** appropriate for a business having multiple offices within a region and a limited recovery budget?

A. A hot site maintained by the business
B. A commercial cold site
C. A reciprocal arrangement between its offices
D. A third-party hot site

C is the correct answer.

Justification:

A. A hot site maintained by the business would be a costly solution but would provide a high degree of confidence.
B. Multiple cold sites leased for the multiple offices would lead to an ineffective solution with poor availability.
C. For a business having many offices within a region, a reciprocal arrangement among its offices would be most appropriate. Each office could be designated as a recovery site for some other office. This would be the least expensive approach and would provide an acceptable level of confidence.
D. A third-party facility for recovery is provided by a traditional hot site. This would be a costly approach providing a high degree of confidence.

A4-7 During an application audit, an IS auditor is asked to provide assurance of the database referential integrity. Which of the following should be reviewed?

A. Field definition
B. Master table definition
C. Composite keys
D. Foreign key structure

D is the correct answer.

Justification:
A. Field definitions describe the layout of the table but are not directly related to referential integrity.
B. Master table definition describes the structure of the database but is not directly related to referential integrity.
C. Composite keys describe how the keys are created but are not directly related to referential integrity.
D. Referential integrity in a relational database refers to consistency between coupled (linked) tables. Referential integrity is usually enforced by the combination of a primary key or candidate key (alternate key) and a foreign key. For referential integrity to hold, any field in a table that is declared a foreign key should contain only values from a parent table's primary key or a candidate key.

A4-8 An IS auditor is reviewing database security for an organization. Which of the following is the **MOST** important consideration for database hardening?

A. The default configurations are changed.
B. All tables in the database are denormalized.
C. Stored procedures and triggers are encrypted.
D. The service port used by the database server is changed.

A is the correct answer.

Justification:
A. Default database configurations, such as default passwords and services, need to be changed; otherwise, the database could be easily compromised by malicious code and by intruders.
B. The denormalization of a database is related more to performance than to security.
C. Limiting access to stored procedures is a valid security consideration but not as critical as changing default configurations.
D. Changing the service port used by the database is a component of the configuration changes that could be made to the database, but there are other more critical configuration changes that should be made first.

A4-9 In auditing a database environment, an IS auditor will be **MOST** concerned if the database administrator is performing which of the following functions?

A. Performing database changes according to change management procedures
B. Installing patches or upgrades to the operating system
C. Sizing table space and consulting on table join limitations
D. Performing backup and recovery procedures

B is the correct answer.

Justification:
A. Performing database changes according to change management procedures would be a normal function of the database administrator (DBA) and would be compliant with the procedures of the organization.
B. Installing patches or upgrades to the operating system is a function that should be performed by a systems administrator, not by a DBA. If a DBA were performing this function, there would be a risk based on inappropriate segregation of duties.
C. A DBA is expected to support the business through helping design, create and maintain databases and the interfaces to the databases.
D. The DBA often performs or supports database backup and recovery procedures.

A4-10 Which of the following is the **MOST** reasonable option for recovering a non-critical system?

A. Warm site
B. Mobile site
C. Hot site
D. Cold site

D is the correct answer.

Justification:
A. A warm site is generally available at a medium cost, requires less time to become operational and is suitable for sensitive operations that should be recovered in a moderate amount of time.
B. A mobile site is a vehicle ready with all necessary computer equipment that can be moved to any location, depending upon the need. The need for a mobile site depends upon the scale of operations.
C. A hot site is contracted for a shorter time period at a higher cost, and it is better suited for recovery of vital and critical applications.
D. Generally, a cold site is contracted for a longer period at a lower cost. Because it requires more time to make a cold site operational, it is generally used for noncritical applications.

A4-11 An IS auditor is evaluating the effectiveness of the change management process in an organization. What is the **MOST** important control that the IS auditor should look for to ensure system availability?

A. Changes are authorized by IT managers at all times.
B. User acceptance testing is performed and properly documented.
C. Test plans and procedures exist and are closely followed.
D. Capacity planning is performed as part of each development project.

C is the correct answer.

Justification:
A. Changes are usually required to be signed off by a business analyst, member of the change control board or other authorized representative, not necessarily by IT management.
B. User acceptance testing is important but not a critical element of change control and would not usually address the topic of availability as asked in the question.
C. The most important control for ensuring system availability is to implement a sound test plan and procedures that are followed consistently.
D. While capacity planning should be considered in each development project, it will not ensure system availability, nor is it part of the change control process.

A4-12 Data flow diagrams are used by IS auditors to:

A. identify key controls.
B. highlight high-level data definitions.
C. graphically summarize data paths and storage.
D. portray step-by-step details of data generation.

C is the correct answer.

Justification:
A. Identifying key controls is not the focus of data flow diagrams. The focus is as the name states—flow of data.
B. A data dictionary may be used to document data definitions, but the data flow diagram is used to document how data move through a process.
C. Data flow diagrams are used as aids to graph or chart data flow and storage. They trace data from their origination to destination, highlighting the paths and storage of data.
D. The purpose of a data flow diagram is to track the movement of data through a process and is not primarily to document or indicate how data are generated.

A4-13 Which of the following statements is useful while drafting a disaster recovery plan?

A. Downtime costs decrease as the recovery point objective increases.
B. Downtime costs increase with time.
C. Recovery costs are independent of time.
D. Recovery costs can only be controlled on a short-term basis.

B is the correct answer.

Justification:
A. Downtime costs are not related to the recovery point objective (RPO). The RPO defines the data backup strategy, which is related to recovery costs rather than to downtime costs.
B. Downtime costs—such as loss of sales, idle resources, salaries—increase with time. A disaster recovery plan should be drawn to achieve the lowest downtime costs possible.
C. Recovery costs decrease with the time allowed for recovery. For example, recovery costs to recover business operations within two days will be higher than the cost to recover business within seven days. The essence of an effective DRP is to minimize uncertainty and increase predictability.
D. With good planning, recovery costs can be predicted and contained.

A4-14 Although management has stated otherwise, an IS auditor has reasons to believe that the organization is using software that is not licensed. In this situation, the IS auditor should **FIRST**:

A. include the statement from management in the audit report.
B. verify the software is in use through testing.
C. include the item in the audit report.
D. discuss the issue with senior management because it could have a negative impact on the organization.

B is the correct answer.

Justification:
A. The statement from management may be included in the audit report, but the auditor should independently validate the statements made by management to ensure completeness and accuracy.
B. When there is an indication that an organization might be using unlicensed software, the IS auditor should obtain sufficient evidence before including it in report.
C. With respect to this matter, representations obtained from management cannot be independently verified.
D. If the organization is using software that is not licensed, the IS auditor, to maintain objectivity and independence, must include this in the report, but the IS auditor should verify that this is in fact the case before presenting it to senior management.

A4-15 An advantage of using unshielded twisted-pair (UTP) cable for data communication over other copper-based cables is that UTP cable:

A. reduces crosstalk between pairs.
B. provides protection against wiretapping.
C. can be used in long-distance networks.
D. is simple to install.

A is the correct answer.

Justification:
A. The use of unshielded twisted-pair (UTP) in copper will reduce the likelihood of crosstalk.
B. While the twisted nature of the media will reduce sensitivity to electromagnetic disturbances, an unshielded copper wire does not provide adequate protection against wiretapping.
C. Attenuation sets in if copper twisted-pair cable is used for longer than 100 meters, necessitating the use of a repeater.
D. The tools and techniques to install UTP are not simpler or easier than other copper-based cables.

A4-16 Which of the following is the **MOST** critical element to effectively execute a disaster recovery plan?

A. Offsite storage of backup data
B. Up-to-date list of key disaster recovery contacts
C. Availability of a replacement data center
D. Clearly defined recovery time objective

A is the correct answer.

Justification:
A. Remote storage of backups is the most critical disaster recovery plan (DRP) element of the items listed because access to backup data is required to restore systems.
B. Having a list of key contacts is important but not as important as having adequate data backup.
C. A DRP may use a replacement data center or some other solution such as a mobile site, reciprocal agreement or outsourcing agreement.
D. Having a clearly defined recovery time objective is especially important for business continuity planning, but the core element of disaster recovery (the recovery of IT infrastructure and capability) is data backup.

A4-17 While reviewing the process for continuous monitoring of the capacity and performance of IT resources, an IS auditor should **PRIMARILY** ensure that the process is focused on:

A. adequately monitoring service levels of IT resources and services.
B. providing data to enable timely planning for capacity and performance requirements.
C. providing accurate feedback on IT resource capacity.
D. properly forecasting performance, capacity and throughput of IT resources.

C is the correct answer.

Justification:

A. Continuous monitoring helps to ensure that service level agreements (SLAs) are met, but this would not be the primary focus of monitoring. It is possible that even if a system were offline, it would meet the requirements of an SLA. Therefore, accurate availability monitoring is more important.
B. While data gained from capacity and performance monitoring would be an input to the planning process, the primary focus would be to monitor availability.
C. Accurate capacity monitoring of IT resources would be the most critical element of a continuous monitoring process.
D. While continuous monitoring would help management to predict likely IT resource capabilities, the more critical issue would be that availability monitoring is accurate.

A4-18 Which of the following groups is the **BEST** source of information for determining the criticality of application systems as part of a business impact analysis?

A. Business processes owners
B. IT management
C. Senior business management
D. Industry experts

A is the correct answer.

Justification:

A. Business process owners have the most relevant information to contribute because the business impact analysis (BIA) is designed to evaluate criticality and recovery time lines, based on business needs.
B. While IT management must be involved, they may not be fully aware of the business processes that need to be protected.
C. While senior management must be involved, they may not be fully aware of the criticality of applications that need to be protected.
D. The BIA is dependent on the unique business needs of the organization and the advice of industry experts is of limited value.

A4-19 An IS auditor is reviewing an organization's disaster recovery plan (DRP) implementation. The project was completed on time and on budget. During the review, the auditor uncovers several areas of concern. Which of the following presents the **GREATEST** risk?

A. Testing of the DRP has not been performed.
B. The disaster recovery strategy does not specify use of a hot site.
C. The business impact analysis was conducted, but the results were not used.
D. The disaster recovery project manager for the implementation has recently left the organization.

C is the correct answer.

Justification:

A. Although testing a disaster recovery plan (DRP) is a critical component of a successful disaster recovery strategy, this is not the biggest risk; the biggest risk comes from a plan that is not properly designed.
B. Use of a hot site is a strategic determination based on tolerable downtime, cost and other factors. Although using a hot site may be considered a good practice, this is a very costly solution that may not be required for the organization.
C. The risk of not using the results of the business impact analysis (BIA) for disaster recovery planning means that the DRP may not be designed to recover the most critical assets in the correct order. As a result, the plan may not be adequate to allow the organization to recover from a disaster.
D. If the DRP is designed and documented properly, the loss of an experienced project manager should have minimal impact. The risk of a poorly designed plan that may not meet the requirements of the business is much more significant than the risk posed by loss of the project manager.

A4-20 A vendor has released several critical security patches over the past few months and this has put a strain on the ability of the administrators to keep the patches tested and deployed in a timely manner. The administrators have asked if they could reduce the testing of the patches. What approach should the organization take?

A. Continue the current process of testing and applying patches.
B. Reduce testing and ensure that an adequate backout plan is in place.
C. Delay patching until resources for testing are available.
D. Rely on the vendor's testing of the patches.

A is the correct answer.

Justification:

A. Applying security software patches promptly is critical to maintain the security of the servers; further, testing the patches is important because the patches may affect other systems and business operations. Because the vendor has recently released several critical patches in a short time, it can be hoped that this is a temporary problem and does not need a revision to policy or procedures.
B. Reduced testing increases the risk of business operation disruption due to a faulty or incompatible patch. While a backout plan does help mitigate this risk, a thorough testing up front would be the more appropriate option.
C. Applying security software patches promptly is critical to maintain the security of the servers. Delaying patching would increase the risk of a security breach due to system vulnerability.
D. The testing done by the vendor may not be applicable to the systems and environment of the organization that needs to deploy the patches.

A4-21 Which of the following issues should be a **MAJOR** concern to an IS auditor who is reviewing a service level agreement (SLA)?

A. A service adjustment resulting from an exception report took a day to implement.
B. The complexity of application logs used for service monitoring made the review difficult.
C. Service measures were not included in the SLA.
D. The document is updated on an annual basis.

C is the correct answer.

Justification:
A. Resolving issues related to exception reports is an operational issue that should be addressed in the service level agreement (SLA); however, a response time of one day may be acceptable depending on the terms of the SLA.
B. The complexity of application logs is an operational issue, which is not related to the SLA.
C. Lack of service measures will make it difficult to gauge the efficiency and effectiveness of the IT services being provided.
D. While it is important that the document be current, depending on the term of the agreement, it may not be necessary to change the document more frequently than annually.

A4-22 During an IS audit of the disaster recovery plan of a global enterprise, the auditor observes that some remote offices have very limited local IT resources. Which of the following observations would be the **MOST** critical for the IS auditor?

A. A test has not been made to ensure that local resources could maintain security and service standards when recovering from a disaster or incident.
B. The corporate business continuity plan does not accurately document the systems that exist at remote offices.
C. Corporate security measures have not been incorporated into the test plan.
D. A test has not been made to ensure that tape backups from the remote offices are usable.

A is the correct answer.

Justification:
A. Regardless of the capability of local IT resources, the most critical risk would be the lack of testing, which would identify quality issues in the recovery process.
B. The corporate business continuity plan may not include disaster recovery plan (DRP) details for remote offices. It is important to ensure that the local plans have been tested.
C. Security is an important issue because many controls may be missing during a disaster. However, not having a tested plan is more important.
D. The backups cannot be trusted until they have been tested. However, this should be done as part of the overall tests of the DRP.

A4-23 Which of the following reports should an IS auditor use to check compliance with a service level agreement's requirement for uptime?

A. Utilization reports
B. Hardware error reports
C. System logs
D. Availability reports

D is the correct answer.

Justification:

A. Utilization reports document the use of computer equipment, and can be used by management to predict how, where and/or when resources are required.
B. Hardware error reports provide information to aid in detecting hardware failures and initiating corrective action. These error reports may not indicate actual system uptime.
C. System logs are used for recording the system's activities. They may not indicate availability.
D. IS inactivity, such as downtime, is addressed by availability reports. These reports provide the time periods during which the computer was available for utilization by users or other processes.

A4-24 Which of the following would an IS auditor use to determine if unauthorized modifications were made to production programs?

A. System log analysis
B. Compliance testing
C. Forensic analysis
D. Analytical review

B is the correct answer.

Justification:

A. System log analysis would identify changes and activity on a system but would not identify whether the change was authorized unless conducted as a part of a compliance test.
B. Determining that only authorized modifications are made to production programs would require the change management process be reviewed to evaluate the existence of a trail of documentary evidence. Compliance testing would help to verify that the change management process has been applied consistently.
C. Forensic analysis is a specialized technique for criminal investigation.
D. An analytical review assesses the general control environment of an organization.

A4-25 During a change control audit of a production system, an IS auditor finds that the change management process is not formally documented and that some migration procedures failed. What should the IS auditor do next?

A. Recommend redesigning the change management process.
B. Gain more assurance on the findings through root cause analysis.
C. Recommend that program migration be stopped until the change process is documented.
D. Document the finding and present it to management.

B is the correct answer.

Justification:

A. While it may be necessary to redesign the change management process, this cannot be done until a root cause analysis is conducted to determine why the current process is not being followed.
B. A change management process is critical to IT production systems. Before recommending that the organization take any other action (e.g., stopping migrations, redesigning the change management process), the IS auditor should gain assurance that the incidents reported are related to deficiencies in the change management process and not caused by some process other than change management.
C. A business relies on being able to make changes when necessary, and security patches must often be deployed promptly. It would not be feasible to halt all changes until a new process is developed.
D. The results of the audit including the findings of noncompliance will be delivered to management once a root cause analysis of the issue has been completed.

A4-26 An IS auditor evaluating the resilience of a high-availability network should be **MOST** concerned if:

A. the setup is geographically dispersed.
B. the servers are clustered in one site.
C. a hot site is ready for activation.
D. diverse routing is implemented for the network.

B is the correct answer.

Justification:

A. Dispersed geographic locations provide backup if a site has been destroyed.
B. A clustered setup in one site makes the entire network vulnerable to natural disasters or other disruptive events.
C. A hot site would also be a good alternative for a single point-of-failure site.
D. Diverse routing provides telecommunications backup if a network is not available.

A4-27 Management considered two projections for its disaster recovery plan: plan A with two months to fully recover and plan B with eight months to fully recover. The recovery point objectives are the same in both plans. It is reasonable to expect that plan B projected higher:

A. downtime costs.
B. resumption costs.
C. recovery costs.
D. walk-through costs.

A is the correct answer.

Justification:

A. Because management considered a longer time window for recovery in plan B, downtime costs included in the plan are likely to be higher.
B. Because the recovery time for plan B is longer, resumption costs can be expected to be lower.
C. Because the recovery time for plan B is longer, recovery costs can be expected to be lower.
D. Walk-through costs are not a part of disaster recovery.

A4-28 Which of the following would an IS auditor consider to be **MOST** helpful when evaluating the effectiveness and adequacy of a preventive computer maintenance program?

A. A system downtime log
B. Vendors' reliability figures
C. Regularly scheduled maintenance log
D. A written preventive maintenance schedule

A is the correct answer.

Justification:

A. A system downtime log provides evidence regarding the effectiveness and adequacy of computer preventive maintenance programs. The log is a detective control, but because it is validating the effectiveness of the maintenance program, it is validating a preventive control.
B. Vendor's reliability figures are not an effective measure of a preventive maintenance program.
C. Reviewing the log is a good detective control to ensure that maintenance is being done; however, only the system downtime will indicate whether the preventive maintenance is actually working well.
D. A schedule is a good control to ensure that maintenance is scheduled and that no items are missed in the maintenance schedule; however, it is not a guarantee that the work is actually being done.

A4-29 An organization has implemented an online customer help desk application using a software as a service (SaaS) operating model. An IS auditor is asked to recommend the best control to monitor the service level agreement (SLA) with the SaaS vendor as it relates to availability. What is the **BEST** recommendation that the IS auditor can provide?

A. Ask the SaaS vendor to provide a weekly report on application uptime.
B. Implement an online polling tool to monitor the application and record outages.
C. Log all application outages reported by users and aggregate the outage time weekly.
D. Contract an independent third party to provide weekly reports on application uptime.

B is the correct answer.

Justification:

A. Weekly application availability reports are useful, but these reports represent only the vendor's perspective. While monitoring these reports, the organization can raise concerns of inaccuracy; however, without internal monitoring, such concerns cannot be substantiated.
B. Implementing an online polling tool to monitor and record application outages is the best option for an organization to monitor the software as a service application availability. Comparing internal reports with the vendor's service level agreement (SLA) reports would ensure that the vendor's monitoring of the SLA is accurate and that all conflicts are appropriately resolved.
C. Logging the outage times reported by users is helpful but does not give a true picture of all outages of the online application. Some outages may go unreported, especially if the outages are intermittent.
D. Contracting a third party to implement availability monitoring is not a cost-effective option. Additionally, this results in a shift from monitoring the SaaS vendor to monitoring the third party.

A4-30 Applying a retention date on a file will ensure that:

A. data cannot be read until the date is set.
B. data will not be deleted before that date.
C. backup copies are not retained after that date.
D. datasets having the same name are differentiated.

B is the correct answer.

Justification:

A. The retention date will not affect the ability to read the file.
B. A retention date will ensure that a file cannot be overwritten or deleted before that date has passed.
C. Backup copies would be expected to have a different retention date and, therefore, may be retained after the file has been overwritten.
D. The creation date, not the retention date, will differentiate files with the same name.

A4-31 Which of the following is a network diagnostic tool that monitors and records network information?

A. Online monitor
B. Downtime report
C. Help desk report
D. Protocol analyzer

D is the correct answer.

Justification:

A. Online monitors measure telecommunication transmissions and determine whether transmissions were accurate and complete.
B. Downtime reports track the availability of telecommunication lines and circuits.
C. Help desk reports are prepared by the help desk, which is staffed or supported by IS technical support personnel trained to handle problems occurring during the course of IS operations.
D. Protocol analyzers are network diagnostic tools that monitor and record network information from packets traveling in the link to which the analyzer is attached.

A4-32 An IS auditor needs to review the procedures used to restore a software application to its state prior to an upgrade. Therefore, the auditor needs to assess:

A. problem management procedures.
B. software development procedures.
C. backout procedures.
D. incident management procedures.

C is the correct answer.

Justification:

A. Problem management procedures are used to track user feedback and issues related to the operation of an application for trend analysis and problem resolution.
B. Software development procedures such as the software development life cycle (SDLC) are used to manage the creation or acquisition of new or modified software.
C. Backout procedures are used to restore a system to a previous state and are an important element of the change control process. The other choices are not related to the change control process—a process which specifies what procedures should be followed when software is being upgraded but the upgrade does not work and requires a fallback to its former state.
D. Incident management procedures are used to manage errors or problems with system operation. They are usually used by a help desk. One of the incident management procedures may be how to follow a fallback plan.

A4-33 Which of the following is a **MAJOR** concern during a review of help desk activities?

A. Certain calls could not be resolved by the help desk team.
B. A dedicated line is not assigned to the help desk team.
C. Resolved incidents are closed without reference to end users.
D. The help desk instant messaging has been down for more than six months.

C is the correct answer.

Justification:

A. Although this is of concern, it should be expected. A problem escalation procedure should be developed to handle such scenarios.
B. Ideally, a help desk team should have dedicated lines, but this exception is not as serious as the technical team unilaterally closing an incident.
C. The help desk function is a service-oriented unit. The end users must be advised before an incident can be regarded as closed.
D. Instant messaging is an add-on to improve the effectiveness of the help desk team. Its absence cannot be seen as a major concern as long as calls can still be made.

A4-34 The **MAIN** purpose for periodically testing offsite disaster recovery facilities is to:

A. protect the integrity of the data in the database.
B. eliminate the need to develop detailed contingency plans.
C. ensure the continued compatibility of the contingency facilities.
D. ensure that program and system documentation remains current.

C is the correct answer.

Justification:

A. The testing of an offsite facility does nothing to protect the integrity of the database. It may test the validity of backups but does not protect their integrity.
B. Testing an offsite location validates the value of the contingency plans and is not used to eliminate detailed plans.
C. The main purpose of offsite hardware testing is to ensure the continued compatibility of the contingency facilities so that assurance can be gained that the contingency plans would work in an actual disaster.
D. Program and system documentation should be reviewed continuously for currency. A test of an offsite facility may ensure that the documentation for that site is current, but this is not the purpose of testing an offsite facility.

A4-35 A large chain of shops with electronic funds transfer at point-of-sale devices has a central communications processor for connecting to the banking network. Which of the following is the **BEST** disaster recovery plan for the communications processor?

A. Offsite storage of daily backups
B. Alternative standby processor onsite
C. Installation of duplex communication links
D. Alternative standby processor at another network node

D is the correct answer.

Justification:

A. Offsite storage of backups would not help, because electronic funds transfer tends to be an online process and offsite storage will not replace the dysfunctional processor.
B. The provision of an alternate processor onsite would be fine if it were an equipment problem but would not help in the case of a power outage and may require technical expertise to cutover to the alternate equipment.
C. Installation of duplex communication links would be most appropriate if it were only the communication link that failed.
D. Having an alternative standby processor at another network node would be the best solution. The unavailability of the central communications processor would disrupt all access to the banking network, resulting in the disruption of operations for all of the shops. This could be caused by failure of equipment, power or communications.

A4-36 The database administrator suggests that database efficiency can be improved by denormalizing some tables. This would result in:

A. loss of confidentiality.
B. increased redundancy.
C. unauthorized accesses.
D. application malfunctions.

B is the correct answer.

Justification:

A. Denormalization should not cause loss of confidentiality even though confidential data may be involved. The database administrator should ensure that access controls to the databases remain effective.
B. Normalization is a design or optimization process for a relational database that increases redundancy. Redundancy, which is usually considered positive when it is a question of resource availability, is negative in a database environment because it demands additional and otherwise unnecessary data handling efforts. Denormalization is sometimes advisable for functional reasons.
C. Denormalization pertains to the structure of the database, not the access controls. It should not result in unauthorized access.
D. Denormalization may require some changes to the calls between databases and applications but should not cause application malfunctions.

A4-37 An IS auditor has been assigned to conduct a test that compares job run logs to computer job schedules. Which of the following observations would be of the **GREATEST** concern to the IS auditor?

A. There are a growing number of emergency changes.
B. There were instances when some jobs were not completed on time.
C. There were instances when some jobs were overridden by computer operators.
D. Evidence shows that only scheduled jobs were run.

C is the correct answer.

Justification:
A. Emergency changes are acceptable as long as they are properly documented as part of the process.
B. Instances of jobs not being completed on time is a potential issue and should be investigated, but it is not the greatest concern.
C. The overriding of computer processing jobs by computer operators could lead to unauthorized changes to data or programs. This is a control concern; thus, it is always critical.
D. The audit should find that all scheduled jobs were run and that any exceptions were documented. This would not be a violation.

A4-38 A new business requirement required changing database vendors. Which of the following areas should the IS auditor **PRIMARILY** examine in relation to this implementation?

A. Integrity of the data
B. Timing of the cutover
C. Authorization level of users
D. Normalization of the data

A is the correct answer.

Justification:
A. A critical issue when migrating data from one database to another is the integrity of the data and ensuring that the data are migrated completely and correctly.
B. The timing of the cutover is important, but because the data are being migrated to a new database, duplication should not be an issue.
C. The authorization of the users is not as relevant as the authorization of the application because the users will interface with the database through an application, and the users will not directly interface with the database.
D. Normalization is used to design the database and is not necessarily related to database migration.

A4-39 The objective of concurrency control in a database system is to:

A. restrict updating of the database to authorized users.
B. ensure integrity when two processes attempt to update the same data at the same time.
C. prevent inadvertent or unauthorized disclosure of data in the database.
D. ensure the accuracy, completeness and consistency of data.

B is the correct answer.

Justification:
A. Access controls restrict updating of the database to authorized users.
B. Concurrency controls prevent data integrity problems, which can arise when two update processes access the same data item at the same time.
C. Controls such as passwords prevent the inadvertent or unauthorized disclosure of data from the database.
D. Quality controls such as edits ensure the accuracy, completeness and consistency of data maintained in the database.

A4-40 Which of the following controls would provide the **GREATEST** assurance of database integrity?

A. Audit log procedures
B. Table link/reference checks
C. Query/table access time checks
D. Rollback and rollforward database features

B is the correct answer.

Justification:

A. Audit log procedures enable recording of all events that have been identified and help in tracing the events. However, they only point to the event and do not ensure completeness or accuracy of the database contents.

B. Performing table link/reference checks serves to detect table linking errors (such as completeness and accuracy of the contents of the database), and thus provides the greatest assurance of database integrity.

C. Querying/monitoring table access time checks helps designers improve database performance but not integrity.

D. Rollback and rollforward database features ensure recovery from an abnormal disruption. They assure the integrity of the transaction that was being processed at the time of disruption, but do not provide assurance on the integrity of the contents of the database.

A4-41 Which of the following is widely accepted as one of the critical components in networking management?

A. Configuration and change management
B. Topological mappings
C. Application of monitoring tools
D. Proxy server troubleshooting

A is the correct answer.

Justification:

A. Configuration management is widely accepted as one of the key components of any network because it establishes how the network will function internally and externally. It also deals with the management of configuration and monitoring performance. Change management ensures that the setup and management of the network is done properly, including managing changes to the configuration, removal of default passwords and possibly hardening the network by disabling unneeded services.

B. Topological mappings provide outlines of the components of the network and its connectivity. This is important to address issues such as single points of failure and proper network isolation but is not the most critical component of network management.

C. Application monitoring is not a critical part of network management.

D. Proxy server troubleshooting is used for troubleshooting purposes, and managing a proxy is only a small part of network management.

A4-42 In evaluating programmed controls over password management, which of the following is the IS auditor **MOST** likely to rely on?

A. A size check
B. A hash total
C. A validity check
D. A field check

C is the correct answer.

Justification:

A. A size check is useful because passwords should have a minimum length, but it is not as strong of a control as validity.
B. Passwords are not typically entered in a batch mode, so a hash total would not be effective. More important, a system should not accept incorrect values of a password, so a hash total as a control will not indicate any weak passwords, errors or omissions.
C. A validity check would be the most useful for the verification of passwords because it would verify that the required format has been used—for example, not using a dictionary word, including non-alphabetical characters, etc. An effective password must have several different types of characters: alphabetical, numeric and special.
D. The implementation of a field check would not be as effective as a validity check that verifies that all password criteria have been met.

A4-43 Which of the following represents the **GREATEST** risk created by a reciprocal agreement for disaster recovery made between two companies?

A. Developments may result in hardware and software incompatibility.
B. Resources may not be available when needed.
C. The recovery plan cannot be live tested.
D. The security infrastructures in each company may be different.

A is the correct answer.

Justification:

A. If one organization updates its hardware and software configuration, it may mean that it is no longer compatible with the systems of the other party in the agreement. This may mean that each company is unable to use the facilities at the other company to recover their processing following a disaster.
B. Resources being unavailable when needed are an intrinsic risk in any reciprocal agreement, but this is a contractual matter and is not the greatest risk.
C. The plan can be tested by paper-based walk-throughs and possibly by agreement between the companies.
D. The difference in security infrastructures, while a risk, is not insurmountable.

A4-44 Which of the following is **MOST** directly affected by network performance monitoring tools?

A. Integrity
B. Availability
C. Completeness
D. Confidentiality

B is the correct answer.

Justification:

A. Network monitoring tools can be used to detect errors that are propagating through a network, but their primary focus is on network reliability so that the network is available when required.
B. Network monitoring tools allow observation of network performance and problems. This allows the administrator to take corrective action when network problems are observed. Therefore, the characteristic that is most directly affected by network monitoring is availability.
C. Network monitoring tools will not measure completeness of the communication. This is measured by the end points in the communication.
D. A network monitoring tool can violate confidentiality by allowing a network administrator to observe non-encrypted traffic. This requires careful protection and policies regarding the use of network monitoring tools.

A4-45 When auditing the onsite archiving process of emails, the IS auditor should pay the **MOST** attention to:

A. the existence of a data retention policy.
B. the storage capacity of the archiving solution.
C. the level of user awareness concerning email use.
D. the support and stability of the archiving solution manufacturer.

A is the correct answer.

Justification:

A. Without a data retention policy that is aligned to the company's business and compliance requirements, the email archive may not preserve and reproduce the correct information when required.
B. The storage capacity of the archiving solution would be irrelevant if the proper email messages have not been properly preserved and others have been deleted.
C. The level of user awareness concerning email use would not directly affect the completeness and accuracy of the archived email.
D. The support and stability of the archiving solution manufacturer is secondary to the need to ensure a retention policy. Vendor support would not directly affect the completeness and accuracy of the archived email.

A4-46 Vendors have released patches fixing security flaws in their software. Which of the following should an IS auditor recommend in this situation?

A. Assess the impact of patches prior to installation.
B. Ask the vendors for a new software version with all fixes included.
C. Install the security patch immediately.
D. Decline to deal with these vendors in the future.

A is the correct answer.

Justification:

A. The effect of installing the patch should be immediately evaluated and installation should occur based on the results of the evaluation. There are numerous cases where a patch from one vendor has affected other systems; therefore, it is necessary to test the patches as much as possible before rolling them out to the entire organization.
B. New software versions with all fixes included are not always available and a full installation could be time consuming.
C. To install the patch without knowing what it might affect could easily cause problems. The installation of a patch may also affect system availability; therefore, the patch should be rolled out at a time that is acceptable to the business.
D. Declining to deal with vendors does not take care of the flaw and may severely limit service options.

A4-47 Which of the following controls would be **MOST** effective in ensuring that production source code and object code are synchronized?

A. Release-to-release source and object comparison reports
B. Library control software restricting changes to source code
C. Restricted access to source code and object code
D. Date and time-stamp reviews of source and object code

D is the correct answer.

Justification:

A. Using version control software and comparing source and object code is a good practice but may not detect a problem where the source code is a different version than the object code.
B. All production libraries should be protected with access controls, and this may protect source code from tampering. However, this will not ensure that source and object codes are based on the same version.
C. It is a good practice to protect all source and object code—even in development. However, this will not ensure the synchronization of source and object code.
D. Date and time-stamp reviews of source and object code would ensure that source code, which has been compiled, matches the production object code. This is the most effective way to ensure that the approved production source code is compiled and is the one being used.

A4-48 A database administrator (DBA) who needs to make emergency changes to a database after normal working hours should log in:

A. with their named account to make the changes.
B. with the shared DBA account to make the changes.
C. to the server administrative account to make the changes.
D. to the user's account to make the changes.

A is the correct answer.

Justification:

A. Logging in using the named user account before using the database administrator (DBA) account provides accountability by noting the person making the changes.

B. The DBA account is typically a shared user account. The shared account makes it difficult to establish the identity of the support user who is performing the database update.

C. The server administrative accounts are shared and may be used by multiple support users. In addition, the server privilege accounts may not have the ability to perform database changes.

D. The use of a normal user account would not have sufficient privileges to make changes on the database.

A4-49 During an assessment of software development practices, an IS auditor finds that open source software components were used in an application designed for a client. What is the **GREATEST** concern the auditor would have about the use of open source software?

A. The client did not pay for the open source software components.
B. The organization and client must comply with open source software license terms.
C. Open source software has security vulnerabilities.
D. Open source software is unreliable for commercial use.

B is the correct answer.

Justification:

A. A major benefit of using open source software is that it is free. The client is not required to pay for the open source software components; however, both the developing organization and the client should be concerned about the licensing terms and conditions of the open source software components that are being used.

B. There are many types of open source software licenses and each has different terms and conditions. Some open source software licensing allows use of the open source software component freely but requires that the completed software product must also allow the same rights. This is known as viral licensing, and if the development organization is not careful, its products could violate licensing terms by selling the product for profit. The IS auditor should be most concerned with open source software licensing compliance to avoid unintended intellectual property risk or legal consequences.

C. Open source software, just like any software code, should be tested for security flaws and should be part of the normal system development life cycle (SDLC) process. This is not more of a concern than licensing compliance.

D. Open source software does not inherently lack quality. Like any software code, it should be tested for reliability and should be part of the normal SDLC process. This is not more of a concern than licensing compliance.

A4-50 An IS auditor reviewing database controls discovered that changes to the database during normal working hours were handled through a standard set of procedures. However, changes made after normal hours required only an abbreviated number of steps. In this situation, which of the following would be considered an adequate set of compensating controls?

A. Allow changes to be made only with the database administrator (DBA) user account
B. Make changes to the database after granting access to a normal user account
C. Use the DBA user account to make changes, log the changes and review the change log the following day
D. Use the normal user account to make changes, log the changes and review the change log the following day

C is the correct answer.

Justification:

A. The use of the database administrator (DBA) user account without logging would permit uncontrolled changes to be made to databases after access to the account was obtained.
B. A normal user account should not have access to a database. This would permit uncontrolled changes to any of the databases.
C. The use of a DBA user account is normally set up to log all changes made and is most appropriate for changes made outside of normal hours. The use of a log, which records the changes, allows changes to be reviewed. Because an abbreviated number of steps are used, this represents an adequate set of compensating controls.
D. Users should not be able to make changes. Logging would only provide information on changes made but would not limit changes to only those who were authorized.

A4-51 Which of the following tests performed by an IS auditor would be the **MOST** effective in determining compliance with change control procedures in an organization?

A. Review software migration records and verify approvals.
B. Identify changes that have occurred and verify approvals.
C. Review change control documentation and verify approvals.
D. Ensure that only appropriate staff can migrate changes into production.

B is the correct answer.

Justification:

A. Software migration records may not have all changes listed—changes could have been made that were not included in the migration records.
B. The most effective method is to determine what changes have been made (check logs and modified dates) and then verify that they have been approved.
C. Change control records may not have all changes listed.
D. Ensuring that only appropriate staff can migrate changes into production is a key control process but, in itself, does not verify compliance.

A4-52 When an organization's disaster recovery plan has a reciprocal agreement, which of the following risk treatment approaches is being applied?

A. Transfer
B. Mitigation
C. Avoidance
D. Acceptance

B is the correct answer.

Justification:

A. Risk transfer is the transference of risk to a third party (e.g., buying insurance for activities that pose a risk).

B. A reciprocal agreement in which two organizations agree to provide computing resources to each other in the event of a disaster is a form of risk mitigation. This usually works well if both organizations have similar information processing facilities. Because the intended effect of reciprocal agreements is to have a functional disaster recovery plan, it is a risk mitigation strategy.

C. Risk avoidance is the decision to cease operations or activities that give rise to a risk. For example, a company may stop accepting credit card payments to avoid the risk of credit card information disclosure.

D. Risk acceptance occurs when an organization decides to accept the risk as it is and to do nothing to mitigate or transfer it.

A4-53 A programmer maliciously modified a production program to change data and then restored it back to the original code. Which of the following would **MOST** effectively detect the malicious activity?

A. Comparing source code
B. Reviewing system log files
C. Comparing object code
D. Reviewing executable and source code integrity

B is the correct answer.

Justification:

A. Source code comparisons are ineffective because the original programs were restored, and the changed program does not exist.

B. Reviewing system log files is the only trail that may provide information about the unauthorized activities in the production library.

C. Object code comparisons are ineffective because the original programs were restored, and the changed program does not exist.

D. Reviewing executable and source code integrity is an ineffective control, because the source code was changed back to the original and will agree with the current executable.

A4-54 An IS auditor is reviewing an organization's recovery from a disaster in which not all the critical data needed to resume business operations were retained. Which of the following was incorrectly defined?

A. The interruption window
B. The recovery time objective
C. The service delivery objective
D. The recovery point objective

D is the correct answer.

Justification:

A. The interruption window is defined as the amount of time during which the organization is unable to maintain operations from the point of failure to the time that the critical services/applications are restored.
B. The recovery time objective is determined based on the acceptable downtime in the case of a disruption of operations.
C. The service delivery objective (SDO) is directly related to the business needs. SDO is the level of services to be reached during the alternate process mode until the normal situation is restored.
D. The recovery point objective (RPO) is determined based on the acceptable data loss in the case of a disruption of operations. RPO defines the point in time from which it is necessary to recover the data and quantifies, in terms of time, the permissible amount of data loss in the case of interruption.

A4-55 The **PRIMARY** benefit of an IT manager monitoring technical capacity is to:

A. identify the need for new hardware and storage procurement.
B. determine the future capacity need based on usage.
C. ensure that the service level requirements are met.
D. ensure that systems operate at optimal capacity.

C is the correct answer.

Justification:

A. This is one benefit of monitoring technical capacity because it can help forecast future demands, not just react to system failures. However, the primary responsibility of the IT manager is to meet the overall requirement to ensure that IT is meeting the service level expectations of the business.
B. Determining future capacity is one definite benefit of technical capability monitoring.
C. Capacity monitoring has multiple objectives; however, the primary objective is to ensure compliance with the internal service level agreement between the business and IT.
D. IT management is interested in ensuring that systems are operating at optimal capacity, but their primary obligation is to ensure that IT is meeting the service level requirements of the business.

A4-56 An IS auditor reviewing an organization's disaster recovery plan should **PRIMARILY** verify that it is:

A. tested every six months.
B. regularly reviewed and updated.
C. approved by the chief executive officer.
D. communicated to every department head in the organization.

B is the correct answer.

Justification:

A. The plan must be subjected to regular testing, but the period between tests will depend on the nature of the organization, the amount of change in the organization and the relative importance of IS. Three months, or even annually, may be appropriate in different circumstances.

B. The plan should be reviewed at appropriate intervals, depending on the nature of the business and the rate of change of systems and personnel. Otherwise, it may become out of date and may no longer be effective.

C. Although the disaster recovery plan should receive the approval of senior management, it need not be the chief executive officer if another executive officer is equally or more appropriate. For a purely IS-related plan, the executive responsible for technology may have approved the plan.

D. Although a business continuity plan is likely to be circulated throughout an organization, the IS disaster recovery plan will usually be a technical document and only relevant to IS and communication staff.

A4-57 There are several methods of providing telecommunication continuity. The method of routing traffic through split-cable or duplicate-cable facilities is called:

A. alternative routing.
B. diverse routing.
C. long-haul network diversity.
D. last-mile circuit protection.

B is the correct answer.

Justification:

A. Alternative routing is a method of routing information via an alternate medium such as copper cable or fiber optics. This involves the use of different networks, circuits or end points should the normal network be unavailable.

B. Diverse routing routes traffic through split-cable facilities or duplicate-cable facilities. This can be accomplished with different and/or duplicate cable sheaths. If different cable sheaths are used, the cable may be in the same conduit and, therefore, subject to the same interruptions as the cable it is backing up. The communication service subscriber can duplicate the facilities by having alternate routes, although the entrance to and from the customer premises may be in the same conduit. The subscriber can obtain diverse routing and alternate routing from the local carrier, including dual-entrance facilities. This type of access is time consuming and costly.

C. Long-haul network diversity is a diverse, long-distance network using different packet switching circuits among the major long-distance carriers. It ensures long-distance access should any carrier experience a network failure.

D. Last-mile circuit protection is a redundant combination of local carrier T-1s (E-1s in Europe), microwave and/or coaxial cable access to the local communications loop. This enables the facility to have access during a local carrier communication disaster. Alternate local-carrier routing is also used.

A4-58 Recovery procedures for an information processing facility are **BEST** based on:

A. recovery time objective.
B. recovery point objective.
C. maximum tolerable outage.
D. information security policy.

A is the correct answer.

Justification:

A. The recovery time objective (RTO). is the amount of timè allowed for the recovery of a business function or resource after a disaster occurs; the RTO is the desired recovery time frame based on maximum tolerable outage (MTO) and available recovery alternatives.

B. The recovery point objective (RPO) has the greatest influence on the recovery strategies for given data. It is determined based on the acceptable data loss in case of a disruption of operations. The RPO effectively quantifies the permissible amount of data loss in case of interruption.

C. MTO is the amount of time allowed for the recovery of a business function or resource after a disaster occurs; it represents the time by which the service must be restored before the organization is faced with the threat of collapse.

D. An information security policy does not address recovery procedures.

A4-59 An IS auditor is performing an audit in the data center when the fire alarm begins sounding. The audit scope includes disaster recovery, so the auditor observes the data center staff respond to the alarm. Which of the following is the **MOST** important action for the data center staff to complete in this scenario?

A. Notify the local fire department of the alarm condition.
B. Prepare to activate the fire suppression system.
C. Ensure all persons in the data center are evacuated.
D. Remove all backups from the data center.

C is the correct answer.

Justification:

A. Life safety is always the first priority, and notifying the fire department of the alarm is not typically necessary because most data center alarms are configured to automatically report to the local authorities.

B. Fire suppression systems are designed to operate automatically, and activating the system when staff are not yet evacuated could create confusion and panic, leading to injuries or even fatalities. Manual triggering of the system could be necessary under certain conditions, but only after all other data center personnel are safely evacuated.

C. In an emergency, safety of life is always the first priority; therefore, the complete and orderly evacuation of the facility staff would be the most important activity.

D. Removal of backups from the data center is not an appropriate action because it could delay the evacuation of personnel. Most companies would have copies of backups in offsite storage to mitigate the risk of data loss for this type of disaster.

A4-60 An IS auditor discovers that the disaster recovery plan (DRP) for a company does not include a critical application hosted in the cloud. Management's response states that the cloud vendor is responsible for disaster recovery (DR) and DR-related testing. What is the **NEXT** course of action for the IS auditor to pursue?

A. Plan an audit of the cloud vendor.
B. Review the vendor contract to determine its DR capabilities.
C. Review an independent auditor's report of the cloud vendor.
D. Request a copy of the DRP from the cloud vendor.

B is the correct answer.

Justification:

A. Auditing the cloud vendor would be useful; however, this would only be useful if the vendor is contractually required to provide disaster recovery (DR) services.
B. DR services can only be expected from the vendor when explicitly listed in the contract with well-defined recovery time objectives and recovery point objectives. Without the contractual language, the vendor is not required to provide DR services.
C. An independent auditor's report, such as Statements on Standards for Attestation Engagements 16, on DR capabilities can be reviewed to ascertain the vendor's DR capabilities; however, this will only be fruitful if the vendor is contractually required to provide DR services.
D. A copy of DR policies can be requested to review their adequacy; however, this will only be useful if the vendor is contractually required to provide DR services.

A4-61 An IS auditor is performing a review of the disaster recovery hot site used by a financial institution. Which of the following would be the **GREATEST** concern?

A. System administrators use shared accounts which never expire at the hot site.
B. Disk space utilization data are not kept current.
C. Physical security controls at the hot site are less robust than at the main site.
D. Servers at the hot site do not have the same specifications as at the main site.

B is the correct answer.

Justification:

A. While it is not a good practice for security administrators to share accounts that do not expire, the greater risk in this scenario would be running out of disk space.
B. Not knowing how much disk space is in use and, therefore, how much is needed at the disaster recovery site could create major issues in the case of a disaster.
C. Physical security controls are important, and this would be a concern, but the more important concern would be running out of disk space. The particular physical characteristic of the disaster recovery site may call for different controls that may appear to be less robust than the main site; however, such a risk could be addressed through policy and procedures or by adding additional personnel if needed.
D. As long as the servers at the hot site are capable of running the programs that are required in a disaster recovery situation, the precise capabilities of the servers at the hot site is not a major risk. It is necessary to ensure that software configuration and settings match the servers at the main site, but it is not unusual for newer and more powerful servers to exist at the main site for everyday production use while the standby servers are less powerful.

A4-62 When reviewing system parameters, an IS auditor's **PRIMARY** concern should be that:

A. they are set to meet both security and performance requirements.
B. changes are recorded in an audit trail and periodically reviewed.
C. changes are authorized and supported by appropriate documents.
D. access to parameters in the system is restricted.

A is the correct answer.

Justification:

A. The primary concern is to find the balance between security and performance. Recording changes in an audit trail and periodically reviewing them is a detective control; however, if parameters are not set according to business rules, monitoring of changes may not be an effective control.
B. Reviewing changes to ensure that they are supported by appropriate documents is also a detective control.
C. If parameters are set incorrectly, the related documentation and the fact that these are authorized does not reduce the impact.
D. Restriction of access to parameters ensures that only authorized staff can access the parameters; however, if the parameters are set incorrectly, restricting access will still have an adverse impact.

A4-63 An offsite information processing facility with electrical wiring, air conditioning and flooring, but no computer or communications equipment, is a:

A. cold site.
B. warm site.
C. dial-up site.
D. duplicate processing facility.

A is the correct answer.

Justification:

A. A cold site is ready to receive equipment but does not offer any components at the site in advance of the need.
B. A warm site is an offsite backup facility that is partially configured with network connections and selected peripheral equipment—such as disk and tape units, controllers and central processing units—to operate an information processing facility.
C. A dial-up site is used for remote access, but not for offsite information processing.
D. A duplicate information processing facility is a dedicated, fully-developed recovery site that can back up critical applications.

A4-64 An optimized disaster recovery plan for an organization should:

A. reduce the length of the recovery time and the cost of recovery.
B. increase the length of the recovery time and the cost of recovery.
C. reduce the duration of the recovery time and increase the cost of recovery.
D. not affect the recovery time or the cost of recovery.

A is the correct answer.

Justification:

A. One of the objectives of a disaster recovery plan (DRP) is to reduce the duration and cost of recovering from a disaster.
B. A DRP would increase the cost of operations before and after the disaster occurs.
C. A DRP should reduce the time to return to normal operations.
D. A DRP should reduce the cost that could result from a disaster.

A4-65 A disaster recovery plan for an organization's financial system specifies that the recovery point objective is zero and the recovery time objective is 72 hours. Which of the following is the **MOST** cost-effective solution?

A. A hot site that can be operational in eight hours with asynchronous backup of the transaction logs
B. Distributed database systems in multiple locations updated asynchronously
C. Synchronous updates of the data and standby active systems in a hot site
D. Synchronous remote copy of the data in a warm site that can be operational in 48 hours

D is the correct answer.

Justification:
A. A hot site would meet the recovery time objective (RTO) but would incur higher costs than necessary.
B. Asynchronous updates of the database in distributed locations do not meet the recovery point objective (RPO).
C. Synchronous updates of the data and standby active systems in a hot site meet the RPO and RTO requirements but are costlier than a warm site solution.
D. The synchronous copy of the data storage achieves the RPO, and a warm site operational in 48 hours meets the required RTO.

A4-66 A financial institution that processes millions of transactions each day has a central communications processor (switch) for connecting to automated teller machines. Which of the following would be the **BEST** contingency plan for the communications processor?

A. Reciprocal agreement with another organization
B. Alternate processor in the same location
C. Alternate processor at another network node
D. Duplex communication links

C is the correct answer.

Justification:
A. Reciprocal agreements make an organization dependent on the other organization and raise privacy, competition and regulatory issues.
B. Having an alternate processor in the same location resolves the equipment problem but would not be effective if the failure was caused by environmental conditions (i.e., power disruption).
C. The unavailability of the central communications processor would disrupt all access to the banking network. This could be caused by an equipment, power or communications failure. Having a duplicate processor in another location that could be used for alternate processing is the best solution.
D. The installation of duplex communication links would only be appropriate if the failure were limited to the communication link.

A4-67 Which of the following provides the **BEST** evidence of an organization's disaster recovery capability readiness?

A. A disaster recovery plan (DRP)
B. Customer references for the alternate site provider
C. Processes for maintaining the DRP
D. Results of tests and exercises

D is the correct answer.

Justification:

A. Having a plan is important, but a plan cannot be considered effective until it has been tested.
B. Customer references may aid in choosing an alternate site provider but will not ensure the effectiveness of the plan.
C. A disaster recovery plan must be kept up to date through a regular maintenance and review schedule, but this is not as important as testing.
D. Only tests and exercises demonstrate the adequacy of the plans and provide reasonable assurance of an organization's disaster recovery capability readiness.

A4-68 An IS auditor finds that database administrators (DBAs) have access to the log location on the database server and the ability to purge logs from the system. What is the **BEST** audit recommendation to ensure that DBA activity is effectively monitored?

A. Change permissions to prevent DBAs from purging logs.
B. Forward database logs to a centralized log server to which the DBAs do not have access.
C. Require that critical changes to the database are formally approved.
D. Back up database logs to tape.

B is the correct answer.

Justification:

A. Changing the database administrator (DBA) permissions to prevent DBAs from purging logs may not be feasible and does not adequately protect the availability and integrity of the database logs.
B. To protect the availability and integrity of the database logs, it is most feasible to forward the database logs to a centralized log server to which the DBAs do not have access.
C. Requiring that critical changes to the database are formally approved does not adequately protect the availability and integrity of the database logs.
D. Backing up database logs to tape does not adequately protect the availability and integrity of the database logs.

A4-69 While performing a review of a critical third-party application, an IS auditor would be **MOST** concerned with discovering:

A. inadequate procedures for ensuring adequate system portability.
B. inadequate operational documentation for the system.
C. an inadequate alternate service provider listing.
D. an inadequate software escrow agreement.

D is the correct answer.

Justification:
A. Procedures to ensure that systems are developed so that they can be ported to other system platforms will help ensure that the system can still continue functioning without affecting the business process if changes to the infrastructure occur. This is less important than availability of the software.
B. Inadequate operational documentation is a risk but would be less significant than the risk of unavailability of the software.
C. While alternate service providers could be used if a vendor goes out of business, having access to the source code via a software escrow agreement is more important.
D. The inclusion of a clause in the agreement that requires software code to be placed in escrow helps to ensure that the customer can continue to use the software and/or obtain technical support if a vendor were to go out of business.

A4-70 Which of the following activities should the business continuity manager perform **FIRST** after the replacement of hardware at the primary information processing facility?

A. Verify compatibility with the hot site
B. Review the implementation report
C. Perform a walk-through of the disaster recovery plan
D. Update the IT assets inventory

D is the correct answer.

Justification:
A. Before validating that the new hardware is compatible with the recovery site, the business continuity manager should update the listing of all equipment and IT assets included in the business continuity plan.
B. The implementation report will be of limited value to the business continuity manager because the equipment has been installed.
C. The walk-through of the plan should only be done after the asset inventory has been updated.
D. An IT assets inventory is the basic input for the business continuity/disaster recovery plan, and the plan must be updated to reflect changes in the IT infrastructure.

A4-71 Which of the following would an IS auditor consider to be the **MOST** important to review when conducting a disaster recovery audit?

A. A hot site is contracted for and available as needed.
B. A business continuity manual is available and current.
C. Insurance coverage is adequate and premiums are current.
D. Data backups are performed timely and stored offsite.

D is the correct answer.

Justification:
A. A hot site is important, but it is of no use if there are no data backups for it.
B. A business continuity manual is advisable but not most important in a disaster recovery audit.
C. Insurance coverage should be adequate to cover costs but is not as important as having the data backup.
D. Without data to process, all other components of the recovery effort are in vain. Even in the absence of a plan, recovery efforts of any type would not be practical without data to process.

A4-72 Which of the following should the IS auditor review to ensure that servers are optimally configured to support processing requirements?

A. Benchmark test results
B. Server logs
C. Downtime reports
D. Server utilization data

D is the correct answer.

Justification:
A. Benchmark tests are designed to compare system performance using standardized criteria; however, benchmark testing does not provide the best data to ensure the optimal configuration of servers in an organization.
B. A server log contains data showing activities performed on the server but does not contain the utilization data required to ensure the optimal configuration of servers.
C. A downtime report identifies the elapsed time when a computer is not operating correctly because of machine failure but is not useful in determining optimal server configurations.
D. Monitoring server utilization identifies underutilized servers and monitors overall server utilization. Underutilized servers do not provide the business with optimal cost-effectiveness. By monitoring server usage, IT management can take appropriate measures to raise the utilization ratio and provide the most effective return on investment.

A4-73 Which of the following is a continuity plan test that simulates a system crash and uses actual resources to cost-effectively obtain evidence about the plan's effectiveness?

A. Paper test
B. Post-test
C. Preparedness test
D. Walk-through

C is the correct answer.

Justification:

A. A paper test is a walk-through of the plan, involving major players, who attempt to determine what might happen in a particular type of service disruption in the plan's execution. A paper test usually precedes the preparedness test.

B. A post-test is actually a test phase and is comprised of a group of activities such as returning all resources to their proper place, disconnecting equipment, returning personnel and deleting all company data from third-party systems.

C. A preparedness test is a localized version of a full test, wherein resources are expended in the simulation of a system crash. This test is performed regularly on different aspects of the plan and can be a cost-effective way to gradually obtain evidence about the plan's effectiveness. It also provides a means to improve the plan in increments.

D. A walk-through is a test involving a simulated disaster situation that tests the preparedness and understanding of management and staff rather than the actual resources.

A4-74 While designing the business continuity plan (BCP) for an airline reservation system, the **MOST** appropriate method of data transfer/backup at an offsite location would be:

A. shadow file processing.
B. electronic vaulting.
C. hard-disk mirroring.
D. hot-site provisioning.

A is the correct answer.

Justification:

A. In shadow file processing, exact duplicates of the files are maintained at the same site or at a remote site. The two files are processed concurrently. This is used for critical data files such as airline booking systems.

B. Electronic vaulting electronically transmits data either to direct access storage, an optical disc or another storage medium; this is a method used by banks. This is not usually in real time as much as a shadow file system is.

C. Hard-disk mirroring provides redundancy in case the primary hard disk fails. All transactions and operations occur on two hard disks in the same server.

D. A hot site is an alternate site ready to take over business operations within a few hours of any business interruption and is not a method for backing up data.

A4-75 Which of the following is the **BEST** method for determining the criticality of each application system in the production environment?

A. Interview the application programmers.
B. Perform a gap analysis.
C. Review the most recent application audits.
D. Perform a business impact analysis.

D is the correct answer.

Justification:

A. Interviews with the application programmers will provide limited information related to the criticality of the systems.
B. A gap analysis is relevant to system development and project management but does not determine application criticality.
C. The audits may not contain the required information about application criticality or may not have been done recently.
D. A business impact analysis (BIA) will give the impact of the loss of each application. A BIA is conducted with representatives of the business that can accurately describe the criticality of a system and its importance to the business.

A4-76 Code erroneously excluded from a production release was subsequently moved into the production environment, bypassing normal change procedures. Which of the following choices is of **MOST** concern to the IS auditor performing a post-implementation review?

A. The code was missed during the initial implementation.
B. The change did not have change management approval.
C. The error was discovered during the postimplementation review.
D. The release team used the same change order number.

B is the correct answer.

Justification:

A. Although missing a component of a release is indicative of a process deficiency, it is of more concern that the missed change was promoted into the production environment without management approval.
B. Change management approval of changes mitigates the risk of unauthorized changes being introduced to the production environment. Unauthorized changes might result in disruption of systems or fraud. It is, therefore, imperative to ensure that each change has appropriate change management approval.
C. Most release/change control errors are discovered during postimplementation review. It is of greater concern that the change was promoted without management approval after it was discovered.
D. Using the same change order number is not a relevant concern.

A4-77 A hot site should be implemented as a recovery strategy when the:

A. disaster downtime tolerance is low.
B. recovery point objective is high.
C. recovery time objective is high.
D. maximum tolerable downtime is long.

A is the correct answer.

Justification:

A. Disaster downtime tolerance is the time gap during which the business can accept non-availability of IT facilities. If this time gap is low, recovery strategies that can be implemented within a short period of time, such as a hot site, should be used.

B. The recovery point objective (RPO) is the earliest point in time at which it is possible to recover the data. A high RPO means that the process would result in greater losses of data.

C. A high recovery time objective means that additional time would be available for the recovery strategy, thus making other recovery alternatives—such as warm or cold sites—viable alternatives.

D. If the maximum tolerable downtime is long, then a warm or cold site is a more cost-effective solution.

A4-78 In which of the following situations is it **MOST** appropriate to implement data mirroring as the recovery strategy?

A. Disaster tolerance is high.
B. The recovery time objective is high.
C. The recovery point objective is low.
D. The recovery point objective is high.

C is the correct answer.

Justification:

A. Data mirroring is a data recovery technique, and disaster tolerance addresses the allowable time for an outage of the business.

B. The recovery time objective (RTO) is an indicator of the disaster tolerance. Data mirroring addresses data loss, not the RTO.

C. The recovery point objective (RPO) indicates the latest point in time at which it is possible to recover the data. This determines how often the data must be backed up to minimize data loss. If the RPO is low, then the organization does not want to lose much data and must use a process such as data mirroring to prevent data loss.

D. If the RPO is high, then a less expensive backup strategy can be used; data mirroring should not be implemented as the data recovery strategy.

A4-79 Which of the following stakeholders is the **MOST** important in terms of developing a business continuity plan?

A. Process owners
B. Application owners
C. The board of directors
D. IT management

A is the correct answer.

Justification:

A. **Process owners are essential in identifying the critical business functions, recovery times and resources needed.**
B. A business continuity plan (BCP) is concerned with the continuity of business processes, while applications may or may not support critical business processes.
C. The board of directors might approve the plan, but they are typically not involved in the details of developing the BCP.
D. IT management will identify the IT resources, servers and infrastructure needed to support the critical business functions as defined by the business process owners.

A4-80 Which of the following is the **MOST** efficient and sufficiently reliable way to test the design effectiveness of a change control process?

A. Test a sample population of change requests
B. Test a sample of authorized changes
C. Interview personnel in charge of the change control process
D. Perform an end-to-end walk-through of the process

D is the correct answer.

Justification:

A. Testing a sample population of changes is a test of compliance and operating effectiveness to ensure that users submitted the proper documentation/requests. It does not test the effectiveness of the design.
B. Testing changes that have been authorized may not provide sufficient assurance of the entire process because it does not test the elements of the process related to authorization or detect changes that bypassed the controls.
C. Interviewing personnel in charge of the change control process is not as effective as a walk-through of the change controls process because people may know the process but not follow it.
D. **Observation is the best and most effective method to test changes to ensure that the process is effectively designed.**

A4-81 During fieldwork, an IS auditor experienced a system crash caused by a security patch installation. To provide reasonable assurance that this event will not recur, the IS auditor should ensure that:

A. only systems administrators perform the patch process.
B. the client's change management process is adequate.
C. patches are validated using parallel testing in production.
D. an approval process of the patch, including a risk assessment, is developed.

B is the correct answer.

Justification:

A. While system administrators would normally install patches, it is more important that changes be made according to a formal procedure that includes testing and implementing the change during nonproduction times.
B. The change management process, which would include procedures regarding implementing changes during production hours, helps to ensure that this type of event does not recur. An IS auditor should review the change management process, including patch management procedures, to verify that the process has adequate controls and to make suggestions accordingly.
C. While patches would normally undergo testing, it is often impossible to test all patches thoroughly. It is more important that changes be made during nonproduction times, and that a backout plan is in place in case of problems.
D. An approval process alone could not directly prevent this type of incident from happening. There should be a complete change management process that includes testing, scheduling and approval.

A4-82 A batch transaction job failed in production; however, the same job returned no issues during user acceptance testing (UAT). Analysis of the production batch job indicates that it was altered after UAT. Which of the following ways would be the **BEST** to mitigate this risk in the future?

A. Improve regression test cases.
B. Activate audit trails for a limited period after release.
C. Conduct an application user access review.
D. Ensure that developers do not have access to code after testing.

D is the correct answer.

Justification:

A. Improving the quality of the testing would not be applicable in this case because the more important issue is that developers have access to the production environment.
B. Activating audit trails or performing additional logging may be useful; however, the more important issue is that developers have access to the production environment.
C. Conducting an application user access review would not identify developers' access to code because they would not be included in this review.
D. To ensure proper segregation of duties, developers should be restricted to the development environment only. If code needs to be modified after user acceptance testing, the process must be restarted in development.

A4-83 An organization completed a business impact analysis as part of business continuity planning. The **NEXT** step in the process is to develop:

A. a business continuity strategy.
B. a test and exercise plan.
C. a user training program.
D. the business continuity plan (BCP).

A is the correct answer.

Justification:

A. A business continuity strategy is the next phase because it identifies the best way to recover. The criticality of the business process, the cost, the time required to recover, and security must be considered during this phase.
B. The recovery strategy and plan development precede the test plan.
C. Training can only be developed once the business continuity plan (BCP) is in place.
D. A strategy must be determined before the BCP is developed.

A4-84 An IS auditor performing an application maintenance audit would review the log of program changes for the:

A. Authorization of program changes.
B. Creation date of a current object module.
C. Number of program changes actually made.
D. Creation date of a current source program.

A is the correct answer.

Justification:

A. The auditor wants to ensure that only authorized changes have been made to the application. The auditor would therefore review the log of program changes to verify that all changes have been approved.
B. The creation date of the current object module will not indicate earlier changes to the application.
C. The auditor will review the system to notice the number of changes actually made but then will verify that all the changes were authorized.
D. The creation date of the current source program will not identify earlier changes.

A4-85 Which of the following assures an enterprise of the existence and effectiveness of internal controls relative to the service provided by a third party?

A. The current service level agreement
B. A recent independent third-party audit report
C. The current business continuity plan procedures
D. A recent disaster recovery plan test report

B is the correct answer.

Justification:

A. A service level agreement defines the contracted level of service; however, it would not provide assurance related to internal controls.
B. An independent third-party audit report such as a Statements on Standards for Attestation Engagements 16 would provide assurance of the existence and effectiveness of internal controls at the third party.
C. While a business continuity plan is essential, it would not provide assurance related to internal controls.
D. While a disaster recovery plan is essential, it would not provide assurance related to internal controls.

A4-86 When reviewing a disaster recovery plan, an IS auditor should be **MOST** concerned with the lack of:

A. process owner involvement.
B. well-documented testing procedures.
C. an alternate processing facility.
D. a well-documented data classification scheme.

A is the correct answer.

Justification:

A. Process owner involvement is a critical part of the business impact analysis (BIA), which is used to create the disaster recovery plan. If the IS auditor determined that process owners were not involved, this would be a significant concern.

B. While well-documented testing procedures are important, unless process owners are involved there is no way to know whether the priorities and critical elements of the plan are valid.

C. An alternate processing facility may be a requirement to meet the needs of the business; however, such a decision needs to be based on the BIA.

D. A data classification scheme is important to ensure that controls over data are appropriate; however, this is a lesser concern than a lack of process owner involvement.

A4-87 An organization has outsourced its help desk function. Which of the following indicators would be the **BEST** to include in the service level agreement?

A. Overall number of users supported
B. First call resolution rate
C. Number of incidents reported to the help desk
D. Number of agents answering the phones

B is the correct answer.

Justification:

A. The contract price will usually be based on the number of users supported, but the performance metrics should be based on the ability to provide effective support and address user problems rapidly.

B. Because it is about service level (performance) indicators, the percentage of incidents solved on the first call is a good way to measure the effectiveness of the supporting organization.

C. The number of reported incidents cannot be controlled by the outsource supplier; therefore, that cannot be an effective measure.

D. The efficiency and effectiveness of the people answering the calls and being able to address problems rapidly are more important than the number of people answering the calls.

A4-88 Which of the following activities performed by a database administrator should be performed by a different person?

A. Deleting database activity logs
B. Implementing database optimization tools
C. Monitoring database usage
D. Defining backup and recovery procedures

A is the correct answer.

Justification:

A. Because database activity logs record activities performed by the database administrator (DBA), deleting them should be performed by an individual other than the DBA. This is a compensating control to aid in ensuring an appropriate segregation of duties and is associated with the DBA's role.
B. Implementing database optimization tools is part of the DBA's normal job function.
C. Monitoring database usage is part of the DBA's normal job function.
D. Defining backup and recovery procedures is part of the DBA's normal job function.

A4-89 Which of the following is the **BEST** reason for integrating the testing of non-critical systems in disaster recovery plans (DRPs) with business continuity plans (BCPs)?

A. To ensure that DRPs are aligned to the business impact analysis.
B. Infrastructure recovery personnel can be assisted by business subject matter experts.
C. BCPs may assume the existence of capabilities that are not in DRPs.
D. To provide business executives with knowledge of disaster recovery capabilities.

C is the correct answer.

Justification:

A. Disaster recovery plans (DRPs) should be aligned with the business impact analysis; however, this has no impact on integrating the testing of noncritical systems in DRPs with business continuity plans (BCPs).
B. Infrastructure personnel will be focused on restoring the various platforms that make up the infrastructure, and it is not necessary for business subject matter experts to be involved.
C. BCPs may assume the existence of capabilities that are not part of the DRPs, such as allowing employees to work from home during the disaster; however, IT may not have made sufficient provisions for these capabilities (e.g., they cannot support a large number of employees working from home). While the noncritical systems are important, it is possible that they are not part of the DRPs. For example, an organization may use an online system that does not interface with the internal systems. If the business function using the system is a critical process, the system should be tested, and it may not be part of the DRP. Therefore, DRP and BCP testing should be integrated.
D. While business executives may be interested in the benefits of disaster recovery, testing is not the best way to accomplish this task.

A4-90 An IS auditor finds out-of-range data in some tables of a database. Which of the following controls should the IS auditor recommend to avoid this situation?

A. Log all table update transactions.
B. Implement before-and-after image reporting.
C. Use tracing and tagging.
D. Implement integrity constraints in the database.

D is the correct answer.

Justification:

A. Logging all table update transactions is a detective control that would not help avoid invalid data entry.
B. Implementing before-and-after image reporting is a detective control that would not help avoid the situation.
C. Tracing and tagging are used to test application systems and controls and could not prevent out-of-range data.
D. Implementing integrity constraints in the database is a preventive control because data are checked against predefined tables or rules, preventing any undefined data from being entered.

A4-91 An IS auditor discovers that some users have installed personal software on their PCs. This is not explicitly forbidden by the security policy. Of the following, the **BEST** approach for an IS auditor is to recommend that the:

A. IT department implement control mechanisms to prevent unauthorized software installation.
B. Security policy be updated to include the specific language regarding unauthorized software.
C. IT department prohibit the download of unauthorized software.
D. Users obtain approval from an IS manager before installing nonstandard software.

B is the correct answer.

Justification:

A. An IS auditor's obligation is to report on observations noted and make the best recommendation, which is to address the situation through policy. The IT department cannot implement controls in the absence of the authority provided through policy.
B. Lack of specific language addressing unauthorized software in the acceptable use policy is a weakness in administrative controls. The policy should be reviewed and updated to address the issue—and provide authority for the IT department to implement technical controls.
C. Preventing downloads of unauthorized software is not the complete solution. Unauthorized software can be also introduced through compact discs (CDs) and universal serial bus (USB) drives.
D. Requiring approval from the IS manager before installation of the nonstandard software is an exception handling control. It would not be effective unless a preventive control to prohibit user installation of unauthorized software is established first.

A4-92 The purpose of code signing is to provide assurance that:

A. the software has not been subsequently modified.
B. the application can safely interface with another signed application.
C. the signer of the application is trusted.
D. the private key of the signer has not been compromised.

A is the correct answer.

Justification:

A. Code signing ensures that the executable code came from a reputable source and has not been modified after being signed.
B. The signing of code will not ensure that it will integrate with other applications.
C. Code signing will provide assurance of the source but will not ensure that the source is trusted. The code signing will, however, ensure that the code has not been modified.
D. The compromise of the sender's private key would result in a loss of trust and is not the purpose of code signing.

A4-93 An IS auditor analyzing the audit log of a database management system (DBMS) finds that some transactions were partially executed as a result of an error and have not been rolled back. Which of the following transaction processing features has been violated?

A. Consistency
B. Isolation
C. Durability
D. Atomicity

D is the correct answer.

Justification:

A. Consistency ensures that the database is in a proper state when the transaction begins and ends and that the transaction has not violated integrity rules.
B. Isolation means that, while in an intermediate state, the transaction data are invisible to external operations. This prevents two transactions from attempting to access the same data at the same time.
C. Durability guarantees that a successful transaction will persist and cannot be undone.
D. Atomicity guarantees that either the entire transaction is processed or none of it is.

A4-94 Responsibility and reporting lines cannot always be established when auditing automated systems because:

A. diversified control makes ownership irrelevant.
B. staff traditionally changes jobs with greater frequency.
C. ownership is difficult to establish where resources are shared.
D. duties change frequently in the rapid development of technology.

C is the correct answer.

Justification:

A. Ownership is required to ensure that someone has responsibility for the secure and proper operation of a system and the protection of data.
B. The movement of staff is not a serious issue because the responsibility should be linked to a job description, not an individual.
C. The actual data and/or application owner may be hard to establish because of the complex nature of both data and application systems and many systems support more than one business department.
D. Duties may change frequently, but that does not absolve the organization of having a declared owner for systems and data.

A4-95 Which of the following distinguishes a business impact analysis from a risk assessment?

A. An inventory of critical assets
B. An identification of vulnerabilities
C. A listing of threats
D. A determination of acceptable downtime

D is the correct answer.

Justification:
A. An inventory of critical assets is completed in both a risk assessment and a business impact analysis (BIA).
B. An identification of vulnerabilities is relevant in both a risk assessment and a BIA.
C. A listing of threats is relevant both in a risk assessment and a BIA.
D. A determination of acceptable downtime is made only in a BIA.

A4-96 When reviewing a hardware maintenance program, an IS auditor should assess whether:

A. the schedule of all unplanned maintenance is maintained.
B. it is in line with historical trends.
C. it has been approved by the IS steering committee.
D. the program is validated against vendor specifications.

D is the correct answer.

Justification:
A. Unplanned maintenance cannot be scheduled.
B. Hardware maintenance programs do not necessarily need to be in line with historic trends.
C. Maintenance schedules normally are not approved by the steering committee.
D. Although maintenance requirements vary based on complexity and performance workloads, a hardware maintenance schedule should be validated against the vendor-provided specifications.

A4-97 An IS auditor should recommend the use of library control software to provide reasonable assurance that:

A. program changes have been authorized.
B. only thoroughly tested programs are released.
C. modified programs are automatically moved to production.
D. source and executable code integrity is maintained.

A is the correct answer.

Justification:
A. Library control software should be used to separate test from production libraries in mainframe and/or client server environments. The main objective of library control software is to provide assurance that program changes have been authorized.
B. Library control software is concerned with authorized program changes and cannot determine whether programs have been thoroughly tested.
C. Programs should not be moved automatically into production without proper authorization.
D. Library control software provides reasonable assurance that the source code and executable code are matched at the time a source code is moved to production. Access control will ensure the integrity of the software, but the most important benefit of version control software is to ensure that all changes are authorized.

A4-98 Which of the following would help to ensure the portability of an application connected to a database?

A. Verification of database import and export procedures
B. Usage of a Structured Query Language
C. Analysis of stored procedures/triggers
D. Synchronization of the entity-relation model with the database physical schema

B is the correct answer.

Justification:

A. Verification of import and export procedures with other systems ensures better interfacing with other systems but does not contribute to the portability of an application connecting to a database.
B. The use of Structured Query Language facilitates portability because it is an industry standard used by many systems.
C. Analyzing stored procedures/triggers ensures proper access/performance but does not contribute to the portability of an application connecting to a database.
D. Reviewing the design entity-relation model will be helpful but does not contribute to the portability of an application connecting to a database.

A4-99 Business units are concerned about the performance of a newly implemented system. Which of the following should an IS auditor recommend?

A. Develop a baseline and monitor system usage.
B. Define alternate processing procedures.
C. Prepare the maintenance manual.
D. Implement the changes users have suggested.

A is the correct answer.

Justification:

A. An IS auditor should recommend the development of a performance baseline and monitor the system's performance against the baseline to develop empirical data upon which decisions for modifying the system can be made.
B. Alternate processing procedures will not alter a system's performance, and no changes should be made until the reported issue has been examined more thoroughly.
C. A maintenance manual will not alter a system's performance or address the user concerns.
D. Implementing changes without knowledge of the cause(s) for the perceived poor performance may not result in a more efficient system.

A4-100 The **PRIMARY** objective of service-level management is to:

A. define, agree on, record and manage the required levels of service.
B. ensure that services are managed to deliver the highest achievable level of availability.
C. keep the costs associated with any service at a minimum.
D. monitor and report any legal noncompliance to business management.

A is the correct answer.

Justification:

A. The objective of service-level management (SLM) is to negotiate, document and manage (i.e., provide and monitor) the services in the manner in which the customer requires those services.
B. SLM does not necessarily ensure that services are delivered at the highest achievable level of availability (e.g., redundancy and clustering). Although maximizing availability might be necessary for some critical services, it cannot be applied as a general rule of thumb.
C. SLM cannot ensure that costs for all services will be kept at a low or minimum level because costs associated with a service will directly reflect the customer's requirements.
D. Monitoring and reporting legal noncompliance is not a primary objective of SLM.

A4-101 Which of the following should be a **MAJOR** concern for an IS auditor reviewing a business continuity plan?

A. The plan is approved by the chief information officer.
B. The plan contact lists have not been updated.
C. Test results are not adequately documented.
D. The training schedule for recovery personnel is not included.

C is the correct answer.

Justification:

A. Ideally, the board of directors should approve the plan to ensure acceptability, but it is possible to delegate approval authority to the chief information officer. Pragmatically, lack of documenting test results could have more significant consequences.
B. The contact lists are an important part of the business continuity plan (BCP); however, they are not as important as documenting the test results.
C. The effectiveness of a BCP can best be determined through tests. If results of tests are not documented, then there is no basis for feedback, updates, etc.
D. If test results are documented, a need for training will be identified and the BCP will be updated.

A4-102 Which of the following processes will be **MOST** effective in reducing the risk that unauthorized software on a backup server is distributed to the production server?

A. Manually copy files to accomplish replication.
B. Review changes in the software version control system.
C. Ensure that developers do not have access to the backup server.
D. Review the access control log of the backup server.

B is the correct answer.

Justification:

A. Even if replication is be conducted manually with due care, there still remains a risk to copying unauthorized software from one server to another.
B. It is common practice for software changes to be tracked and controlled using version control software. An IS auditor should review reports or logs from this system to identify the software that is promoted to production. Only moving the versions on the version control system program will prevent the transfer of development or earlier versions.
C. If unauthorized code was introduced onto the backup server by developers, controls on the production server and the software version control system should mitigate this risk.
D. Review of the access log will identify staff access or the operations performed; however, it may not provide enough information to detect the release of unauthorized software.

A4-103 An organization has recently installed a security patch, which crashed the production server. To minimize the probability of this occurring again, an IS auditor should:

A. apply the patch according to the patch's release notes.
B. ensure that a good change management process is in place.
C. thoroughly test the patch before sending it to production.
D. approve the patch after doing a risk assessment.

B is the correct answer.

Justification:
A. The IS auditor should not apply the patch. That is an administrator responsibility.
B. An IS auditor must review the change management process, including patch management procedures, and verify that the process has adequate controls and make suggestions accordingly.
C. The testing of the patch is the responsibility of the development or production support team, not the auditor.
D. The IS auditor is not authorized to approve a patch. That is a responsibility of a steering committee.

A4-104 During maintenance of a relational database, several values of the foreign key in a transaction table have been corrupted. The consequence is that:

A. the detail of involved transactions may no longer be associated with master data, causing errors when these transactions are processed.
B. there is no way of reconstructing the lost information, except by deleting the dangling tuples and reentering the transactions.
C. the database will immediately stop execution and lose more information.
D. the database will no longer accept input data.

A is the correct answer.

Justification:
A. When the external key of a transaction is corrupted or lost, the application system will normally be incapable of directly attaching the master data to the transaction data. Normally, this will cause the system to undertake a sequential search and slow down the processing. If the concerned files are big, this slowdown will be unacceptable. This is a violation of referential integrity.
B. A system can recover the corrupted external key by re-indexing the table.
C. The corruption of a foreign key will not stop program execution.
D. The corruption of a foreign key will not affect database input.

A4-105 In a relational database with referential integrity, the use of which of the following keys would prevent deletion of a row from a customer table as long as the customer number of that row is stored with live orders on the orders table?

A. Foreign key
B. Primary key
C. Secondary key
D. Public key

A is the correct answer.

Justification:
A. In a relational database with referential integrity, the use of foreign keys would prevent events such as primary key changes and record deletions, resulting in orphaned relations within the database.
B. It should not be possible to delete a row from a customer table when the customer number (primary key) of that row is stored with live orders on the orders table (the foreign key to the customer table). A primary key works in one table so it is not able to provide/ensure referential integrity by itself.
C. Secondary keys that are not foreign keys are not subject to referential integrity checks.
D. A public key is related to encryption and not linked in any way to referential integrity.

A4-106 The **PRIMARY** objective of testing a business continuity plan is to:

A. familiarize employees with the business continuity plan.
B. ensure that all residual risk is addressed.
C. exercise all possible disaster scenarios.
D. identify limitations of the business continuity plan.

D is the correct answer.

Justification:
A. Familiarizing employees with the business continuity plan is a secondary benefit of a test.
B. It is not cost-effective to address all residual risk in a business continuity plan.
C. It is not practical to test all possible disaster scenarios.
D. Testing the business continuity plan provides the best evidence of any limitations that may exist.

A4-107 An IS auditor examining the security configuration of an operating system should review the:

A. transaction logs.
B. authorization tables.
C. parameter settings.
D. routing tables.

C is the correct answer.

Justification:

A. Transaction logs are used to track and analyze transactions related to an application or system interface, but that is not the primary source of audit evidence in an operating system audit.
B. Authorization tables are used to verify implementation of logical access controls and will not be of much help when reviewing control features of an operating system.
C. Configuration parameters allow a standard piece of software to be customized for diverse environments and are important in determining how a system runs. The parameter settings should be appropriate to an organization's workload and control environment. Improper implementation and/or monitoring of operating systems can result in undetected errors and corruption of the data being processed, as well as lead to unauthorized access and inaccurate logging of system usage.
D. Routing tables do not contain information about the operating system and, therefore, provide no information to aid in the evaluation of controls.

A4-108 During a data center audit, an IS auditor observes that some parameters in the tape management system are set to bypass or ignore tape header records. Which of the following is the **MOST** effective compensating control for this weakness?

A. Staging and job setup
B. Supervisory review of logs
C. Regular backup of tapes
D. Offsite storage of tapes

A is the correct answer.

Justification:

A. If the IS auditor finds that there are effective staging and job setup processes, this can be accepted as a compensating control. Not reading header records may otherwise result in loading the wrong tape and deleting or accessing data on the loaded tape.
B. Supervisory review of logs is a detective control that would not prevent loading of the wrong tapes.
C. Regular tape backup is not related to bypassing tape header records.
D. Offsite storage of tapes would not prevent loading the wrong tape because of bypassing header records.

A4-109 While reviewing the IT infrastructure, an IS auditor notices that storage resources are continuously being added. The IS auditor should:

A. recommend the use of disk mirroring.
B. review the adequacy of offsite storage.
C. review the capacity management process.
D. recommend the use of a compression algorithm.

C is the correct answer.

Justification:
A. A disk mirroring solution would increase storage requirements. This would not be advisable until a proper capacity management plan is in place.
B. Offsite storage is unrelated to the problem.
C. Capacity management is the planning and monitoring of computer resources to ensure that available IT resources are used efficiently and effectively. This will look at capacity from a strategic viewpoint and allow a plan to forecast and purchase additional equipment in a planned manner.
D. Though data compression may save disk space, it could affect system performance. This is not the first choice—the auditor should recommend more investigation into the increased demand for storage before providing any recommended solutions.

A4-110 Which of the following is the **GREATEST** risk of an organization using reciprocal agreements for disaster recovery between two business units?

A. The documents contain legal deficiencies.
B. Both entities are vulnerable to the same incident.
C. IT systems are not identical.
D. One party has more frequent disruptions than the other.

B is the correct answer.

Justification:
A. Inadequate agreements between two business units is a risk, but generally a lesser one than the risk that both organizations will suffer a disaster at the same time.
B. The use of reciprocal disaster recovery is based on the probability that both organizations will not suffer a disaster at the same time.
C. While incompatible IT systems could create problems, it is a less significant risk than both organizations suffering from the same disaster at the same time.
D. While one party may use the other's resources more frequently, this can be addressed by contractual provisions and is not a major risk.

A4-111 In determining the acceptable time period for the resumption of critical business processes:

A. only downtime costs need to be considered.
B. recovery operations should be analyzed.
C. both downtime costs and recovery costs need to be evaluated.
D. indirect downtime costs should be ignored.

C is the correct answer.

Justification:

A. Downtime costs cannot be looked at in isolation. The quicker information assets can be restored and business processing resumed, the smaller the downtime costs. However, the expenditure needed to have the redundant capability required to rapidly recover information resources might be prohibitive for nonessential business processes.
B. Recovery operations alone do not determine the acceptable time period for the resumption of critical business processes, and indirect downtime costs should be considered in addition to the direct cash outflows incurred due to business disruption.
C. Both downtime costs and recovery costs need to be evaluated in determining the acceptable time period before the resumption of critical business processes. The outcome of the business impact analysis should be a recovery strategy that represents the optimal balance.
D. The indirect costs of a serious disruption to normal business activity (e.g., loss of customer and supplier goodwill and loss of market share) may actually be more significant than direct costs over time, thus reaching the point where business viability is threatened.

A4-112 To verify that the correct version of a data file was used for a production run, an IS auditor should review:

A. operator problem reports.
B. operator work schedules.
C. system logs.
D. output distribution reports.

C is the correct answer.

Justification:

A. Operator problem reports are used by operators to log computer operation problems.
B. Operator work schedules are maintained to assist in human resource planning.
C. System logs are automated reports which identify most of the activities performed on the computer. Programs that analyze the system log have been developed to report on specifically defined items. The IS auditor can then carry out tests to ensure that the correct file version was used for a production run.
D. Output distribution reports identify all application reports generated and their distribution.

A4-113 The **BEST** audit procedure to determine if unauthorized changes have been made to production code is to:

A. examine the change control system records and trace them forward to object code files.
B. review access control permissions operating within the production program libraries.
C. examine object code to find instances of changes and trace them back to change control records.
D. review change approved designations established within the change control system.

C is the correct answer.

Justification:
A. Checking the change control system will not detect changes that were not recorded in the control system.
B. Reviewing access control permissions will not identify unauthorized changes made previously.
C. The procedure of examining object code files to establish instances of code changes and tracing these back to change control system records is a substantive test that directly addresses the risk of unauthorized code changes.
D. Reviewing change approved designations will not identify unauthorized changes.

A4-114 When performing a database review, an IS auditor notices that some tables in the database are not normalized. The IS auditor should next:

A. recommend that the database be normalized.
B. review the conceptual data model.
C. review the stored procedures.
D. review the justification.

D is the correct answer.

Justification:
A. The IS auditor should not recommend normalizing the database until further investigation takes place.
B. Reviewing the conceptual data model will not provide information about normalization or the justification for the level of normalization.
C. Reviewing the stored procedures will not provide information about normalization.
D. If the database is not normalized, the IS auditor should review the justification because, in some situations, denormalization is recommended for performance reasons.

A4-115 Which of the following would be **MOST** important for an IS auditor to verify while conducting a business continuity audit?

A. Data backups are performed on a timely basis.
B. A recovery site is contracted for and available as needed.
C. Human safety procedures are in place.
D. Insurance coverage is adequate and premiums are current.

C is the correct answer.

Justification:
A. Performing data backups is necessary for a business continuity plan, but the IS auditor will always be most concerned with human safety.
B. A recovery site is important for business continuity, but life safety is always the first priority.
C. The most important element in any business continuity process is the protection of human life. This takes precedence over all other aspects of the plan.
D. Insurance coverage is not as important as life safety.

A4-116 The application systems of an organization using open-source software have no single recognized developer producing patches. Which of the following would be the **MOST** secure way of updating open-source software?

A. Rewrite the patches and apply them.
B. Review the code and application of available patches.
C. Develop in-house patches.
D. Identify and test suitable patches before applying them.

D is the correct answer.

Justification:
A. Rewriting the patches and applying them would require skilled resources and time to rewrite the patches.
B. Code review could be possible, but tests need to be performed before applying the patches.
C. Because the system was developed outside the organization, the IT department may not have the necessary skills and resources to develop patches.
D. Suitable patches from the existing developers should be selected and tested before applying them.

A4-117 During the audit of a database server, which of the following would be considered the **GREATEST** exposure?

A. The password on the administrator account does not expire.
B. Default global security settings for the database remain unchanged.
C. Old data have not been purged.
D. Database activity is not fully logged.

B is the correct answer.

Justification:
A. A nonexpiring password is a risk and an exposure but not as serious a risk as a weak password or the continued use of default settings.
B. Default security settings for the database could allow issues such as blank user passwords or passwords that were the same as the username.
C. Failure to purge old data may present a performance issue but is not an immediate security concern.
D. Logging all database activity is a potential risk but not as serious a risk as default settings.

A4-118 An IS auditor discovers that developers have operator access to the command line of a production environment operating system. Which of the following controls would **BEST** mitigate the risk of undetected and unauthorized program changes to the production environment?

A. Commands typed on the command line are logged.
B. Hash keys are calculated periodically for programs and matched against hash keys calculated for the most recent authorized versions of the programs.
C. Access to the operating system command line is granted through an access restriction tool with preapproved rights.
D. Software development tools and compilers have been removed from the production environment.

B is the correct answer.

Justification:
A. Having a log is not a control; reviewing the log is a control.
B. The matching of hash keys over time would allow detection of changes to files.
C. Because the access was already granted at the command line level, it will be possible for the developers to bypass the control.
D. Removing the tools from the production environment will not mitigate the risk of unauthorized activity by the developers.

A4-119 A new application has been purchased from a vendor and is about to be implemented. Which of the following choices is a key consideration when implementing the application?

A. Preventing the compromise of the source code during the implementation process
B. Ensuring that vendor default accounts and passwords have been disabled
C. Removing the old copies of the program from escrow to avoid confusion
D. Verifying that the vendor is meeting support and maintenance agreements

B is the correct answer.

Justification:
A. The source code may not even be available to the purchasing organization, and it is the executable or object code that must be protected during implementation.
B. Disabling vendor default accounts and passwords is a critical part of implementing a new application.
C. Because this is a new application, there should not be any problem with older versions in escrow.
D. It is not possible to ensure that the vendor is meeting support and maintenance requirements until the system is operating.

A4-120 The **MAIN** criterion for determining the severity level of a service disruption incident is:

A. cost of recovery.
B. negative public opinion.
C. geographic location.
D. downtime.

D is the correct answer.

Justification:
A. The cost of recovery could be minimal, yet the service downtime could have a major impact.
B. Negative public opinion is a symptom of an incident; it is a factor in determining impact but not the most important one.
C. Geographic location does not determine the severity of the incident.
D. The longer the period of time a client cannot be serviced, the greater the severity (impact) of the incident.

A4-121 Doing which of the following during peak production hours could result in unexpected downtime?

A. Performing data migration or tape backup
B. Performing preventive maintenance on electrical systems
C. Promoting applications from development to the staging environment
D. Reconfiguring a standby router in the data center

B is the correct answer.

Justification:
A. Performing data migration may impact performance but would not cause downtime.
B. Preventive maintenance activities should be scheduled for non-peak times of the day, and preferably during a maintenance window time period. A mishap or incident caused by a maintenance worker could result in unplanned downtime.
C. Promoting applications into a staging environment (not production) should not affect systems operations in any significant manner.
D. Reconfiguring a standby router should not cause unexpected downtime because the router is not operational and any problems should not affect network traffic.

A4-122 During a human resources (HR) audit, an IS auditor is informed that there is a verbal agreement between the IT and HR departments as to the level of IT services expected. In this situation, what should the IS auditor do **FIRST**?

A. Postpone the audit until the agreement is documented.
B. Report the existence of the undocumented agreement to senior management.
C. Confirm the content of the agreement with both departments.
D. Draft a service level agreement for the two departments.

C is the correct answer.

Justification:

A. There is no reason to postpone an audit because a service agreement is not documented, unless that is all that is being audited. The agreement can be documented after it has been established that there is an agreement in place.
B. Reporting to senior management is not necessary at this stage of the audit because this is not a serious immediate vulnerability.
C. An IS auditor should first confirm and understand the current practice before making any recommendations. Part of this will be to ensure that both parties agree with the terms of the agreement.
D. Drafting a service level agreement is not the IS auditor's responsibility.

A4-123 A database administrator has detected a performance problem with some tables, which could be solved through denormalization. This situation will increase the risk of:

A. concurrent access.
B. deadlocks.
C. unauthorized access to data.
D. a loss of data integrity.

D is the correct answer.

Justification:

A. Denormalization will have no effect on concurrent access to data in a database; concurrent access is resolved through locking.
B. Deadlocks are a result of locking of records. This is not related to normalization.
C. Access to data is controlled by defining user rights to information and is not affected by denormalization.
D. Normalization is the removal of redundant data elements from the database structure. Disabling normalization in relational databases will create redundancy and a risk of not maintaining consistency of data, with the consequent loss of data integrity.

A4-124 Which of the following processes should an IS auditor recommend to assist in the recording of baselines for software releases?

A. Change management
B. Backup and recovery
C. Incident management
D. Configuration management

D is the correct answer.

Justification:

A. Change management is important to control changes to the configuration, but the baseline itself refers to a standard configuration.
B. Backup and recovery of the configuration are important, but not used to create the baseline.
C. Incident management will determine how to respond to an adverse event but is not related to recording baseline configurations.
D. The configuration management process may include automated tools that will provide an automated recording of software release baselines. Should the new release fail, the baseline will provide a point to which to return.

A4-125 An IS auditor notes that patches for the operating system used by an organization are deployed by the IT department as advised by the vendor. The **MOST** significant concern an IS auditor should have with this practice is that IT has **NOT** considered:

A. the training needs for users after applying the patch.
B. any beneficial impact of the patch on the operational systems.
C. delaying deployment until testing the impact of the patch.
D. the necessity of advising end users of new patches.

C is the correct answer.

Justification:

A. Normally, there is no need for training users when a new operating system patch has been installed.
B. Any beneficial impact is less important than the risk of unavailability, which could be avoided with proper testing.
C. Deploying patches without testing exposes an organization to the risk of system disruption or failure.
D. Normally, there is no need for advising users when a new operating system patch has been installed except to ensure that the patch is applied at a time that will have minimal impact on operations.

A4-126 The **BEST** method for assessing the effectiveness of a business continuity plan is to review the:

A. plans and compare them to appropriate standards.
B. results from previous tests.
C. emergency procedures and employee training.
D. offsite storage and environmental controls.

B is the correct answer.

Justification:

A. Comparisons to standards will give some assurance that the plan addresses the critical aspects of a business continuity plan but will not reveal anything about its effectiveness.
B. Previous test results will provide evidence of the effectiveness of the business continuity plan.
C. Reviewing emergency procedures would provide insight into some aspects of the plan but would fall short of providing assurance of the plan's overall effectiveness.
D. Reviewing offsite storage and environmental controls would provide insight into some aspects of the plan but would fall short of providing assurance of the plan's overall effectiveness.

A4-127 With respect to business continuity strategies, an IS auditor interviews key stakeholders in an organization to determine whether they understand their roles and responsibilities. The IS auditor is attempting to evaluate the:

A. clarity and simplicity of the business continuity plans.
B. adequacy of the business continuity plans.
C. effectiveness of the business continuity plans.
D. ability of IS and end-user personnel to respond effectively in emergencies.

A is the correct answer.

Justification:

A. The IS auditor should interview key stakeholders to evaluate how well they understand their roles and responsibilities. When all stakeholders have a detailed understanding of their roles and responsibilities in the event of a disaster, an IS auditor can deem the business continuity plan to be clear and simple.

B. To evaluate adequacy, the IS auditor should review the plans and compare them to appropriate standards and the results of tests of the plan.

C. To evaluate effectiveness, the IS auditor should review the results from previous tests or incidents. This is the best determination for the evaluation of effectiveness. An understanding of roles and responsibilities by key stakeholders will assist in ensuring the business continuity plan is effective.

D. To evaluate the response, the IS auditor should review results of continuity tests. This will provide the IS auditor with assurance that target and recovery times are met. Emergency procedures and employee training need to be reviewed to determine whether the organization has implemented plans to allow for an effective response.

A4-128 During the design of a business continuity plan, the business impact analysis identifies critical processes and supporting applications. This will **PRIMARILY** influence the:

A. responsibility for maintaining the business continuity plan.
B. criteria for selecting a recovery site provider.
C. recovery strategy.
D. responsibilities of key personnel.

C is the correct answer.

Justification:

A. The responsibility for maintaining the business continuity plan is decided after the selection or design of the appropriate recovery strategy and development of the plan.

B. The criteria for selecting a recovery site provider are decided after the selection or design of the appropriate recovery strategy.

C. The most appropriate strategy is selected based on the relative risk level, time lines and criticality identified in the business impact analysis.

D. The responsibilities of key personnel are decided after the selection or design of the appropriate recovery strategy during the plan development phase.

A4-129 During a review of a business continuity plan, an IS auditor noticed that the point at which a situation is declared to be a crisis has not been defined. The **MAJOR** risk associated with this is that:

A. assessment of the situation may be delayed.
B. execution of the disaster recovery plan could be impacted.
C. notification of the teams might not occur.
D. potential crisis recognition might be delayed.

B is the correct answer.

Justification:

A. Problem and severity assessment would provide information necessary in declaring a disaster, but the lack of a crisis declaration point would not delay the assessment.
B. Execution of the business continuity and disaster recovery plans would be impacted if the organization does not know when to declare a crisis.
C. After a potential crisis is recognized, the teams responsible for crisis management need to be notified. Delaying the declaration of a disaster would impact or negate the effect of having response teams, but this is only one part of the larger impact.
D. Potential crisis recognition is the first step in recognizing or responding to a disaster and would occur prior to the declaration of a disaster.

A4-130 An organization has just completed its annual risk assessment. Regarding the business continuity plan, what should an IS auditor recommend as the next step for the organization?

A. Review and evaluate the business continuity plan for adequacy
B. Perform a full simulation of the business continuity plan
C. Train and educate employees regarding the business continuity plan
D. Notify critical contacts in the business continuity plan

A is the correct answer.

Justification:

A. The business continuity plan should be reviewed every time a risk assessment is completed for the organization.
B. Performing a simulation should be completed after the business continuity plan has been deemed adequate for the organization.
C. Training of the employees should be performed after the business continuity plan has been deemed adequate for the organization.
D. There is no reason to notify the business continuity plan contacts at this time.

A4-131 Which of the following database controls would ensure that the integrity of transactions is maintained in an online transaction processing system's database?

A. Authentication controls
B. Data normalization controls
C. Read/write access log controls
D. Commitment and rollback controls

D is the correct answer.

Justification:

A. Authentication controls would ensure that only authorized personnel can make changes but would not ensure the integrity of the changes.
B. Data normalization is not used to protect the integrity of online transactions.
C. Log controls are a detective control but will not ensure the integrity of the data in the database.
D. Commitment and rollback controls are directly relevant to integrity. These controls ensure that database operations that form a logical transaction unit will be completed entirely or not at all (i.e., if, for some reason, a transaction cannot be fully completed, then incomplete inserts/updates/deletes are rolled back so that the database returns to its pretransition state).

A4-132 An IS auditor finds that the data warehouse query performance decreases significantly at certain times of the day. Which of the following controls would be **MOST** relevant for the IS auditor to review?

A. Permanent table-space allocation
B. Commitment and rollback controls
C. User spool and database limit controls
D. Read/write access log controls

C is the correct answer.

Justification:

A. Table-space allocation will not affect performance at different times of the day.
B. Commitment and rollback will only apply to errors or failures and will not affect performance at different times of the day.
C. User spool limits restrict the space available for running user queries. This prevents poorly formed queries from consuming excessive system resources and impacting general query performance. Limiting the space available to users in their own databases prevents them from building excessively large tables. This helps to control space utilization which itself acts to help performance by maintaining a buffer between the actual data volume stored and the physical device capacity. Additionally, it prevents users from consuming excessive resources in ad hoc table builds (as opposed to scheduled production loads that often can run overnight and are optimized for performance purposes). In a data warehouse, because you are not running online transactions, commitment and rollback does not have an impact on performance.
D. Read/write access log controls will not affect performance at different times of the day.

A4-133 In a small organization, developers may release emergency changes directly to production. Which of the following will **BEST** control the risk in this situation?

A. Approve and document the change the next business day.
B. Limit developer access to production to a specific time frame.
C. Obtain secondary approval before releasing to production.
D. Disable the compiler option in the production machine.

A is the correct answer.

Justification:

A. It may be appropriate to allow programmers to make emergency changes as long as they are documented and approved after the fact.
B. Restricting release time frame may help somewhat; however, it would not apply to emergency changes and cannot prevent unauthorized release of the programs.
C. Obtaining secondary approval before releasing to production is not relevant in an emergency situation.
D. Disabling the compiler option in the production machine is not relevant in an emergency situation.

A4-134 Of the following alternatives, the **FIRST** approach to developing a disaster recovery strategy would be to assess whether:

A. all threats can be completely removed.
B. a cost-effective, built-in resilience can be implemented.
C. the recovery time objective can be optimized.
D. the cost of recovery can be minimized.

B is the correct answer.

Justification:

A. It is impossible to remove all existing and future threats.
B. It is critical to initially identify information assets that can be made more resilient to disasters (e.g., diverse routing, alternate paths or multiple communication carriers). Preventing a problem is always better than planning to address a problem when it happens.
C. The optimization of the recovery time objective comes later in the development of the disaster recovery strategy.
D. Efforts to minimize the cost of recovery come later in the development of the disaster recovery strategy.

A4-135 An IS auditor determined that the IT manager recently changed the vendor that is responsible for performing maintenance on critical computer systems to cut costs. While the new vendor is less expensive, the new maintenance contract specifies a change in incident resolution time specified by the original vendor. Which of the following should be the **GREATEST** concern to the IS auditor?

A. Disaster recovery plans may be invalid and need to be revised.
B. Transactional business data may be lost in the event of system failure.
C. The new maintenance vendor is not familiar with the organization's policies.
D. Application owners were not informed of the change.

D is the correct answer.

Justification:

A. Disaster recovery plans (DRPs) must support the needs of the business, but the greater risk is that application owners are not aware of the change in resolution time.
B. Transactional business data loss is determined by data backup frequency and, consequently, the backup schedule.
C. The vendor must abide by the terms of the contract and those should include compliance with the privacy policies of the organization, but the lack of application owner involvement is the most important concern.
D. The greatest risk of making a change to the maintenance of critical systems is that the change could have an adverse impact on a critical business process. While there is a benefit in selecting a less expensive maintenance vendor, the resolution time must be aligned with the needs of the business.

A4-136 In the event of a data center disaster, which of the following would be the **MOST** appropriate strategy to enable a complete recovery of a critical database?

A. Daily data backup to tape and storage at a remote site
B. Real-time replication to a remote site
C. Hard disk mirroring to a local server
D. Real-time data backup to the local storage area network

B is the correct answer.

Justification:

A. Daily tape backup recovery could result in a loss of a day's work of data.
B. With real-time replication to a remote site, data are updated simultaneously in two separate locations; therefore, a disaster in one site would not damage the information located in the remote site. This assumes that both sites were not affected by the same disaster.
C. Hard disk mirroring to a local server takes place in the same data center and could possibly be affected by the same disaster.
D. Real-time data backup to the local storage area network takes place in the same data center and could possibly be affected by the same disaster.

A4-137 If the recovery time objective increases:

A. the disaster tolerance increases.
B. the cost of recovery increases.
C. a cold site cannot be used.
D. the data backup frequency increases.

A is the correct answer.

Justification:

A. The longer the recovery time objective (RTO), the higher disaster tolerance. The disaster tolerance is the amount of time the business can afford to be disrupted before resuming critical operations.
B. The longer the RTO, the lower the recovery cost.
C. It cannot be concluded that a cold site is inappropriate; with a longer RTO the use of a cold site may become feasible.
D. RTO is not related to the frequency of data backups—that is related to recovery point objective.

A4-138 Due to changes in IT, the disaster recovery plan of a large organization has been changed. What is the **PRIMARY** risk if the new plan is not tested?

A. Catastrophic service interruption
B. High consumption of resources
C. Total cost of the recovery may not be minimized
D. Users and recovery teams may face severe difficulties when activating the plan

A is the correct answer.

Justification:

A. If a new disaster recovery plan (DRP) is not tested, the possibility of a catastrophic service interruption that the organization cannot recover from is the most critical of all risk.
B. A DRP that has not been tested may lead to a higher consumption of resources than expected, but that is not the most critical risk.
C. An untested DRP may be inefficient and lead to extraordinary costs, but the most serious risk is the failure of critical services.
D. Testing educates users and recovery teams so that they can effectively execute the DRP, but the most critical risk is the failure of core business services.

A4-139 When developing a disaster recovery plan, the criteria for determining the acceptable downtime should be the:

A. annual loss expectancy.
B. service delivery objective.
C. quantity of orphan data.
D. maximum tolerable outage.

D is the correct answer.

Justification:

A. The acceptable downtime would not be determined by the annual loss expectancy (ALE); ALE is related to risk management calculations, not disaster recovery.
B. The service delivery objective is relevant to business continuity, but it is not determined by acceptable downtime.
C. The quantity of orphan data is relevant to business continuity, but it is not determined by acceptable downtime.
D. Recovery time objective is determined based on the acceptable downtime in case of a disruption of operations. It indicates the maximum tolerable outage that an organization considers to be acceptable before a system or process must resume following a disaster.

A4-140 During the review of an enterprise's preventive maintenance process for systems at a data center, the IS auditor has determined that adequate maintenance is being performed on all critical computing, power and cooling systems. Additionally, it is **MOST** important for the IS auditor to ensure that the organization:

A. has performed background checks on all service personnel.
B. escorts service personnel at all times when performing their work.
C. performs maintenance during noncritical processing times.
D. independently verifies that maintenance is being performed.

C is the correct answer.

Justification:

A. While the trustworthiness of the service personnel is important, it is normal practice for these individuals to be escorted and supervised by the data center personnel. It is also expected that the service provider would perform this background check, not the customer.
B. Escorting service personnel is common and a good practice, but the greater risk in this case would be if work were performed during critical processing times.
C. The biggest risk to normal operations in a data center would be if an incident or mishap were to happen during critical peak processing times; therefore, it would be prudent to ensure that no type of system maintenance be performed at these critical times.
D. It is possible that the service provider is performing inadequate maintenance; therefore, this issue may need to be investigated; however, the bigger risk is maintenance being performed at critical processing times.

A4-141 Which of the following backup techniques is the **MOST** appropriate when an organization requires extremely granular data restore points, as defined in the recovery point objective?

A. Virtual tape libraries
B. Disk-based snapshots
C. Continuous data backup
D. Disk-to-tape backup

C is the correct answer.

Justification:

A. Virtual tape libraries would require time to complete the backup, while continuous data backup happens online (in real time).
B. Disk-based snapshots would require time to complete the backup and would lose some data between the times of the backup and the failure, while continuous data backup happens online (in real time).
C. Recovery point objective (RPO) is based on the acceptable data loss in the case of a disruption. In this scenario the organization needs a short RPO and continuous data backup is the best option.
D. Disk-to-tape backup would require time to complete the backup, while continuous data backup happens online (in real time).

A4-142 A lower recovery time objective results in:

A. higher disaster tolerance.
B. higher cost.
C. wider interruption windows.
D. more permissive data loss.

B is the correct answer.

Justification:

A. Disaster tolerance relates the length of time that critical business processes can be interrupted. A higher disaster tolerance allows for a longer outage and, therefore, longer recovery time.
B. Recovery time objective (RTO) is based on the acceptable down time in case of a disruption of operations. The lower the RTO, the higher the cost of recovery strategies.
C. The lower the disaster tolerance, the narrower the interruption windows. The interruption window is the length of the outage of critical processes.
D. Permissive data loss relates to recovery point objective, not disaster tolerance.

A4-143 During an implementation review of a recent application deployment, it was determined that several incidents were assigned incorrect priorities and, because of this, failed to meet the business service level agreement (SLA). What is the **GREATEST** concern?

A. The support model was not approved by senior management.
B. The incident resolution time specified in the SLA is not realistic.
C. There are inadequate resources to support the applications.
D. The support model was not properly developed and implemented.

D is the correct answer.

Justification:

A. While senior management involvement is important, the more critical issue is whether the support model was not properly developed and implemented.
B. While the incident resolution time specified in the service level agreement may not always be attainable, the more critical issue is whether the support model was not properly developed and implemented.
C. While adequate support resources are important, the more critical issue is whether the support model was not properly developed and implemented.
D. The greatest concern for the IS auditor is that the support model was not developed and implemented correctly to prevent or react to potential outages. Incidents could cost the business a significant amount of money and a support model should be implemented with the project. This should be a step within the system development life cycle and procedures and, if it is missed on one project, it may be a symptom of an overall breakdown in process.

A4-144 What is the **BEST** backup strategy for a large database with data supporting online sales?

A. Weekly full backup with daily incremental backup
B. Daily full backup
C. Clustered servers
D. Mirrored hard disks

D is the correct answer.

Justification:

A. Weekly full backup and daily incremental backup is a poor backup strategy for online transactions. Because this system supports online sales it can be difficult to recreate lost data and this solution may result in a loss of up to one day's worth of data.
B. A full backup normally requires a couple of hours, and therefore, it can be impractical to conduct a full backup every day.
C. Clustered servers provide a redundant processing capability but are not a backup.
D. Mirrored hard disks will ensure that all data are backed up to more than one disk so that a failure of one disk will not result in loss of data.

A4-145 An IS auditor notes during an audit that an organization's business continuity plan (BCP) does not adequately address information confidentiality during the recovery process. The IS auditor should recommend that the plan be modified to include:

A. the level of information security required when business recovery procedures are invoked.
B. information security roles and responsibilities in the crisis management structure.
C. information security resource requirements.
D. change management procedures for information security that could affect business continuity arrangements.

A is the correct answer.

Justification:

A. Business should consider whether information security levels required during recovery should be the same, lower or higher than when business is operating normally. In particular, any special rules for access to confidential data during a crisis need to be identified.
B. During a time of crisis, the security needs of the organization may increase because many usual controls such as separation of duties are missing. Having security roles in the crisis management plan is important, but that is not the best answer to this scenario.
C. Identifying the resource requirements for information security, as part of the business continuity plan (BCP), is important, but it is more important to set out the security levels that would be required for protected information.
D. Change management procedures can help keep a BCP up to date but are not relevant to this scenario.

A4-146 During a disaster recovery test, an IS auditor observes that the performance of the disaster recovery site's server is slow. To find the root cause of this, the IS auditor should **FIRST** review the:

A. event error log generated at the disaster recovery site.
B. disaster recovery test plan.
C. disaster recovery plan.
D. configurations and alignment of the primary and disaster recovery sites.

D is the correct answer.

Justification:

A. If the issue cannot be clarified, the IS auditor should then review the event error log.
B. The disaster recovery test plan would not identify any issues related to system performance unless the test was poorly designed and inefficient, but that would come after checking the configuration.
C. Reviewing the disaster recovery plan would be unlikely to provide any information about system performance issues.
D. Because the configuration of the system is the most probable cause, the IS auditor should review that first.

A4-147 Which of the following is the **GREATEST** risk when storage growth in a critical file server is not managed properly?

A. Backup time would steadily increase.
B. Backup operational costs would significantly increase.
C. Storage operational costs would significantly increase.
D. Server recovery work may not meet the recovery time objective.

D is the correct answer.

Justification:

A. Backup time may increase, but that can be managed. The most important issue is the time taken to recover the data.
B. The backup cost issues are not as significant as not meeting the recovery time objective (RTO).
C. The storage cost issues are not as significant as not meeting the RTO.
D. In case of a crash, recovering a server with an extensive amount of data could require a significant amount of time. If the recovery cannot meet the RTO, there will be a discrepancy in IT strategies. It is important to ensure that server restoration can meet the RTO.

A4-148 An organization has a business process with a recovery time objective equal to zero and a recovery point objective close to one minute. This implies that the process can tolerate:

A. a data loss of up to one minute, but the processing must be continuous.
B. a one-minute processing interruption but cannot tolerate any data loss.
C. a processing interruption of one minute or more.
D. both a data loss and a processing interruption longer than one minute.

A is the correct answer.

Justification:

A. Recovery time objective (RTO) measures an organization's tolerance for downtime and recovery point objective (RPO) measures how much data loss can be accepted.
B. A processing interruption of one minute would exceed the zero RTO set by the organization.
C. A processing interruption of one minute or more would exceed the continuous availability requirements of an RTO of zero.
D. An RPO of one minute would only allow data loss of one minute.

A4-149 Which of the following issues should be the **GREATEST** concern to the IS auditor when reviewing an IT disaster recovery test?

A. Due to the limited test time window, only the most essential systems were tested. The other systems were tested separately during the rest of the year.
B. During the test, some of the backup systems were defective or not working, causing the test of these systems to fail.
C. The procedures to shut down and secure the original production site before starting the backup site required far more time than planned.
D. Every year, the same employees perform the test. The recovery plan documents are not used because every step is well known by all participants.

B is the correct answer.

Justification:
A. This is not a concern because over the course of the year, all the systems were tested.
B. The purpose of the test is to test the backup plan. When the backup systems are not working then the plan cannot be counted on in a real disaster. This is the most serious problem.
C. In a real disaster, there is no need for a clean shutdown of the original production environment because the first priority is to bring the backup site up.
D. A disaster recovery test should test the plan, processes, people and IT systems. Therefore, if the plan is not used, its accuracy and adequacy cannot be verified. Disaster recovery should not rely on key staff because a disaster can occur when they are not available. However, the fact that the test works is less serious than the failure of the systems and infrastructure that the recovery plan counts on. Good practice would rotate different people through the test and ensure that the plan itself is followed and tested.

A4-150 The frequent updating of which of the following is key to the continued effectiveness of a disaster recovery plan?

A. Contact information of key personnel
B. Server inventory documentation
C. Individual roles and responsibilities
D. Procedures for declaring a disaster

A is the correct answer.

Justification:
A. In the event of a disaster, it is important to have a current updated list of personnel who are key to the operation of the plan.
B. Asset inventory is important and should be linked to the change management process of the organization but having access to key people may compensate for outdated records.
C. Individual roles and responsibilities are important, but in a disaster many people could fill different roles depending on their experience.
D. The procedures for declaring a disaster are important because this can affect response, customer perception and regulatory issues, but not as important as having the right people there when needed.

A4-151 A live test of a mutual agreement for IT system recovery has been carried out, including a four-hour test of intensive usage by the business units. The test has been successful, but gives only partial assurance that the:

A. system and the IT operations team can sustain operations in the emergency environment.
B. resources and the environment could sustain the transaction load.
C. connectivity to the applications at the remote site meets response time requirements.
D. workflow of actual business operations can use the emergency system in case of a disaster.

A is the correct answer.

Justification:

A. The applications have been operated intensively, but the capability of the system and the IT operations team to sustain and support this environment (ancillary operations, batch closing, error corrections, output distribution, etc.) is only partially tested.
B. Because the test involved intensive usage, the backup would seem to be able to handle the transaction load.
C. Because users were able to connect to and use the system, the response time must have been satisfactory.
D. The intensive tests by the business indicated that the workflow systems worked correctly. Changes to the environment could pose a problem in the future, but it is working correctly now.

A4-152 Which of the following is the **MOST** important consideration when defining recovery point objectives?

A. Minimum operating requirements
B. Acceptable data loss
C. Mean time between failures
D. Acceptable time for recovery

B is the correct answer.

Justification:

A. Minimum operating requirements help define recovery strategies.
B. Recovery point objectives are the level of data loss/reworking an organization is willing to accept.
C. Mean time between failures helps define likelihood of system failure.
D. Recovery time objectives are the acceptable time delay in availability of business operations.

A4-153 To address an organization's disaster recovery requirements, backup intervals should not exceed the:

A. service level objective.
B. recovery time objective.
C. recovery point objective.
D. maximum acceptable outage.

C is the correct answer.

Justification:

A. Organizations will try to set service level objective to meet established business targets. The resulting time for the service level agreement relates to recovery of services, not to recovery of data.
B. Recovery time objective (RTO) defines the time period after the disaster in which normal business functionality needs to be restored.
C. Recovery point objective defines the point in time to which data must be restored after a disaster to resume processing transactions. Backups should be performed in a way that the latest backup is no older than this maximum time frame. If the backups are not done frequently enough, then too many data are likely to be lost.
D. Maximum acceptable outage (MAO) is the maximum amount of system downtime that is tolerable. It can be used as a synonym for maximum tolerable period of disruption or maximum allowable downtime. However, the RTO denotes an objective/target, while the MAO constitutes a vital necessity for an organization's survival.

A4-154 The **FIRST** step in the execution of a problem management mechanism should be:

A. issue analysis.
B. exception ranking.
C. exception reporting.
D. root cause analysis.

C is the correct answer.

Justification:

A. Analysis and resolution are performed after logging and triage have been performed.
B. Exception ranking can only be performed once the exceptions have been reported.
C. The reporting of operational issues is normally the first step in tracking problems.
D. Root cause analysis is performed once the exceptions have been identified and is not normally the first part of problem management.

A4-155 Which of the following would **BEST** support 24/7 availability?

A. Daily backup
B. Offsite storage
C. Mirroring
D. Periodic testing

C is the correct answer.

Justification:

A. Daily backup implies that it is reasonable for restoration to take place within a number of hours but not immediately.
B. Offsite storage does not, itself, support continuous availability.
C. Mirroring of critical elements is a tool that facilitates immediate (failover) recoverability.
D. Periodic testing of systems does not, itself, support continuous availability.

A4-156 The **PRIMARY** purpose of implementing Redundant Array of Inexpensive Disks (RAID) level 1 in a file server is to:

A. achieve performance improvement.
B. provide user authentication.
C. ensure availability of data.
D. ensure the confidentiality of data.

C is the correct answer.

Justification:
A. Redundant Array of Inexpensive Disks (RAID) level 1 does not improve performance. It writes the data to two separate disk drives.
B. RAID level 1 has no relevance to authentication.
C. RAID level 1 provides disk mirroring. Data written to one disk are also written to another disk. Users in the network access data in the first disk; if disk one fails, the second disk takes over. This redundancy ensures the availability of data.
D. RAID level 1 does nothing to provide for data confidentiality.

A4-157 Which of the following is the **MOST** important criterion when selecting a location for an offsite storage facility for IS backup files? The offsite facility must be:

A. physically separated from the data center and not subject to the same risk.
B. given the same level of protection as that of the computer data center.
C. outsourced to a reliable third party.
D. equipped with surveillance capabilities.

A is the correct answer.

Justification:
A. It is important that there is an offsite storage location for IS files and that it is in a location not subject to the same risk as the primary data center.
B. The offsite location may be shared with other companies and, therefore, have an even higher level of protection than the primary data center.
C. An offsite location may be owned by a third party or by the organization itself.
D. Physical protection is important but not as important as not being affected by the same crisis.

A4-158 If a database is restored using before-image dumps, where should the process begin following an interruption?

A. Before the last transaction
B. After the last transaction
C. As the first transaction after the latest checkpoint
D. As the last transaction before the latest checkpoint

A is the correct answer.

Justification:
A. If before images are used, the last transaction in the dump will not have updated the database prior to the dump being taken.
B. The last transaction will not have updated the database and must be reprocessed.
C. Program checkpoints are irrelevant in this situation. Checkpoints are used in application failures.
D. Program checkpoints are irrelevant in this situation. Checkpoints are used in application failures.

A4-159 In addition to the backup considerations for all systems, which of the following is an important consideration in providing backup for online systems?

A. Maintaining system software parameters
B. Ensuring periodic dumps of transaction logs
C. Ensuring grandfather-father-son file backups
D. Maintaining important data at an offsite location

B is the correct answer.

Justification:

A. Maintaining system software parameters is important for all systems, not just online systems.
B. Ensuring periodic dumps of transaction logs is the only safe way of preserving timely historic data. Because online systems do not have a paper trail that can be used to recreate data, maintaining transaction logs is critically important to prevent data loss. The volume of activity usually associated with an online system may make other more traditional methods of backup impractical.
C. Having generations of backups is a good practice for all systems.
D. All backups should consider offsite storage at a location that is accessible but not likely to be affected by the same disaster.

A4-160 Which of the following disaster recovery testing techniques is the **MOST** efficient way to determine the effectiveness of the plan?

A. Preparedness tests
B. Paper tests
C. Full operational tests
D. Actual service disruption

A is the correct answer.

Justification:

A. Preparedness tests involve simulation of the entire environment (in phases) at relatively low cost and help the team to better understand and prepare for the actual test scenario.
B. Paper tests in a walk-through test the entire plan, but there is no simulation and less is learned. It also is difficult to obtain evidence that the team has understood the test plan.
C. Full operational tests would require approval from management, are not easy or practical to test in most scenarios and may trigger a real disaster.
D. An actual service disruption is not recommended in most cases unless required by regulation or policy.

A4-161 Online banking transactions are being posted to the database when processing suddenly comes to a halt. The integrity of the transaction processing is **BEST** ensured by:

A. database integrity checks.
B. validation checks.
C. input controls.
D. database commits and rollbacks.

D is the correct answer.

Justification:

A. Database integrity checks are important to ensure database consistency and accuracy. These include isolation, concurrency and durability controls, but the most important issue here is atomicity—the requirement for transactions to complete entirely and commit or else roll back to the last known good point.
B. Validation checks will prevent introduction of corrupt data but will not address system failure.
C. Input controls are important to protect the integrity of input data but will not address system failure.
D. Database commits ensure that the data are saved after the transaction processing is completed. Rollback ensures that the processing that has been partially completed as part of the transaction is reversed back and not saved if the entire transaction does not complete successfully.

A4-162 Which of the following security measures **BEST** ensures the integrity of information stored in a data warehouse?

A. Validated daily backups
B. Change management procedures
C. Data dictionary maintenance
D. A read-only restriction

D is the correct answer.

Justification:

A. Backups address availability, not integrity. Validated backups ensure that the backup will work when needed.
B. Adequate change management procedures protect the data warehouse and the systems with which the data warehouse interfaces from unauthorized changes but are not usually concerned with the data.
C. Data dictionary maintenance procedures provide for the definition and structure of data that are input to the data warehouse. This will not affect the integrity of data already stored.
D. Because most data in a data warehouse are historic and do not need to be changed, applying read-only restrictions prevents data manipulation.

A4-163 Which of the following ensures the availability of transactions in the event of a disaster?

A. Send hourly tapes containing transactions offsite.
B. Send daily tapes containing transactions offsite.
C. Capture transactions to multiple storage devices.
D. Transmit transactions offsite in real time.

D is the correct answer.

Justification:

A. Sending hourly tapes containing transactions offsite is not in real time and, therefore, would possibly result in the loss of one hour's worth of transactional data.
B. Sending daily tapes containing transactions offsite is not in real time and, therefore, could result in the loss of one day's worth of transactional data.
C. Capturing transactions to multiple storage devices does not ensure availability at an offsite location.
D. The only way to ensure availability of all transactions is to perform a real-time transmission to an offsite facility.

A4-164 IT management has decided to install a level 1 Redundant Array of Inexpensive Disks (RAID) system in all servers to compensate for the elimination of offsite backups. The IS auditor should recommend:

A. upgrading to a level 5 RAID.
B. increasing the frequency of onsite backups.
C. reinstating the offsite backups.
D. establishing a cold site in a secure location.

C is the correct answer.

Justification:

A. Upgrading to level 5 Redundant Array of Inexpensive Disks (RAID) will not address the problem of catastrophic failure of the data center housing all the data.
B. Increasing the frequency of onsite backups is not relevant to RAID 1 because all data are being mirrored already.
C. A RAID system, at any level, will not protect against a natural disaster. The problem will not be alleviated without offsite backups.
D. A cold site is an offsite recovery location but will not provide for data recovery because a cold site is not used to store data.

A4-165 In a contract with a hot, warm or cold site, contractual provisions should **PRIMARILY** cover which of the following considerations?

A. Physical security measures
B. Total number of subscribers
C. Number of subscribers permitted to use a site at one time
D. References by other users

C is the correct answer.

Justification:

A. Physical security measures are not always part of the contract, although they are an important consideration when choosing a third-party site.
B. The total number of subscribers is a consideration, but more important is whether the agreement limits the number of subscribers in a building or in a specific area. It is also good to know if other subscribers are competitors.
C. The contract should specify the number of subscribers permitted to use the site at any one time. The contract can be written to give preference to certain subscribers.
D. The references that other users can provide are a consideration taken before signing the contract; it is by no means part of the contractual provisions.

A4-166 Which of the following reports is the **MOST** appropriate source of information for an IS auditor to validate that an Internet service provider (ISP) has been complying with an enterprise service level agreement for the availability of outsourced telecommunication services?

A. Downtime reports on the telecommunication services generated by the ISP
B. A utilization report of automatic failover services generated by the enterprise
C. A bandwidth utilization report provided by the ISP
D. Downtime reports on the telecommunication services generated by the enterprise

D is the correct answer.

Justification:
A. The Internet service provider (ISP)-generated downtime reports are produced by the same entity that is being monitored. As a result, it will be necessary to review these reports for possible bias and/or errors against other data.
B. The information provided by these reports is indirect evidence of the extent that the backup telecommunication services were used. These reports may not indicate compliance with the service level agreement, just that the failover systems had been used.
C. Utilization reports are used to measure the usage of bandwidth, not uptime.
D. The enterprise should use internally generated downtime reports to monitor the service provided by the ISP and, as available, to compare with the reports provided by the ISP.

A4-167 Integrating the business continuity plan into IT project management aids in:

A. the testing of the business continuity requirements.
B. the development of a more comprehensive set of requirements.
C. the development of a transaction flowchart.
D. ensuring the application meets the user's needs.

B is the correct answer.

Justification:
A. Testing the business continuity plan's (BCP) requirements is not related to IT project management.
B. Integrating the BCP into the development process ensures complete coverage of the requirements through each phase of the project.
C. A transaction flowchart aids in analyzing an application's controls but does not affect business continuity.
D. A BCP will not directly address the detailed processing needs of the users.

A4-168 An enterprise uses privileged accounts to process configuration changes for mission-critical applications. Which of the following would be the **BEST** and appropriate control to limit the risk in such a situation?

A. Ensure that audit trails are accurate and specific.
B. Ensure that personnel have adequate training.
C. Ensure that personnel background checks are performed for critical personnel.
D. Ensure that supervisory approval and review are performed for critical changes.

D is the correct answer.

Justification:

A. Audit trails are a detective control and, in many cases, can be altered by those with privileged access.
B. Staff proficiency is important and good training may be somewhat of a deterrent, but supervisory approval and review is the best choice.
C. Performing background checks is a very basic control and will not effectively prevent or detect errors or malfeasance.
D. Supervisory approval and review of critical changes by the accountable managers in the enterprise are required to avoid and detect any unauthorized change. In addition to authorization, supervision enforces a separation of duties and prevents an unauthorized attempt by any single employee.

A4-169 An IS auditor observed that multiple applications are hosted on the same server. The recovery time objective (RTO) for the server will be:

A. based on the application with the longest RTO.
B. based on the application with the shortest RTO.
C. based on the mean of each application's RTO.
D. independent of the RTO and based on the criticality of the application.

B is the correct answer.

Justification:

A. The longest recovery time objective (RTO) will be determined for noncritical applications, which will not help in meeting the objectives for critical systems.
B. When several applications are hosted on a server, the server's RTO must be determined by taking the RTO of the most critical application, which is the shortest RTO.
C. The mean value will be higher than the RTO for a critical application.
D. Critical applications usually have the shortest RTOs. The RTO of the server cannot be independent of the application RTO.

A4-170 During an application audit, the IS auditor finds several problems related to corrupt data in the database. Which of the following is a corrective control that the IS auditor should recommend?

A. Define the standards, and closely monitor them for compliance.
B. Ensure that only authorized personnel can update the database.
C. Establish controls to handle concurrent access problems.
D. Proceed with restore procedures.

D is the correct answer.

Justification:
A. Establishing standards is a preventive control, and monitoring for compliance is a detective control.
B. Ensuring that only authorized personnel can update the database is a preventive control.
C. Establishing controls to handle concurrent access problems is a preventive control.
D. Proceeding with restore procedures is a corrective control. Restore procedures can be used to recover databases to their last-known archived version.

A4-171 Which of the following scenarios provides the **BEST** disaster recovery plan to implement for critical applications?

A. Daily data backups that are stored offsite and a hot site located 140 kilometers from the main data center
B. Daily data backups that are stored onsite in a fireproof safe
C. Real-time data replication between the main data center and the hot site located 500 meters from the main site
D. Daily data backups that are stored offsite with a warm site located 70 kilometers from the main data center

A is the correct answer.

Justification:
A. Of the given choices, this is the most suitable answer. The disaster recovery plan includes a hot site that is located sufficiently away from the main data center and will allow recovery in the event of a major disaster. Not having real-time backups may be a problem depending on recovery point objective (RPO).
B. Having data backups is necessary, but not having a replication site would be insufficient for the critical application.
C. Depending on the type of disaster, a hot site should normally be located more than 500 meters from the main facility. Having real-time backups may be the best option though, depending on the data RPO.
D. A warm site may take days to recover, and therefore, it may not be a suitable solution.

A4-172 Which of the following is the **BEST** indicator of the effectiveness of backup and restore procedures while restoring data after a disaster?

A. Members of the recovery team were available.
B. Recovery time objectives were met.
C. Inventory of backup tapes was properly maintained.
D. Backup tapes were completely restored at an alternate site.

B is the correct answer.

Justification:
A. The availability of key personnel does not ensure that backup and restore procedures will work effectively.
B. The effectiveness of backup and restore procedures is best ensured by recovery time objectives (RTOs) being met because these are the requirements that are critically defined during the business impact analysis stage, with the inputs and involvement of all business process owners.
C. The inventory of the backup tapes is only one element of the successful recovery.
D. The restoration of backup tapes is a critical success, but only if they were able to be restored within the time frames set by the RTO.

A4-173 Which of the following would be the **MOST** appropriate recovery strategy for a sensitive system with a high recovery time objective (RTO)?

A. Warm site
B. Hot site
C. Cold site
D. Mobile recovery site

C is the correct answer.

Justification:

A. While a warm site may be a good solution, it would not be the most appropriate because it is more expensive than a cold site.
B. A hot site is used for those systems classified as critical that have a low recovery time objective (RTO).
C. Sensitive systems having a high RTO can be performed manually at a tolerable cost for an extended period of time. The cold site would be the most cost-effective solution for such a system.
D. A mobile recovery site would not be as cost-effective as a cold site and would not be appropriate for systems with high RTOs.

A4-174 Which of the following should an incident response team address **FIRST** after a major incident in an information processing facility?

A. Restoration at the facility
B. Documentation of the facility
C. Containment at the facility
D. Monitoring of the facility

C is the correct answer.

Justification:

A. Restoration ensures that the affected systems or services are restored to a condition specified in the restore point objective. This action will be possible only after containment of the damage.
B. Documentation of the facility should be prepared to inform management of the incident; however, damage must be contained first.
C. The first priority (after addressing life safety) is the containment of the incident at the facility so that spread of the damage is minimized. The incident team must gain control of the situation.
D. Monitoring of the facility is important, although containment must take priority to avoid spread of the damage.

A4-175 An IS auditor discovers that some hard drives disposed of by an enterprise were not sanitized in a manner that would reasonably ensure the data could not be recovered. In addition, the enterprise does not have a written policy on data disposal. The IS auditor should **FIRST**:

A. draft an audit finding and discuss it with the auditor in charge.
B. determine the sensitivity of the information on the hard drives.
C. discuss with the IT manager good practices in data disposal.
D. develop an appropriate data disposal policy for the enterprise.

B is the correct answer.

Justification:

A. Drafting a finding without a quantified risk would be premature.
B. Even though a policy is not available, the IS auditor should determine the nature of the information on the hard drives to quantify, as much as possible, the risk.
C. It would be premature to discuss good practices with the IT manager until the extent of the incident has been quantified.
D. An IS auditor should not develop policies.

A4-176 An IS auditor is assessing services provided by an Internet service provider (ISP) during an IS compliance audit of a nationwide corporation that operates a governmental program. Which of the following is **MOST** important?

A. Review the request for proposal.
B. Review monthly performance reports generated by the ISP.
C. Review the service level agreement.
D. Research other clients of the ISP.

C is the correct answer.

Justification:
A. Because the request for proposal is not the contracted agreement, it is more relevant to review the terms of the SLA.
B. The reports from the Internet service provider (ISP) are indirect evidence that may require further review to ensure accuracy and completeness.
C. A service level agreement provides the basis for an adequate assessment of the degree to which the provider is meeting the level of agreed-on service.
D. The services provided to other clients of the ISP are irrelevant to the IS auditor.

A4-177 During an audit of a small enterprise, the IS auditor noted that the IS director has superuser-privilege access that allows the director to process requests for changes to the application access roles (access types). Which of the following should the IS auditor recommend?

A. Implement a properly documented process for application role change requests.
B. Hire additional staff to provide a segregation of duties for application role changes.
C. Implement an automated process for changing application roles.
D. Document the current procedure in detail and make it available on the enterprise intranet.

A is the correct answer.

Justification:
A. The IS auditor should recommend implementation of processes that could prevent or detect improper changes from being made to the major application roles. The application role change request process should start and be approved by the business owner; then, the IS director can make the changes to the application.
B. While it is preferred that a strict segregation of duties be adhered to and that additional staff be recruited, this practice is not always possible in small enterprises. The IS auditor must look at recommended alternative processes.
C. An automated process for managing application roles may not be practical to prevent improper changes being made by the IS director, who also has the most privileged access to the application.
D. Making the existing process available on the enterprise intranet would not provide any value to protect the system.

A4-178 While observing a full simulation of the business continuity plan, an IS auditor notices that the notification systems within the organizational facilities could be severely impacted by infrastructure damage. The **BEST** recommendation the IS auditor can provide to the organization is to ensure:

A. the salvage team is trained to use the notification system.
B. the notification system provides for the recovery of the backup.
C. redundancies are built into the notification system.
D. the notification systems are stored in a vault.

C is the correct answer.

Justification:
A. The salvage team would not be able to use a severely damaged notification system, even if they are trained to use it.
B. The recovery of the backups has no bearing on the notification system.
C. If the notification system has been severely impacted by the damage, redundancy would be the best control.
D. Storing the notification system in a vault would be of little value if the building is damaged.

A4-179 To ensure structured disaster recovery, it is **MOST** important that the business continuity plan and disaster recovery plan are:

A. stored at an alternate location.
B. communicated to all users.
C. tested regularly.
D. updated regularly.

C is the correct answer.

Justification:
A. Storing the business continuity plan (BCP) at an alternate location is useful in the case of complete site outage; however, the BCP is not useful during a disaster without adequate tests.
B. Communicating to users is not of much use without actual tests.
C. If the BCP is tested regularly, the BCP and disaster recovery plan team is adequately aware of the process and that helps in structured disaster recovery.
D. Even if the plan is updated regularly, it is of less use during an actual disaster if it is not adequately tested.

A4-180 The **PRIMARY** purpose of a business impact analysis is to:

A. define recovery strategies.
B. identify the alternate site.
C. improve recovery testing.
D. calculate the annual loss expectancy.

A is the correct answer.

Justification:
A. One of the primary outcomes of a business impact analysis (BIA) is the recovery time objective and the recovery point objective, which help in defining the recovery strategies.
B. A BIA, itself, will not help in identifying the alternate site. That is determined during the recovery strategy phase of the project.
C. A BIA, itself, will not help improve recovery testing. That is done during the implementation and testing phase of the project.
D. The annual loss expectancy of critical business assets and processes is determined during risk assessment and will be reviewed in the BIA, but this is not the primary advantage.

A4-181 Which of the following **BEST** helps define disaster recovery strategies?

A. Annual loss expectancy and exposure factor
B. Maximum tolerable downtime and data loss
C. Existing server and network redundancies
D. Data backup and offsite storage requirements

B is the correct answer.

Justification:
A. Annual loss expectancy and exposure factor are more related to risk in general.
B. One of the key outcomes of the business impact analysis is the recovery time objective (RTO) and recovery point objective (RPO)—maximum tolerable downtime and data loss—that further help in identifying the recovery strategies.
C. Existing server and network redundancies are good to know, but the RTO and RPO are needed to design the right recovery strategies.
D. Data backup and offsite storage requirements are an important aspect of a business continuity plan, but these alone will not help in defining the disaster recovery strategies.

A4-182 After a disaster declaration, the media creation date at a warm recovery site is based on the:

A. recovery point objective.
B. recovery time objective.
C. service delivery objective.
D. maximum tolerable outage.

A is the correct answer.

Justification:
A. The recovery point objective (RPO) is determined based on the acceptable data loss in case of a disruption of operations. It indicates the earliest point in time that is acceptable to recover the data. The RPO effectively quantifies the permissible amount of data loss in case of interruption. The media creation date will reflect the point to which data are to be restored or the RPO.
B. The recovery time objective is the amount of time allowed for the recovery of a business function or resource after a disaster occurs.
C. The service delivery objective is directly related to the business needs and is the level of service to be reached during the alternate process mode until the normal situation is restored.
D. The maximum tolerable outage is the maximum time that an organization can support processing in alternate mode.

A4-183 The activation of an enterprise's business continuity plan should be based on predetermined criteria that address the:

A. duration of the outage.
B. type of outage.
C. probability of the outage.
D. cause of the outage.

A is the correct answer.

Justification:

A. **The initiation of a business continuity plan (action) should primarily be based on the maximum period for which a business function can be disrupted before the disruption threatens the achievement of organizational objectives.**
B. The type of outage is not as important to the activation of the plan as the length or duration of the outage.
C. The probability of the outage would be relevant to the frequency of incidents, not the need to activate the plan. The plan is designed to be activated after an event of a certain duration occurs.
D. The cause of the outage may affect the response plan to be activated, but not the decision to activate the plan. The plan will be activated any time an event of a predetermined duration occurs.

A4-184 During an audit of a small company that provides medical transcription services, an IS auditor observes several issues related to the backup and restore process. Which of the following should be the auditor's **GREATEST** concern?

A. Restoration testing for backup media is not performed; however, all data restore requests have been successful.
B. The policy for data backup and retention has not been reviewed by the business owner for the past three years.
C. The company stores transcription backup tapes offsite using a third-party service provider, which inventories backup tapes annually.
D. Failed backup alerts for the marketing department data files are not followed up on or resolved by the IT administrator.

C is the correct answer.

Justification:

A. Lack of restoration testing does not increase the risk of unauthorized leakage of information. Not performing restoration tests on backup tapes poses a risk; however, this risk is somewhat mitigated because past data restore requests have been successful.
B. Lack of review of the data backup and retention policy may be of a concern if systems and business processes have changed in the past three years. The IS auditor should perform additional procedures to verify the validity of existing procedures. In addition, lack of this control does not introduce a risk of unauthorized leakage of information.
C. **For a company working with confidential patient data, the loss of a backup tape is a significant incident. Privacy laws specify severe penalties for such an event, and the company's reputation could be damaged due to mandated reporting requirements. To gain assurance that tapes are being handled properly, the organization should perform audit tests that include frequent physical inventories and an evaluation of the controls in place at the third-party provider.**
D. Failed backup alerts that are not followed up on and resolved imply that certain data or files are not backed up. This is a concern if the files/data being backed up are critical in nature, but, typically, marketing data files are not regulated in the same way as medical transcription files. Lack of this control does not introduce a risk of unauthorized leakage of sensitive information.

A4-185 Determining the service delivery objective should be based **PRIMARILY** on:

A. the minimum acceptable operational capability.
B. the cost-effectiveness of the restoration process.
C. meeting the recovery time objectives.
D. the allowable interruption window.

A is the correct answer.

Justification:

A. The service delivery objective (SDO) is the level of service to be reached during the alternate process mode until the normal situation is restored. This is directly related to the business needs.
B. The cost-effectiveness of the restoration process is not the main consideration of determining the SDO.
C. Meeting the recovery time objective may be one of the considerations in determining the SDO, but it is a secondary factor.
D. The allowable interruption window may be one of the factors secondary to determining the SDO.

A4-186 An IS auditor reviewing the application change management process for a large multinational company should be **MOST** concerned when:

A. test systems run different configurations than do production systems.
B. change management records are paper based.
C. the configuration management database is not maintained.
D. the test environment is installed on the production server.

C is the correct answer.

Justification:

A. While, ideally, production and test systems should be configured identically, there may be reasons why this does not occur. The more significant concern is whether the configuration management database was not maintained.
B. Paper-based change management records are inefficient to maintain and not easy to review in large volumes; however, they do not present a concern from a control point of view as long as they are properly and diligently maintained.
C. The configuration management database (CMDB) is used to track configuration items (CIs) and the dependencies between them. An out-of-date CMDB in a large multinational company could result in incorrect approvals being obtained or leave out critical dependencies during the test phase.
D. While it is not ideal to have the test environment installed on the production server, it is not a control-related concern. As long as the test and production environments are kept separate, they can be installed on the same physical server(s).

A4-187 An IS auditor can verify that an organization's business continuity plan (BCP) is effective by reviewing the:

A. alignment of the BCP with industry good practices.
B. results of business continuity tests performed by IS and end-user personnel.
C. offsite facility, its contents, security and environmental controls.
D. annual financial cost of the BCP activities versus the expected benefit of the implementation of the plan.

B is the correct answer.

Justification:
A. Alignment of the business continuity plan (BCP) with industry good practices does not provide the assurance of the effectiveness of the BCP.
B. The effectiveness of the BCP can best be evaluated by reviewing the results from previous business continuity tests for thoroughness and accuracy in accomplishing their stated objectives.
C. The offsite facility, its contents, security and environmental controls do not provide the assurance of the effectiveness of the BCP. Only testing will provide an accurate assessment of the effectiveness of the BCP.
D. The annual financial cost of the BCP activities versus the expected benefit of implementation of the plan does not provide the assurance of the effectiveness of the BCP. Only testing will provide an accurate assessment of the effectiveness of the BCP.

A4-188 It is **MOST** appropriate to implement an incremental backup scheme when:

A. there is limited recovery time for critical data.
B. online disk-based media are preferred.
C. there is limited media capacity.
D. a random selection of backup sets is required.

C is the correct answer.

Justification:
A. A full backup or differential backup is preferred in this situation.
B. Incremental backup could be used irrespective of the media adopted.
C. In an incremental backup, after the full backup, only the files that have changed are backed up, thus minimizing media storage.
D. A random selection of backup sets may not be possible with an incremental backup scheme because only fragments of the data are backed up on a daily basis.

A4-189 Which of the following **BEST** mitigates the risk arising from using reciprocal agreements as a recovery alternative?

A. Perform disaster recovery exercises annually.
B. Ensure that partnering organizations are separated geographically.
C. Regularly perform a business impact analysis.
D. Select a partnering organization with similar systems.

B is the correct answer.

Justification:
A. While disaster recovery exercises are important but difficult to perform in a reciprocal agreement, the greater risk is geographic proximity.
B. If the two partnering organizations are in close geographic proximity, this could lead to both organizations being subjected to the same environmental disaster, such as an earthquake.
C. A business impact analysis will help both organizations identify critical applications, but separation is a more important consideration when entering reciprocal agreements.
D. Selecting a partnering organization with similar systems is a good idea, but separation is a more important consideration when entering reciprocal agreements.

A4-190 During the review of an in-house developed application, the **GREATEST** concern to an IS auditor is if a:

A. user raises a change request and tests it in the test environment.
B. programmer codes a change in the development environment and tests it in the test environment.
C. manager approves a change request and then reviews it in production.
D. manager initiates a change request and subsequently approves it.

D is the correct answer.

Justification:
A. Having a user involved in testing changes is common practice.
B. Having a programmer code a change in development and then separately test the change in a test environment is a good practice and preferable over testing in production.
C. Having a manager review a change to make sure it was done correctly is an acceptable practice.
D. Initiating and subsequently approving a change request violates the principle of segregation of duties. A person should not be able to approve their own requests.

A4-191 In a disaster recovery situation, which of the following is the **MOST** important metric to ensure that data are synchronized between critical systems?

A. Recovery point objective
B. Recovery time objective
C. Recovery service resilience
D. Recovery service scalability

A is the correct answer.

Justification:
A. Establishing a common recovery point objective is most critical for ensuring that interdependencies between systems are properly synchronized. It ensures that systems do not contain data from different points in time that may result in accounting transactions that cannot be reconciled and a loss of referential integrity.
B. Recovery time objectives are not as important to synchronize because they normally vary depending on the level of effort and resources required to restore a system.
C. Recovery service resilience measures the fault tolerance due to data exceptions and ability to restart and recover from internal failures.
D. Recovery service scalability refers to the capacity constraints and limitations that a recovery solution may have relative to the original system configuration.

A4-192 Which of the following **BEST** mitigates the risk of backup media containing irreplaceable information being lost or stolen while in transit?

A. Ensure that media are encrypted.
B. Maintain a duplicate copy.
C. Maintain chain of custody.
D. Ensure that personnel are bonded.

B is the correct answer.

Justification:
A. Although strong encryption protects against disclosure, it will not mitigate the loss of irreplaceable data.
B. Sensitive data should always be fully backed up before being transmitted or moved. Backups of sensitive information should be treated with the same control considerations as the actual data.
C. Chain of custody is an important control, but it will not mitigate a loss if a locked area is broken into and media removed or if media are lost while in an individual's custody.
D. Bonded security, although good for preventing theft, will not protect against accidental loss or destruction.

A4-193 An IS auditor is reviewing the change management process for an enterprise resource planning application. Which of the following is the **BEST** method for testing program changes?

A. Select a sample of change tickets and review them for authorization.
B. Perform a walk-through by tracing a program change from start to finish.
C. Trace a sample of modified programs to supporting change tickets.
D. Use query software to analyze all change tickets for missing fields.

C is the correct answer.

Justification:

A. Selecting a sample of change tickets and reviewing them for authorization helps test for authorization controls; however, it does not identify program changes that were made without supporting change tickets.
B. Performing a walk-through assists the IS auditor in understanding the process but does not ensure that all changes adhere to the normal process.
C. Tracing a sample of modified programs to supporting change tickets is the best way to test change management controls. This method is most likely to identify instances in which a change was made without supporting documentation.
D. Using query software to analyze all change tickets for missing fields does not identify program changes that were made without supporting change tickets.

A4-194 Emergency changes that bypass the normal change control process are **MOST** acceptable if:

A. management reviews and approves the changes after they have occurred.
B. the changes are reviewed by a peer at the time of the change.
C. the changes are documented in the change control system by the operations department.
D. management has preapproved all emergency changes.

A is the correct answer.

Justification:

A. Because management cannot always be available when a system failure occurs, it is acceptable for changes to be reviewed and approved within a reasonable time period after they occur.
B. Although peer review provides some accountability, management should review and approve all changes, even if that review and approval must occur after the fact.
C. Documenting the event does not replace the need for a review and approval process to occur.
D. It is not a good control practice for management to ignore its responsibility by preapproving all emergency changes in advance without reviewing them. Unauthorized changes could then be made without management's knowledge.

A4-195 To optimize an organization's business continuity plan, an IS auditor should recommend a business impact analysis to determine:

A. the business processes that generate the most financial value for the organization and, therefore, must be recovered first.
B. the priorities and order for recovery to ensure alignment with the organization's business strategy.
C. the business processes that must be recovered following a disaster to ensure the organization's survival.
D. the priorities and order of recovery, which will recover the greatest number of systems in the shortest time frame.

C is the correct answer.

Justification:

A. It is a common mistake to overemphasize financial value rather than urgency. For example, while the processing of incoming mortgage loan payments is important from a financial perspective, it could be delayed for a few days in the event of a disaster. On the other hand, wiring funds to close on a loan, while not generating direct revenue, is far more critical because of the possibility of regulatory problems, customer complaints and reputation issues.
B. The business strategy (which is often a long-term view) does not have a direct impact at this point in time.
C. To ensure the organization's survival following a disaster, it is important to recover the most critical business processes first.
D. The mere number of recovered systems does not have a direct impact at this point in time. The importance is to recover systems that would impact business survival.

A4-196 Which of the following is the **MOST** efficient strategy for the backup of large quantities of mission-critical data when the systems need to be online to take sales orders 24 hours a day?

A. Implementing a fault-tolerant disk-to-disk backup solution
B. Making a full backup to tape weekly and an incremental backup nightly
C. Creating a duplicate storage area network (SAN) and replicating the data to a second SAN
D. Creating identical server and storage infrastructure at a hot site

A is the correct answer.

Justification:

A. Disk-to-disk backup, also called disk-to-disk-to-tape backup or tape cache, is when the primary backup is written to disk instead of tape. That backup can then be copied, cloned or migrated to tape at a later time (hence the term "disk-to-disk-to-tape"). This technology allows the backup of data to be performed without impacting system performance and allows a large quantity of data to be backed up in a very short backup window. In case of a failure, the fault-tolerant system can transfer immediately to the other disk set.
B. While a backup strategy involving tape drives is valid, because many computer systems must be taken offline so that backups can be performed, there is the need to create a backup window, typically during each night. This would not enable the system to be available 24/7. For a system that must remain online at all times, the only feasible way to back up the data is to either duplicate the data to a server that gets backed up to tape, or deploy a disk-to-disk solution, which is effectively the same thing.
C. While creating a duplicate storage area network (SAN) and replicating the data to a second SAN provides some redundancy and data protection, this is not really a backup solution. If the two systems are at the same site, there is a risk that an incident such as a fire or flood in the data center could lead to data loss.
D. While creating an identical server and storage infrastructure at a hot site provides a great deal of redundancy and availability to enable the system to stay operational, it does not address the need for long-term data storage. There is still the need to create an efficient method of backing up data.

A4-197 Which of the following would **BEST** ensure uninterrupted operations in an organization with IT operation centers in several countries?

A. Distribution of key procedural documentation
B. Reciprocal agreement between business partners
C. Strong senior management leadership
D. Employee training on the business continuity plan

D is the correct answer.

Justification:

A. Procedural documentation should always be up to date and distributed to major locations. However, documents alone are insufficient if employees do not know their role in the plan.
B. A reciprocal agreement is an emergency processing agreement between two or more enterprises with similar equipment or applications. Typically, participants of a reciprocal agreement promise to provide processing time to each other when an emergency arises. While it is integral to business continuity to have a location for business operations, it does not necessarily need to be a reciprocal agreement. For example, in some cases, business operations may be carried out from each employee's home.
C. Senior management may not be readily available to provide leadership during a disaster. Therefore, it is most important that employees fully understand their roles in the business continuity plan (BCP).
D. During a disaster, the chain of command might be interrupted. Therefore, it is important that employees know their roles in the BCP, including where to report and how to perform their job functions. Employee training on the plan is especially important for businesses with offices that are geographically separated because there is a greater chance of communication disruption.

A4-198 Which of the following **BEST** ensures that users have uninterrupted access to a critical, heavily used web-based application?

A. Disk mirroring
B. Redundant Array of Inexpensive Disks
C. Dynamic domain name system
D. Load balancing

D is the correct answer.

Justification:

A. Disk mirroring provides real-time replication of disk drives but does not ensure uninterrupted system availability in the event a server crashes.
B. Redundant Array of Inexpensive Disks technology improves resiliency but does not protect against failure of a network interface card or central processing unit processor failure.
C. Dynamic domain name system is a method used to assign a host name to an Internet Protocol address that is dynamic. This is a useful technology but does not help ensure availability.
D. Load balancing best ensures uninterrupted system availability by distributing traffic across multiple servers. Load balancing helps ensure consistent response time for web applications. Also, if a web server fails, load balancing ensures that traffic will be directed to a different, functional server.

A4-199 Which of the following is the **BEST** method to ensure that critical IT system failures do not recur?

A. Invest in redundant systems.
B. Conduct a follow-up audit.
C. Monitor system performance.
D. Perform root cause analysis.

D is the correct answer.

Justification:

A. Redundancy may be a solution; however, a root cause analysis enables an educated decision to address the origin of the problem instead of simply assuming that system redundancy is the solution.
B. While an audit may discover the root cause of the problem, an audit is not a solution to an operational problem. Identifying the origins of operational failures needs to be part of day-to-day IT processes and owned by the IT department.
C. Use of monitoring tools is a means to gather data and can contribute to root cause analysis, but it does not by itself help prevent an existing problem from recurring.
D. Root cause analysis determines the key reason an incident has occurred and allows for appropriate corrections that will help prevent the incident from recurring.

A4-200 Which of the following is **MOST** important when an operating system patch is to be applied to a production environment?

A. Successful regression testing by the developer
B. Approval from the information asset owner
C. Approval from the security officer
D. Patch installation at alternate sites

B is the correct answer.

Justification:

A. While testing is important for any patch, in this case it should be assumed that the operating system (OS) vendor tested the patch before releasing it. Before this OS patch is put into production, the organization should do system testing to ensure that no issues will occur.
B. It is most important that information owners approve any changes to production systems to ensure that no serious business disruption takes place as the result of the patch release.
C. The security officer does not normally need to approve every OS patch.
D. Security patches need to be deployed consistently across the organization, including alternate sites. However, approval from the information asset owner is still the most important consideration.

A4-201 The IS auditor observes that the latest security-related software patches for a mission-critical system were released two months ago, but IT personnel have not yet installed the patches. The IS auditor should:

A. review the patch management policy and determine the risk associated with this condition.
B. recommend that IT systems personnel test and then install the patches immediately.
C. recommend that patches be applied every month or immediately upon release.
D. take no action, because the IT processes related to patch management appear to be adequate.

A is the correct answer.

Justification:
A. Reviewing the patch management policy and determining whether the IT department is compliant with the policies will detect whether the policies are appropriate and what risk is associated with current practices.
B. While there may be instances in which the patch is an urgent fix for a serious security issue, IT may have made the determination that the risk to system stability is greater than the risk identified by the software vendor who issued the patch. Therefore, the time frame selected by IT may be appropriate.
C. While keeping critical systems properly patched helps to ensure that they are secure, the requirement for a precise timetable to patch systems may create other issues if patches are improperly tested prior to implementation. Therefore, this is not the correct answer.
D. Even if the IS auditor concludes that the patch management process is adequate, the observation related to the time delay in applying patches should be reported.

A4-202 Which of the following **BEST** helps prioritize the recovery of IT assets when planning for a disaster?

A. Incident response plan
B. Business impact analysis
C. Threat and risk analysis
D. Recovery time objective

B is the correct answer.

Justification:
A. An incident response plan is an organized approach to addressing and managing a security breach or attack. The plan defines what constitutes an incident and the process to follow when an incident occurs. It does not prioritize recovery during a disaster.
B. Incorporating the business impact analysis (BIA) into the IT disaster recovery planning process is critical to ensure that IT assets are prioritized to align with the business.
C. Identifying threats and analyzing risk to the business is an important part of disaster planning, but it does not determine the priority of recovery.
D. The recovery time objective is the amount of time allowed for the recovery of a business function or resource after a disaster occurs. This is included as part of the BIA and used to represent the prioritization of recovery.

A4-203 Which of the following is the **MOST** likely reason an organization implements an emergency change to an application using the emergency change control process?

A. The application owner requested new functionality.
B. Changes are developed using an agile methodology.
C. There is a high probability of a significant impact on operations.
D. The operating system vendor has released a security patch.

C is the correct answer.

Justification:

A. Requests for new functionality by the application owner generally follow normal change control procedures, unless they have an impact on the business function.
B. The agile system development methodology breaks down projects into short time-boxed iterations. Each iteration focuses on developing end-to-end functionality from user interface to data storage for the intended architecture. However, the release does not need to follow emergency release procedures unless there is a significant impact on operations.
C. Emergency releases to an application are fixes that require implementation as quickly as possible to prevent significant user downtime. Emergency release procedures are followed in such situations.
D. Operating system security patches are applied after testing, and therefore there is no need for an emergency release.

A4-204 A company with a limited budget has a recovery time objective of 72 hours and a recovery point objective of 24 hours. Which of the following would **BEST** meet the requirements of the business?

A. A hot site
B. A cold site
C. A mirrored site
D. A warm site

D is the correct answer.

Justification:

A. Although a hot site enables the business to meet its recovery point objective (RPO) and recovery time objective (RTO), the cost to maintain a hot site is more than the cost to maintain a warm site, which could also meet the objectives.
B. A cold site, although providing basic infrastructure, lacks the required hardware to meet the business objectives.
C. A mirrored site provides fully redundant facilities with real-time data replication. It can meet the business objectives, but it is not as cost-effective a solution as a warm site.
D. A warm site is the most appropriate solution because it provides basic infrastructure and most of the required IT equipment to affordably meet the business requirements. The remainder of the equipment needed can be provided through vendor agreements within a few days. The RTO is the amount of time allowed for the recovery of a business function or resource after a disaster occurs. The RPO is determined based on the acceptable data loss in case of a disruption of operations. The RPO indicates the earliest point in time that is acceptable to recover the data, and it effectively quantifies the permissible amount of data loss in case of interruption.

A4-205 Which of the following is **MOST** important to determine the recovery point objective for a critical process in an enterprise?

A. Number of hours of acceptable downtime
B. Total cost of recovering critical systems
C. Extent of data loss that is acceptable
D. Acceptable reduction in the level of service

C is the correct answer.

Justification:

A. The recovery time objective is the amount of time allowed for the recovery of a business function or resource after a disaster.
B. The determination of the recovery point objective (RPO) already takes cost into consideration.
C. The RPO is determined based on the acceptable data loss in case of a disruption of operations. It indicates the earliest point in time that is acceptable to recover the data. The RPO effectively quantifies the permissible amount of data loss in case of interruption.
D. The service delivery objective (SDO) is directly related to the business needs. The SDO is the level of services to be reached during the alternate process mode until the normal situation is restored.

A4-206 An IS auditor is assisting in the design of the emergency change control procedures for an organization with a limited budget. Which of the following recommendations **BEST** helps to establish accountability for the system support personnel?

A. Production access is granted to the individual support ID when needed.
B. Developers use a firefighter ID to promote code to production.
C. A dedicated user promotes emergency changes to production.
D. Emergency changes are authorized prior to promotion.

A is the correct answer.

Justification:

A. Production access should be controlled and monitored to ensure segregation of duties. During an emergency change, a user who normally does not have access to production may require access. The best process to ensure accountability within the production system is to have the information security team create a production support group and add the user ID to that group to promote the change. When the change is complete the ID can be removed from the group. This process ensures that activity in production is linked to the specific ID that was used to make the change.
B. Some organizations may use a firefighter ID, which is a generic/shared ID, to promote changes to production. When needed, the developer can use this ID to access production. It may still be difficult to determine who made the change; therefore, although this process is commonly used, the use of a production support ID is a better choice.
C. Having a dedicated user who promotes changes to production in an emergency is ideal but is generally not cost-effective and may not be realistic for emergency changes.
D. Emergency changes are, by definition, unauthorized changes. Approvals usually are obtained following promotion of the change to production. All changes should be auditable, and that can best be accomplished by having a user ID added/removed to the production support group as needed.

A4-207 Segmenting a highly sensitive database results in:

A. reduced exposure.
B. reduced threat.
C. less criticality.
D. less sensitivity.

A is the correct answer.

Justification:
A. **Segmenting data reduces the quantity of data exposed to a particular vulnerability.**
B. The threat may remain constant, but each segment represents a different vector against which it must be directed.
C. Criticality is a data attribute and is not affected by the manner in which it is segmented.
D. Sensitivity is a data attribute and is not affected by the manner in which it is segmented.

A4-208 Which of the following is the **BEST** way to ensure that incident response activities are consistent with the requirements of business continuity?

A. Draft and publish a clear practice for enterprise-level incident response.
B. Establish a cross-departmental working group to share perspectives.
C. Develop a scenario and perform a structured walk-through.
D. Develop a project plan for end-to-end testing of disaster recovery.

C is the correct answer.

Justification:
A. Publishing an enterprise-level incident response plan is effective only if business continuity aligned itself to incident response. Incident response supports business continuity, not the other way around.
B. Sharing perspectives is valuable, but a working group does not necessarily lead to ensuring that the interface between plans is workable.
C. **A structured walk-through including both incident response and business continuity personnel provides the best opportunity to identify gaps or misalignments between the plans.**
D. A project plan developed for disaster recovery will not necessarily address deficiencies in business continuity or incident response.

A4-209 An IS auditor is evaluating network performance for an organization that is considering increasing its Internet bandwidth due to a performance degradation during business hours. Which of the following is **MOST** likely the cause of the performance degradation?

A. Malware on servers
B. Firewall misconfiguration
C. Increased spam received by the email server
D. Unauthorized network activities

D is the correct answer.

Justification:

A. The existence of malware on the organization's server could contribute to network performance issues, but the degraded performance would not likely be restricted to business hours.
B. Firewall misconfiguration could contribute to network performance issues, but the degraded performance would not likely be restricted to business hours.
C. The existence of spam on the organization's email server could contribute to network performance issues, but the degraded performance would not likely be restricted to business hours.
D. Unauthorized network activities—such as employee use of file or music sharing sites or online gambling or personal email containing large files or photos—could contribute to network performance issues. Because the IS auditor found the degraded performance during business hours, this is the most likely cause.

A4-210 Which of the following is the **BEST** method for an IS auditor to verify that critical production servers are running the latest security updates released by the vendor?

A. Ensure that automatic updates are enabled on critical production servers.
B. Verify manually that the patches are applied on a sample of production servers.
C. Review the change management log for critical production servers.
D. Run an automated tool to verify the security patches on production servers.

D is the correct answer.

Justification:

A. Ensuring that automatic updates are enabled on production servers may be a valid way to manage the patching process; however, this would not provide assurance that all servers are being patched appropriately.
B. Verifying patches manually on a sample of production servers will be less effective than automated testing and introduces a significant audit risk. Manual testing is also difficult and time consuming.
C. The change management log may not be updated on time and may not accurately reflect the patch update status on servers. A better testing strategy is to test the server for patches, rather than examining the change management log.
D. An automated tool can immediately provide a report on which patches have been applied and which are missing.

A4-211 An IS auditor is conducting a review of the disaster recovery procedures for a data center. Which of the following indicators **BEST** shows that the procedures meet the requirements?

A. Documented procedures were approved by management.
B. Procedures were reviewed and compared with industry good practices.
C. A tabletop exercise using the procedures was conducted.
D. Recovery teams and their responsibilities are documented.

C is the correct answer.

Justification:
A. Management approval does not necessarily mean that the disaster recovery procedures are sufficient to meet the needs of the business.
B. While it is useful to compare the procedures with documented industry good practices, a tabletop exercise (paper test) is a better indicator that the procedures meet requirements.
C. Conducting a tabletop exercise (paper-based test) of the procedures with all responsible members, best ensures that the procedures meet the requirements. This type of test can identify missing or incorrect procedures because representatives responsible for performing the tasks are present.
D. The documentation of recovery teams and their responsibilities would be part of the procedures and not necessarily validate that the procedures are correct and complete thus meeting requirements.

A4-212 Which of the following choices **BEST** ensures accountability when updating data directly in a production database?

A. Review of audit logs
B. Principle of least privilege
C. Approved validation plan
D. Segregation of duties

A is the correct answer.

Justification:
A. Detailed audit logs that contain the user ID of the individual who performed the change as well as the data before and after the change are the best evidence of database changes. A review of these logs would evidence the individual who changed the data (ensuring accountability) as well as the correctness of the change.
B. Although access to production databases should be controlled by the principle of least privilege, this does not evidence who made the change or if the change was made correctly.
C. Having an approved validation plan evidences that the change was made correctly but does not show who made the change in production. Only a system-generated audit log can prove accountability.
D. Segregation of duties only ensures that the user making the data change is different than the individual who approved the data change. It would not evidence the individual who made the change, nor would it ensure that the data change was correct.

A4-213 An IS auditor has discovered that a new patch is available for an application, but the IT department has decided that the patch is not needed because other security controls are in place. What should the IS auditor recommend?

A. Apply the patch only after it has been thoroughly tested.
B. Implement a host-based intrusion detection system.
C. Modify the firewall rules to further protect the application server.
D. Assess the overall risk, then recommend whether to deploy the patch.

D is the correct answer.

Justification:

A. Applying a patch without first performing a risk assessment might be a waste of resources if it is determined that the application is not mission critical.
B. Implementing a host-based intrusion detection system would be a valid control; however, it may not address vulnerabilities within the application.
C. Modifying the firewall rules may help to mitigate the risk of a security incident; however, first the risk related to the patch would need to be determined.
D. While it is important to ensure that systems are properly patched, a risk assessment needs to be performed to determine the likelihood and probability of the vulnerability being exploited. Therefore, the patch would be applied only if the risk of circumventing the existing security controls is great enough to warrant it.

A4-214 An IS auditor is reviewing the most recent disaster recovery plan of an organization. Which approval is the **MOST** important when determining the availability of system resources required for the plan?

A. Executive management
B. IT management
C. Board of directors
D. Steering committee

B is the correct answer.

Justification:

A. Although executive management's approval is essential, the IT department is responsible for managing system resources and their availability as related to disaster recovery (DR).
B. Because a disaster recovery plan (DRP) is based on the recovery and provisioning of IT services, IT management's approval would be most important to verify that the system resources will be available in the event that a disaster event is triggered.
C. The board of directors may review and approve the DRP, but the IT department is responsible for managing system resources and their availability as related to DR.
D. The steering committee would determine the requirements for disaster recovery (recovery time objective and recovery point objective); however, the IT department is responsible for managing system resources and their availability as related to DR.

A4-215 Which of the following inputs would **PRIMARILY** help in designing the data backup strategy in case of potential natural disasters?

A. Recovery point objective
B. Volume of data to be backed up
C. Available data backup technologies
D. Recovery time objective

A is the correct answer.

Justification:

A. The recovery point objective (RPO) is determined based on the acceptable data loss in case of a disruption of operations. It indicates the earliest point in time that is acceptable to recover the data. The RPO effectively quantifies the acceptable amount of data loss in the case of interruption. Based on the RPO, one can design the data backup strategy for potential disasters using various technologies.

B. While the amount of data to be stored is critical in terms of planning for adequate capacity, the speed of recovery required by the business is the most important factor.

C. While a solid understanding of the capabilities of all types of advanced data backup technologies is necessary, without the knowledge of the RPO one cannot design a backup strategy using these technologies.

D. The recovery time objective is the amount of time allowed for the recovery of a business function or resource after a disaster occurs. This will help in designing disaster site options, but not the data backup strategy in the case of impacting disasters.

A4-216 While conducting an audit on the customer relationship management application, the IS auditor observes that it takes a significantly long time for users to log on to the system during peak business hours as compared with other times of the day. Once logged on, the average response time for the system is within acceptable limits. Which of the following choices should the IS auditor recommend?

A. No action should be taken because the system meets current business requirements.
B. IT should increase the network bandwidth to improve performance.
C. Users should be provided with detailed manuals to use the system properly.
D. Establish performance measurement criteria for the authentication servers.

D is the correct answer.

Justification:

A. The IS auditor should not recommend taking no action because a delayed login process has a negative impact on employee productivity.

B. Network bandwidth may or may not be the root cause of this issue. Performance measurement criteria may help determine the cause, which can then be remediated.

C. Because the problem is related to logging on and not to processing, additional training for users would not be effective in this case.

D. Performance criteria for the authentication servers would help to quantify acceptable thresholds for system performance, which can be measured and remediated.

A4-217 Due to resource constraints, a developer requires full access to production data to support certain problems reported by production users. Which of the following choices would be a good compensating control for controlling unauthorized changes in production?

A. Provide and monitor separate developer login IDs for programming and for production support.
B. Capture activities of the developer in the production environment by enabling detailed audit trails.
C. Back up all affected records before allowing the developer to make production changes.
D. Ensure that all changes are approved by the change manager prior to implementation.

A is the correct answer.

Justification:

A. Providing separate login IDs that would only allow a developer privileged access when required is a good compensating control, but it must also be backed up with monitoring and supervision of the activity of the developer.
B. While capturing activities of the developer via audit trails or logs would be a good practice, the control would not be effective unless these audit trails are reviewed on a periodic basis.
C. Creating a backup of affected records before making the change would allow for rollback in case of an error but would not prevent or detect unauthorized changes.
D. Even though changes are approved by the change manager, a developer with full access can easily circumvent this control.

A4-218 Which of the following choices would **MOST** likely ensure that a disaster recovery effort is successful?

A. The tabletop test was performed.
B. Data restoration was completed.
C. Recovery procedures are approved.
D. Appropriate staff resources are committed.

B is the correct answer.

Justification:

A. Performing a tabletop test is extremely helpful but does not ensure that the recovery process is working properly.
B. The most reliable method to determine whether a backup is valid would be to restore it to a system. A data restore test should be performed at least annually to verify that the process is working properly.
C. Approved recovery procedures will not ensure that data can be successfully restored.
D. While having appropriate staff resources is appropriate, without data the recovery would not be successful.

A4-219 An IS auditor is auditing an IT disaster recovery plan. The IS auditor should **PRIMARILY** ensure that the plan covers:

A. a resilient IT infrastructure.
B. alternate site information.
C. documented disaster recovery test results.
D. analysis and prioritization of business functions.

D is the correct answer.

Justification:

A. A resilient IT infrastructure is typically required to minimize interruptions to IT services; however, if a critical business function does not require high availability of IT, this may not be required for all disaster recovery plan (DRP) elements.
B. While the selection of an alternate site is important, the more critical issue is the prioritization of resources based on impact and recovery time objectives (RTOs) of business functions.
C. Documented DRP test results are helpful when maintaining the DRP; however, the DRP must first and foremost be aligned with business requirements.
D. The DRP must primarily focus on recovering critical business functions in the event of disaster within predefined RTOs; thus, it is necessary to align the recovery of IT services based on the criticality of business functions.

A4-220 An IS auditor observed that users are occasionally granted the authority to change system data. This elevated system access yet is required for smooth functioning of business operations. Which of the following controls would the IS auditor **MOST** likely recommend for long-term resolution?

A. Redesign the controls related to data authorization.
B. Implement additional segregation of duties controls.
C. Review policy to see if a formal exception process is required.
D. Implement additional logging controls.

C is the correct answer.

Justification:

A. Data authorization controls should be driven by the policy. While there may be some technical controls that could be adjusted, if the data changes happen infrequently, then an exception process would be the better choice.
B. While adequate segregation of duties is important, the IS auditor must first review policy to see if there is a formal documented process for this type of temporary access controls to enforce segregation of duties.
C. If the users are granted access to change data in support of the business requirements, and the policy should be followed. If there is no policy for the granting of extraordinary access, then one should be designed to ensure no unauthorized changes are made.
D. Audit trails are needed whenever temporary elevated access is required. However, but this is not the first step the auditor should take in reviewing the overall process.

A4-221 A medium-sized organization, whose IT disaster recovery measures have been in place and regularly tested for years, has just developed a formal business continuity plan (BCP). A basic BCP tabletop exercise has been performed successfully. Which testing should an IS auditor recommend be performed next to verify the adequacy of the new BCP?

A. Full-scale test with relocation of all departments, including IT, to the contingency site
B. Walk-through test of a series of predefined scenarios with all critical personnel involved
C. IT disaster recovery test with business departments involved in testing the critical applications
D. Functional test of a scenario with limited IT involvement

D is the correct answer.

Justification:

A. A full-scale test in the situation described might fail because it would be the first time that the plan is actually exercised, and a number of resources (including IT) and time would be wasted.
B. The walk-through test is a basic type of testing. Its intention is to make key staff familiar with the plan and discuss critical plan elements, rather than verifying its adequacy.
C. The recovery of applications should always be verified and approved by the business instead of being purely IT-driven. The IT plan has been tested repeatedly so a disaster recovery test would not help in verifying the administrative and organizational parts of the BCP, which are not IT-related.
D. After a tabletop exercise has been performed, the next step would be a functional test, which includes the mobilization of staff to exercise the administrative and organizational functions of a recovery. Because the IT part of the recovery has been tested for years, it would be more efficient to verify and optimize the BCP before actually involving IT in a full-scale test. The full-scale test would be the last step of the verification process before entering into a regular annual testing schedule.

A4-222 Which of the following business continuity plan tests involves participation of relevant members of the crisis management/response team to practice proper coordination?

A. Tabletop
B. Functional
C. Full-scale
D. Deskcheck

A is the correct answer.

Justification:

A. The primary purpose of tabletop testing is to practice proper coordination because it involves all or some of the crisis team members and is focused more on coordination and communication issues than on technical process details.
B. Functional testing involves mobilization of personnel and resources at various geographic sites. This is a more in-depth functional test and not primarily focused on coordination and communication.
C. Full-scale testing involves enterprisewide participation and full involvement of external organizations.
D. Deskcheck testing requires the least effort of the options given. Its aim is to ensure the plan is up to date and promote familiarity of the BCP to critical personnel from all areas.

A4-223 Which of the following is the **BEST** method to ensure that the business continuity plan remains up to date?

A. The group walks through the different scenarios of the plan from beginning to end.
B. The group ensures that specific systems can actually perform adequately at the alternate offsite facility.
C. The group is aware of full-interruption test procedures.
D. Interdepartmental communication is promoted to better respond in the case of a disaster.

A is the correct answer.

Justification:

A. A structured walk-through test gathers representatives from each department who will review the plan and identify weaknesses.
B. The ability of the group to ensure that specific systems can actually perform adequately at the alternate offsite facility is a parallel test and does not involve group meetings.
C. Group awareness of full-interruption test procedures is the most intrusive test to regular operations and the business.
D. While improving communication is important, it is not the most valued method to ensure that the plan is up to date.

A4-224 An organization having a number of offices across a wide geographical area has developed a disaster recovery plan. Using actual resources, which of the following is the **MOST** cost-effective test of the disaster recovery plan?

A. Full operational test
B. Preparedness test
C. Paper test
D. Regression test

B is the correct answer.

Justification:

A. A full operational test is conducted after the paper and preparedness test and is quite expensive.
B. A preparedness test is performed by each local office/area to test the adequacy of the preparedness of local operations for disaster recovery.
C. A paper test is a structured walk-through of the disaster recovery plan and should be conducted before a preparedness test, but a paper test (deskcheck) is not sufficient to test the viability of the plan.
D. A regression test is not a disaster recovery plan test and is used in software development and maintenance.

A4-225 An organization's disaster recovery plan should address early recovery of:

A. all information systems processes.
B. all financial processing applications.
C. only those applications designated by the IS manager.
D. processing in priority order, as defined by business management.

D is the correct answer.

Justification:

A. A disaster recovery plan (DRP) will recover most critical systems first according to business priorities.
B. Depending on business priorities, financial systems may or may not be the first to be recovered.
C. The business manager, not the IS manager, will determine priorities for system recovery.
D. Business management should know which systems are critical and what they need to process well in advance of a disaster. It is management's responsibility to develop and maintain the plan. Adequate time will not be available for this determination once the disaster occurs. IS and the information processing facility are service organizations that exist for the purpose of assisting the general user management in successfully performing their jobs.

A4-226 Disaster recovery planning addresses the:

A. technological aspect of business continuity planning (BCP).
B. operational part of BCP.
C. functional aspect of BCP.
D. overall coordination of BCP.

A is the correct answer.

Justification:

A. Disaster recovery planning (DRP) is the technological aspect of business continuity planning (BCP) that focuses on IT systems and operations.
B. Business resumption planning addresses the operational part of BCP.
C. Disaster recovery addresses the technical components of business recovery.
D. The overall coordination of BCP is accomplished through business continuity management and strategic plans. DRP addresses technical aspects of BCP.

A4-227 Which of the following must exist to ensure the viability of a duplicate information processing facility?

A. The site is near the primary site to ensure quick and efficient recovery.
B. The site contains the most advanced hardware available.
C. The workload of the primary site is monitored to ensure adequate backup is available.
D. The hardware is tested when it is installed to ensure it is working properly.

C is the correct answer.

Justification:

A. The site chosen should not be subject to the same natural disaster as the primary site. Being close may be a risk or an advantage, depending on the type of expected disaster.
B. A reasonable compatibility of hardware/software must exist to serve as a basis for backup. The latest or newest hardware may not adequately serve this need.
C. Resource availability must be assured. The workload of the primary site must be monitored to ensure that availability at the alternate site for emergency backup use is sufficient.
D. Testing the hardware when the site is established is essential, but regular testing of the actual backup data is necessary to ensure that the operation will continue to perform as planned.

A4-228 The cost of ongoing operations when a disaster recovery plan (DRP) is in place, compared to not having a DRP, will **MOST** likely:

A. increase.
B. decrease.
C. remain the same.
D. be unpredictable.

A is the correct answer.

Justification:

A. Due to the additional cost of testing, maintaining and implementing disaster recovery plan (DRP) measures, the cost of normal operations for any organization will always increase after a DRP implementation (i.e., the cost of normal operations during a nondisaster period will be more than the cost of operations during a nondisaster period when no DRP was in place).
B. The implementation of a DRP will always result in additional costs to the organization.
C. The implementation of a DRP will always result in additional costs to the organization.
D. The costs of a DRP are fairly predictable and consistent.

A4-229 Which of the following tasks should be performed **FIRST** when preparing a disaster recovery plan?

A. Develop a recovery strategy
B. Perform a business impact analysis
C. Map software systems, hardware and network components
D. Appoint recovery teams with defined personnel, roles and hierarchy

B is the correct answer.

Justification:
A. Developing a recovery strategy will come after performing a business impact analysis (BIA).
B. The first step in any disaster recovery plan is to perform a BIA.
C. The BIA will identify critical business processes and the systems that support those processes. Mapping software systems, hardware and network components will come after performing a BIA.
D. Appointing recovery teams with defined personnel, roles and hierarchy will come after performing a BIA.

A4-230 After completing the business impact analysis, what is the **NEXT** step in the business continuity planning process?

A. Test and maintain the plan.
B. Develop a specific plan.
C. Develop recovery strategies.
D. Implement the plan.

C is the correct answer.

Justification:
A. After selecting a strategy, a specific business continuity plan (BCP) can be developed, tested and implemented.
B. After selecting a strategy, a specific BCP can be developed, tested and implemented.
C. Once the business impact analysis (BIA) is completed, the next phase in the BCP development is to identify the various recovery strategies and select the most appropriate strategy for recovering from a disaster that will meet the time lines and priorities defined through the BIA.
D. After selecting a strategy, a specific BCP can be developed, tested and implemented.

A4-231 Which of the following is an appropriate test method to apply to a business continuity plan?

A. Pilot
B. Paper
C. Unit
D. System

B is the correct answer.

Justification:
A. A pilot test is used for implementing a new process or technology and is not appropriate for a business continuity planning (BCP).
B. A paper test (sometimes called a deskcheck) is appropriate for testing a BCP. It is a walk-through of the entire BCP, or part of the BCP, involving major players in the BCP's execution who reason out what may happen in a particular disaster.
C. A unit test is used to test new software components and is not appropriate for a BCP.
D. A system test is an integrated test used to test a new IT system but is not appropriate for a BCP.

A4-232 As part of the business continuity planning process, which of the following should be identified **FIRST** in the business impact analysis?

A. Risk such as single point-of-failure and infrastructure risk
B. Threats to critical business processes
C. Critical business processes for ascertaining the priority for recovery
D. Resources required for resumption of business

C is the correct answer.

Justification:
A. Risk should be identified after the critical business processes have been identified.
B. The identification of threats to critical business processes can only be determined after the critical business processes have been identified.
C. The identification of critical business processes should be addressed first so that the priorities and time lines for recovery can be documented.
D. Identification of resources required for business resumption will occur after the identification of critical business processes.

A4-233 Which of the following would contribute **MOST** to an effective business continuity plan?

A. The document is circulated to all interested parties.
B. Planning involves all user departments.
C. The plan is approved by senior management.
D. An audit is performed by an external IS auditor.

B is the correct answer.

Justification:
A. The business continuity plan (BCP) circulation will ensure that the BCP document is received by all users. Although essential, this does not contribute significantly to the success of the BCP.
B. The involvement of user departments in the BCP is crucial for the identification of the business processing priorities and the development of an effective plan.
C. A BCP approved by senior management would not necessarily ensure the effectiveness of the BCP.
D. An audit would not necessarily improve the quality of the BCP.

A4-234 The **PRIMARY** objective of business continuity and disaster recovery plans should be to:

A. safeguard critical IS assets.
B. provide for continuity of operations.
C. minimize the loss to an organization.
D. protect human life.

D is the correct answer.

Justification:
A. Safeguarding critical IS assets is a secondary objective of a business continuity and disaster recovery plan. The first priority is always life safety.
B. Providing continuity of operations is a secondary objective of a business continuity and disaster recovery plan. The first priority is always life safety.
C. Minimizing the loss to an organization is a secondary objective of a business continuity and disaster recovery plan. The first priority is always life safety.
D. Because human life is invaluable, the main priority of any business continuity and disaster recovery plan should be to protect people.

A4-235 Depending on the complexity of an organization's business continuity plan (BCP), it may be developed as a set of plans to address various aspects of business continuity and disaster recovery. In such an environment, it is essential that:

A. each plan is consistent with one another.
B. all plans are integrated into a single plan.
C. each plan is dependent on one another.
D. the sequence for implementation of all plans is defined.

A is the correct answer.

Justification:

A. Depending on the complexity of an organization, there could be more than one plan to address various aspects of business continuity and disaster recovery, but the plans must be consistent to be effective.
B. The plans do not necessarily have to be integrated into one single plan.
C. Although each plan may be independent, each plan has to be consistent with other plans to have a viable business continuity planning strategy.
D. It may not be possible to define a sequence in which plans have to be implemented because it may be dependent on the nature of disaster, criticality, recovery time, etc.

A4-236 When developing a business continuity plan, which of the following tools should be used to gain an understanding of the organization's business processes?

A. Business continuity self-audit
B. Resource recovery analysis
C. Risk assessment
D. Gap analysis

C is the correct answer.

Justification:

A. Business continuity self-audit is a tool for evaluating the adequacy of the business continuity plan (BCP) but not for gaining an understanding of the business.
B. Resource recovery analysis is a tool for identifying the components necessary for a business resumption strategy but not for gaining an understanding of the business.
C. Risk assessment and business impact assessment are tools for understanding the business as a part of BCP.
D. The role gap analysis can play in BCP is to identify deficiencies in a plan but not for gaining an understanding of the business.

A4-237 Which of the following should be of **MOST** concern to an IS auditor reviewing the business continuity plan (BCP)?

A. The disaster levels are based on scopes of damaged functions but not on duration.
B. The difference between low-level disaster and software incidents is not clear.
C. The overall BCP is documented, but detailed recovery steps are not specified.
D. The responsibility for declaring a disaster is not identified.

D is the correct answer.

Justification:

A. Although failure to consider duration could be a problem, it is not as significant as scope, and neither is as critical as the need to identify someone with the authority to invoke the business continuity plan (BCP).
B. The difference between incidents and low-level disasters is always unclear and frequently revolves around the amount of time required to correct the damage.
C. The lack of detailed steps should be documented, but their absence does not mean a lack of recovery if, in fact, someone has invoked the BCP.
D. If nobody declares the disaster, the BCP would not be invoked, making all other concerns less important.

A4-238 During an audit of a business continuity plan (BCP), an IS auditor found that, although all departments were housed in the same building, each department had a separate BCP. The IS auditor recommended that the BCPs be reconciled. Which of the following areas should be reconciled **FIRST**?

A. Evacuation plan
B. Recovery priorities
C. Backup storages
D. Call tree

A is the correct answer.

Justification:

A. Protecting human resources during a disaster-related event should be addressed first. Having separate business continuity plans could result in conflicting evacuation plans, thus jeopardizing the safety of staff and clients.
B. Recovery priorities may be unique to each department and could be addressed separately, but still should be reviewed for possible conflicts and/or the possibility of cost reduction, but only after the issue of human safety has been analyzed.
C. Backup strategies are not critical to the integration of the plans for the various departments. Life safety is always the first priority.
D. Communication during a crisis is always a challenge, but the call tree is not as important as ensuring life safety first.

A4-239 For effective implementation after a business continuity plan (BCP) has been developed, it is **MOST** important that the BCP be:

A. stored in a secure, offsite facility.
B. approved by senior management.
C. communicated to appropriate personnel.
D. made available through the enterprise's intranet.

C is the correct answer.

Justification:
A. The business continuity plan (BCP), if kept in a safe place, will not reach the users; users will never implement the BCP and, thus, the BCP will be ineffective.
B. Senior management approval is a prerequisite for designing and approving the BCP but is less important than making sure that the plan is available to all key personnel to ensure that the plan will be effective.
C. The implementation of a BCP will be effective only if appropriate personnel are informed and aware of all the aspects of the BCP.
D. Making a BCP available on an enterprise's intranet does not guarantee that personnel will be able to access, read or understand it.

A4-240 Which of the following is the **PRIMARY** objective of the business continuity plan process?

A. To provide assurance to stakeholders that business operations will continue in the event of disaster
B. To establish an alternate site for IT services to meet predefined recovery time objectives
C. To manage risk while recovering from an event that adversely affected operations
D. To meet the regulatory compliance requirements in the event of natural disaster

C is the correct answer.

Justification:
A. The business continuity plan (BCP) in itself does not provide assurance of continuing operations; however, it helps the organization to respond to disruptions to critical business processes.
B. Establishment of an alternate site is more relevant to disaster recovery than the BCP.
C. The BCP process primarily focuses on managing and mitigating risk during recovery of operations due to an event that affected operations.
D. The regulatory compliance requirements may help establish the recovery time objective (RTO) requirements.

A4-241 Which of the following would **BEST** help to detect errors in data processing?

A. Programmed edit checks
B. Well-designed data entry screens
C. Segregation of duties
D. Hash totals

D is the correct answer.

Justification:
A. Automated controls such as programmed edit checks are preventive controls.
B. Automated controls such as well-designed data entry screens are preventive controls.
C. Enforcing segregation of duties primarily ensures that a single individual does not have the authority to both create and approve a transaction; this is not considered to be a method to detect errors, but a method to help prevent errors.
D. The use of hash totals is an effective method to reliably detect errors in data processing. A hash total would indicate an error in data integrity.

A4-242 Which of the following is the **MOST** critical to the quality of data in a data warehouse?

A. Accuracy of the source data
B. Credibility of the data source
C. Accuracy of the extraction process
D. Accuracy of the data transformation

A is the correct answer.

Justification:

A. Accuracy of source data is a prerequisite for the quality of the data in a data warehouse. Inaccurate source data will corrupt the integrity of the data in the data warehouse.
B. Credibility of the data source is important but would not change inaccurate data into quality (accurate) data.
C. Accurate extraction processes are important but would not change inaccurate data into quality (accurate) data.
D. Accurate transformation routines are important but would not change inaccurate data into quality (accurate) data.

A4-243 A clerk changed the interest rate for a loan on a master file. The rate entered is outside the normal range for such a loan. Which of the following controls is **MOST** effective in providing reasonable assurance that the change was authorized?

A. The system will not process the change until the clerk's manager confirms the change by entering an approval code
B. The system generates a weekly report listing all rate exceptions and the report is reviewed by the clerk's manager
C. The system requires the clerk to enter an approval code
D. The system displays a warning message to the clerk

A is the correct answer.

Justification:

A. Requiring an approval code by a manager would prevent or detect the use of an unauthorized interest rate.
B. A weekly report would inform the manager after the fact that a change was made, thereby making it possible for transactions to use an unauthorized rate prior to management review.
C. Having a clerk enter an approval code would not provide separation of duties and would not prevent the clerk from entering an unauthorized rate change.
D. A warning message would alert the clerk in case the change was being made in error but would not prevent the clerk from entering an unauthorized rate change.

A4-244 The **GREATEST** advantage of using web services for the exchange of information between two systems is:

A. Secure communication
B. Improved performance
C. Efficient interfacing
D. Enhanced documentation

C is the correct answer.

Justification:
A. Communication is not necessarily more secure using web services.
B. The use of web services will not necessarily increase performance.
C. Web services facilitate the interoperable exchange of information between two systems regardless of the operating system or programming language used.
D. There is no documentation benefit in using web services

A4-245 Which of the following is a prevalent risk in the development of end-user computing applications?

A. Applications may not be subject to testing and IT general controls.
B. Development and maintenance costs may be increased.
C. Application development time may be increased.
D. Decision-making may be impaired due to diminished responsiveness to requests for information.

A is the correct answer.

Justification:
A. End-user computing (EUC) is defined as the ability of end users to design and implement their own information system using computer software products. End-user developed applications may not be subjected to an independent outside review by systems analysts and frequently are not created in the context of a formal development methodology. These applications may lack appropriate standards, controls, quality assurance procedures, and documentation. A risk of end-user applications is that management may rely on them as much as traditional applications.
B. EUC systems typically result in reduced application development and maintenance costs.
C. EUC systems typically result in a reduced development cycle time.
D. EUC systems normally increase flexibility and responsiveness to management's information requests because the system is being developed directly by the user community.

A4-246 An IS auditor finds out-of-range data in some tables of a database. Which of the following controls should the IS auditor recommend to avoid this situation?

A. Log all table update transactions
B. Implement integrity constraints in the database
C. Implement before and after image reporting
D. Use tracing and tagging

B is the correct answer.

Justification:
A. Logging all table update transactions provides audit trails and is a detective control but will not prevent the introduction of inaccurate data.
B. Implementing integrity constraints in the database is a preventive control because data are checked against predefined tables or rules, which prevents any undefined data from being entered.
C. Before and after image reporting makes it possible to trace the impact that transactions have on computer records and is a detective control.
D. Tracing and tagging is used to test application systems and controls but is not a preventive control that can avoid out-of-range data.

A4-247 A new database is being set up in an overseas location to provide information to the general public and to increase the speed at which the information is made available. The overseas database is to be housed at a data center and will be updated in real time to mirror the information stored locally. Which of the following areas of operations should be considered as having the **HIGHEST** risk?

A. Confidentiality of the information stored in the database
B. The hardware being used to run the database application
C. Backups of the information in the overseas database
D. Remote access to the backup database

B is the correct answer.

Justification:

A. Confidentiality of the information stored in the database is not a major concern, because the information is intended for public use.
B. The business objective is to make the information available to the public in a timely manner. Because the database is physically located overseas, hardware failures that are left unfixed can reduce the availability of the system to users.
C. Backups of the information in the overseas database are not a major concern, because the overseas database is a mirror of the local database; thus, a backup copy exists locally.
D. Remote access to the backup database does not impact availability.

A4-248 Which of the following is the **MOST** effective when determining the correctness of individual account balances migrated from one database to another?

A. Compare the hash total before and after the migration
B. Verify that the number of records is the same for both databases
C. Perform sample testing of the migrated account balances
D. Compare the control totals of all of the transactions

C is the correct answer.

Justification:

A. The hash total will only validate the data integrity at a batch level rather than at a transaction level.
B. Databases are composed of records that can contain multiple fields. The number of records will not allow an IS auditor to ascertain whether some of these fields have been successfully migrated.
C. Performing sample testing of the migrated account balances will involve the comparison of a selection of individual transactions from the database before and after the migration.
D. Comparing the control totals does not imply that the records are complete or that individual values are accurate.

A4-249 During the review of data file change management controls, which of the following **BEST** helps to decrease the research time needed to investigate exceptions?

A. One-for-one checking
B. Data file security
C. Transaction logs
D. File updating and maintenance authorization

C is the correct answer.

Justification:

A. One-for-one checking is a control procedure in which an individual document agrees with a detailed listing of documents processed by the system. It would take a long time to complete the research using this procedure.
B. Data file security controls prevent access by unauthorized users in their attempt to alter data files. This would not help identify the transactions posted to an account.
C. Transaction logs generate an audit trail by providing a detailed list of date of input, time of input, user ID, terminal location, etc. Research time can be reduced in investigating exceptions because the review can be performed on the logs rather than on the entire transaction file. It also helps to determine which transactions have been posted to an account—by a particular individual during a particular period.
D. File updating and maintenance authorization is a control procedure to update the stored data and ensure accuracy and security of stored data. This does provide evidence regarding the individuals who update the stored data; however, it is not effective in the given situation to determine transactions posted to an account.

A4-250 An IS auditor is reviewing a monthly accounts payable transaction register using audit software. For what purpose would the auditor be interested in using a check digit?

A. To detect data transposition errors
B. To ensure that transactions do not exceed predetermined amounts
C. To ensure that data entered are within reasonable limits
D. To ensure that data entered are within a predetermined range of values

A is the correct answer.

Justification:

A. A check digit is a numeric value added to data to ensure that original data are correct and have not been altered.
B. Ensuring that data have not exceeded a predetermined amount is a limit check.
C. Ensuring that data entered are within predetermined reasonable limits is a reasonableness check.
D. Ensuring that data entered are within a predetermined range of values is a range check.

A4-251 A hard disk containing confidential data was damaged beyond repair. If the goal is to positively prevent access to the data by anyone else, what should be done to the hard disk before it is discarded?

A. Overwriting
B. Low-level formatting
C. Degaussing
D. Destruction

D is the correct answer.

Justification:
A. Rewriting data is impractical because the hard disk is damaged and offers less assurance than physical destruction even when done successfully.
B. Low-level formatting is impractical because the hard disk is damaged and offers less assurance than physical destruction even when done successfully.
C. Degaussing is highly effective but offers less assurance than physical destruction.
D. Physically destroying the hard disk is the most effective way to ensure that data cannot be recovered.

A4-252 Authorizing access to application data is the responsibility of the:

A. data custodian.
B. application administrator.
C. data owner.
D. security administrator.

C is the correct answer.

Justification:
A. Data custodians are responsible only for storing and safeguarding the data according to the direction provided by the data owner.
B. An application administrator is responsible for managing the application itself, not determining who is authorized to access the data that it contains.
C. Data owners have authority to grant or withhold access to the data and applications for which they are responsible.
D. The security administrator may lead investigations and is responsible for implementing and maintaining information security policy, but not for authorizing data access.

A4-253 An IS auditor finds that a database administrator (DBA) has read and write access to production data. The IS auditor should:

A. accept the DBA access as a common practice.
B. assess the controls relevant to the DBA function.
C. recommend the immediate revocation of the DBA access to production data.
D. review user access authorizations approved by the DBA.

B is the correct answer.

Justification:
A. Although granting access to production data to the database administrator (DBA) may be a common practice, the IS auditor should evaluate the relevant controls.
B. When reviewing privileged accounts, the auditor should look for compensating controls that may address a potential exposure.
C. The DBA should have access based on the principle of least privilege; unless care is taken to validate what access is required, revocation may remove access the DBA requires to do his/her job.
D. Granting user authorizations is the responsibility of the data owner, not the DBA, and access to production data is not generally associated with user access authorizations.

A4-254 Which of the following is the **MOST** effective method for disposing of magnetic media that contains confidential information?

A. Degaussing
B. Defragmenting
C. Erasing
D. Destroying

D is the correct answer.

Justification:
A. Degaussing or demagnetizing is a good control, but not sufficient to fully erase highly confidential information from magnetic media.
B. The purpose of defragmentation is to improve efficiency by eliminating fragmentation in file systems; it does not remove information.
C. Erasing or deleting magnetic media does not remove the information; this method simply changes a file's indexing information.
D. Destroying magnetic media is the only way to assure that confidential information cannot be recovered.

A4-255 Which of the following should an IS auditor recommend for the protection of specific sensitive information stored in a data warehouse?

A. Implement column- and row-level permissions
B. Enhance user authentication via strong passwords
C. Organize the data warehouse into subject matter-specific databases
D. Log user access to the data warehouse

A is the correct answer.

Justification:
A. Column- and row-level permissions control what information users can access. Column-level security prevents users from seeing one or more attributes on a table. With row-level security a certain grouping of information on a table is restricted (e.g., if a table held details of employee salaries, then a restriction could be put in place to ensure that, unless specifically authorized, users could not view the salaries of executive staff). Column- and row-level security can be achieved in a relational database by allowing users to access logical representations of data (views) rather than physical tables. This "fine-grained" security model is likely to offer the best balance between information protection while still supporting a wide range of analytical and reporting uses.
B. Enhancing user authentication via strong passwords is a security control that should apply to all users of the data warehouse and does not specifically address protection of specific sensitive data.
C. Organizing a data warehouse into subject-specific databases is a potentially useful practice but, in itself, does not adequately protect sensitive data. Database-level security is normally too "coarse" a level to efficiently and effectively protect information. For example, one database may hold information that needs to be restricted such as employee salary and customer profitability details while other information such as employee department may need to be legitimately accessed by a large number of users. Organizing the data warehouse into subject matter-specific databases is similar to user access in that this control should generally apply. Extra attention could be devoted to reviewing access to tables with sensitive data, but this control is not sufficient without strong preventive controls at the column and row level.
D. Logging user access is important, but it is only a detective control that will not provide adequate protection to sensitive information.

A4-256 The responsibility for authorizing access to a business application system belongs to the:

A. data owner.
B. security administrator.
C. IT security manager.
D. requestor's immediate supervisor.

A is the correct answer.

Justification:

A. When a business application is developed, a good practice is to assign an information or data owner to the application. The information owner should be responsible for authorizing access to the application itself or to back-end databases for queries.

B. The security administrator normally does not have responsibility for authorizing access to business applications.

C. The IT security manager normally does not have responsibility for authorizing access to business applications.

D. The requestor's immediate supervisor may share the responsibility for approving user access to a business application system; however, the final responsibility should go to the information owner.

A4-257 What would be the **MOST** effective control for enforcing accountability among database users accessing sensitive information?

A. Implement a log management process.
B. Implement a two-factor authentication.
C. Use table views to access sensitive data.
D. Separate database and application servers.

A is the correct answer.

Justification:

A. Accountability means knowing what is being done by whom. The best way to enforce the principle is to implement a log management process that would create and store logs with pertinent information such as user name, type of transaction and hour.

B. Implementing a two-factor authentication would prevent unauthorized access to the database but would not record the activity of the user when using the database.

C. Using table views would restrict users from seeing data that they should not be able to see but would not record what users did with data they were allowed to see.

D. Separating database and application servers may help in better administration or even in implementing access controls but does not address the accountability issues.

A4-258 While auditing an ecommerce architecture, an IS auditor notes that customer master data are stored on the web server for six months after the transaction date and then purged due to inactivity. Which of the following would be the **PRIMARY** concern for the IS auditor?

A. Availability of customer data
B. Integrity of customer data
C. Confidentiality of customer data
D. System storage performance

C is the correct answer.

Justification:

A. Availability of customer data may be affected during an Internet connection outage, but this is of a lower concern than confidentiality.

B. Integrity of customer data is affected only if security controls are weak enough to permit unauthorized modifications to the data, and it may be tracked by logging of changes. Confidentiality of data is a larger concern.

C. Due to its exposure to the Internet, storing customer data for six months raises concerns regarding confidentiality of customer data.

D. System storage performance may be a concern due to the volume of data. However, the bigger issue is that the information is protected.

Page intentionally left blank

DOMAIN 5—PROTECTION OF INFORMATION ASSETS (27 %)

A5-1 Web application developers sometimes use hidden fields on web pages to save information about a client session. This technique is used, in some cases, to store session variables that enable persistence across web pages, such as maintaining the contents of a shopping cart on a retail web site application. The **MOST** likely web-based attack due to this practice is:

A. parameter tampering.
B. cross-site scripting.
C. cookie poisoning.
D. stealth commanding.

A is the correct answer.

Justification:

A. **Web application developers sometimes use hidden fields to save information about a client session or to submit hidden parameters, such as the language of the end user, to the underlying application. Because hidden form fields do not display in the browser, developers may feel safe passing unvalidated data in the hidden fields (to be validated later). This practice is not safe because an attacker can intercept, modify and submit requests, which can discover information or perform functions that the web developer never intended. The malicious modification of web application parameters is known as parameter tampering.**

B. Cross-site scripting involves the compromise of the web page to redirect users to content on the attacker web site. The use of hidden fields has no impact on the likelihood of a cross-site scripting attack because these fields are static content that cannot ordinarily be modified to create this type of attack. Web applications use cookies to save session state information on the client machine so that the user does not need to log on every time a page is visited.

C. Cookie poisoning refers to the interception and modification of session cookies to impersonate the user or steal logon credentials. The use of hidden fields has no relation to cookie poisoning.

D. Stealth commanding is the hijacking of a web server by the installation of unauthorized code. While the use of hidden forms may increase the risk of server compromise, the most common server exploits involve vulnerabilities of the server operating system or web server.

A5-2 Which control is the **BEST** way to ensure that the data in a file have not been changed during transmission?

A. Reasonableness check
B. Parity bits
C. Hash values
D. Check digits

C is the correct answer.

Justification:

A. A reasonableness check is used to ensure that input data is within expected values, not to ensure integrity of data transmission. Data can be changed and still pass a reasonableness test.

B. Parity bits are a weak form of data integrity checks used to detect errors in transmission, but they are not as good as using a hash.

C. **Hash values are calculated on the file and are very sensitive to any changes in the data values in the file. Thus, they are the best way to ensure that data has not changed.**

D. Check digits are used to detect an error in a numeric field such as an account number and is usually related to a transposition or transcribing error.

A5-3 The **PRIMARY** purpose of audit trails is to:

A. improve response time for users.
B. establish accountability for processed transactions.
C. improve the operational efficiency of the system.
D. provide information to auditors who wish to track transactions.

B is the correct answer.

Justification:

A. The objective of enabling software to provide audit trails is not to improve system efficiency because it often involves additional processing which may, in fact, reduce response time for users.
B. Enabling audit trails helps in establishing the accountability and responsibility of processed transactions by tracing transactions through the system.
C. Enabling audit trails involves storage and, thus, occupies disk space and may decrease operational efficiency.
D. Audit trails are used to track transactions for various purposes, not just for audit. The use of audit trails for IS auditors is valid; however, it is not the primary reason.

A5-4 Which of the following systems or tools can recognize that a credit card transaction is more likely to have resulted from a stolen credit card than from the holder of the credit card?

A. Intrusion detection systems
B. Data mining techniques
C. Stateful inspection firewalls
D. Packet filtering routers

B is the correct answer.

Justification:

A. An intrusion detection system is effective in detecting network or host-based errors but not effective in measuring fraudulent transactions.
B. Data mining is a technique used to detect trends or patterns of transactions or data. If the historical pattern of charges against a credit card account is changed, then it is a flag that the transaction may have resulted from a fraudulent use of the card.
C. A firewall is an excellent tool for protecting networks and systems but not effective in detecting fraudulent transactions.
D. A packet filtering router operates at a network level and cannot see a transaction.

A5-5 Which of the following **BEST** ensures the integrity of a server's operating system?

A. Protecting the server in a secure location
B. Setting a boot password
C. Hardening the server configuration
D. Implementing activity logging

C is the correct answer.

Justification:

A. Protecting the server in a secure location is a good practice, but it does not ensure that a user will not try to exploit logical vulnerabilities and compromise the operating system (OS).
B. Setting a boot password is a good practice but does not ensure that a user will not try to exploit logical vulnerabilities and compromise the OS.
C. Hardening a system means to configure it in the most secure manner (install latest security patches, properly define access authorization for users and administrators, disable insecure options and uninstall unused services) to prevent nonprivileged users from gaining the right to execute privileged instructions and, thus, take control of the entire machine, jeopardizing the integrity of the OS.
D. Activity logging has two weaknesses in this scenario—it is a detective control (not a preventive one), and the attacker who already gained privileged access can modify logs or disable them.

A5-6 Which of the following network components is **PRIMARILY** set up to serve as a security measure by preventing unauthorized traffic between different segments of the network?

A. Firewalls
B. Routers
C. Layer 2 switches
D. Virtual local area networks

A is the correct answer.

Justification:

A. Firewall systems are the primary tool that enables an organization to prevent unauthorized access between networks. An organization may choose to deploy one or more systems that function as firewalls.
B. Routers can filter packets based on parameters, such as source address but are not primarily a security tool.
C. Based on Media Access Control addresses, layer 2 switches separate traffic without determining whether it is authorized or unauthorized traffic.
D. A virtual local area network is a functionality of some switches that allows them to control traffic between different ports even though they are in the same physical local access network. Nevertheless, they do not effectively deal with authorized versus unauthorized traffic.

A5-7 An IS auditor discovers that the chief information officer (CIO) of an organization is using a wireless broadband modem using global system for mobile communications (GSM) technology. This modem is being used to connect the CIO's laptop to the corporate virtual private network when the CIO travels outside of the office. The IS auditor should:

A. do nothing because the inherent security features of GSM technology are appropriate.
B. recommend that the CIO stop using the laptop computer until encryption is enabled.
C. ensure that media access control address filtering is enabled on the network so unauthorized wireless users cannot connect.
D. suggest that two-factor authentication be used over the wireless link to prevent unauthorized communications.

A is the correct answer.

Justification:

A. The inherent security features of global system for mobile communications (GSM) technology combined with the use of a virtual private network (VPN) are appropriate. The confidentiality of the communication on the GSM radio link is ensured by the use of encryption and the use of a VPN signifies that an encrypted session is established between the laptop and the corporate network. GSM is a global standard for cellular telecommunications that can be used for both voice and data. Currently deployed commercial GSM technology has multiple overlapping security features which prevent eavesdropping, session hijacking or unauthorized use of the GSM carrier network. While other wireless technologies such as 802.11 wireless local area network (LAN) technologies have been designed to allow the user to adjust or even disable security settings, GSM does not allow any devices to connect to the system unless all relevant security features are active and enabled.

B. Because the chief information officer (CIO) is using a VPN it can be assumed that encryption is enabled in addition to the security features in GSM. In addition, VPNs will not allow the transfer of data for storage on the remote device (such as the CIO's laptop).

C. Media access control (MAC) filtering can be used on a wireless LAN but does not apply to a GSM network device.

D. Because the GSM network is being used rather than a wireless LAN, it is not possible to configure settings for two-factor authentication over the wireless link. However, two-factor authentication is recommended as it will better protect against unauthorized access than single factor authentication.

A5-8 Which of the following is the **BEST** way to minimize unauthorized access to unattended end-user PC systems?

A. Enforce use of a password-protected screen saver
B. Implement proximity-based authentication system
C. Terminate user session at predefined intervals
D. Adjust power management settings so the monitor screen is blank

A is the correct answer.

Justification:

A. A password-protected screen saver with a proper time interval is the best measure to prevent unauthorized access to unattended end-user systems. It is important to ensure that users lock the workstation when they step away from the machine, which is something that could be reinforced via awareness training.

B. There are solutions that will lock machines when users step away from their desks, and those would be suitable here; however, those tools are a more expensive solution, which would normally include the use of smart cards and extra hardware. Therefore, the use of a password-protected screen saver would be a better solution.

C. Terminating user sessions is often done for remote login (periodic re-authentication) or after a certain amount of inactivity on a web or server session. There is more risk related to leaving the workstation unlocked; therefore, this is not the correct answer.

D. Switching off the monitor would not be a solution because the monitor could simply be switched on.

A5-9 The implementation of which of the following would **MOST** effectively prevent unauthorized access to a system administration account on a web server?

A. Host intrusion detection software installed on the server
B. Password expiration and lockout policy
C. Password complexity rules
D. Two-factor authentication

D is the correct answer.

Justification:

A. Host intrusion detection software will assist in the detection of unauthorized system access but does not prevent such access.

B. While controls regarding password expiration and lockout from failed login attempts are important, two-factor authentication methods or techniques would most effectively reduce the risk of stolen or compromised credentials. Password-only based authentication may not provide adequate security.

C. While controls regarding password complexity are important, two-factor authentication methods or techniques would most effectively reduce the risk of stolen or compromised credentials.

D. Two-factor authentication requires a user to use a password in combination with another identification factor that is not easily stolen or guessed by an attacker. Types of two-factor authentication include electronic access tokens that show one-time passwords on their display panels or biometric authentication systems.

A5-10 An organization's IT director has approved the installation of a wireless local area network access point in a conference room for a team of consultants to access the Internet with their laptop computers. The **BEST** control to protect the corporate servers from unauthorized access is to ensure that:

A. encryption is enabled on the access point.
B. the conference room network is on a separate virtual local area network (VLAN).
C. antivirus signatures and patch levels are current on the consultants' laptops.
D. default user IDs are disabled and strong passwords are set on the corporate servers.

B is the correct answer.

Justification:

A. Enabling encryption is a good idea to prevent unauthorized network access, but it is more important to isolate the consultants from the rest of the corporate network.
B. The installation of the wireless network device presents risk to the corporate servers from both authorized and unauthorized users. A separate virtual local area network is the best solution because it ensures that both authorized and unauthorized users are prevented from gaining network access to database servers, while allowing Internet access to authorized users.
C. Antivirus signatures and patch levels are good practices but not as critical as preventing network access via access controls for the corporate servers.
D. Protecting the organization's servers through good passwords is good practice, but it is still necessary to isolate the network being used by the consultants. If the consultants can access the rest of the network, they could use password cracking tools against other corporate machines.

A5-11 The IS auditor is reviewing an organization's human resources (HR) database implementation. The IS auditor discovers that the database servers are clustered for high availability, all default database accounts have been removed and database audit logs are kept and reviewed on a weekly basis. What other area should the IS auditor check to ensure that the databases are appropriately secured?

A. Database administrators are restricted from access to HR data.
B. Database logs are encrypted.
C. Database stored procedures are encrypted.
D. Database initialization parameters are appropriate.

D is the correct answer.

Justification:

A. Database administrators would have access to all data on the server, but there is no practical control to prevent that; therefore, this would not be a concern.
B. Database audit logs normally would not contain any confidential data; therefore, encrypting the log files is not required.
C. If a stored procedure contains a security sensitive function such as encrypting data, it can be a requirement to encrypt the stored procedure. However, this is less critical than ensuring initialization parameters are correct.
D. When a database is opened, many of its configuration options are governed by initialization parameters. These parameters are usually governed by a file ("init.ora" in the case of Oracle Database Management System), which contains many settings. The system initialization parameters address many "global" database settings, including authentication, remote access and other critical security areas. To effectively audit a database implementation, the IS auditor must examine the database initialization parameters.

A5-12 An IS auditor has been asked by management to review a potentially fraudulent transaction. The **PRIMARY** focus of an IS auditor while evaluating the transaction should be to:

A. maintain impartiality while evaluating the transaction.
B. ensure that the independence of an IS auditor is maintained.
C. assure that the integrity of the evidence is maintained.
D. assess all relevant evidence for the transaction.

C is the correct answer.

Justification:

A. Although it is important for an IS auditor to be impartial, in this case it is more critical that the evidence be preserved.
B. Although it is important for an IS auditor to maintain independence, in this case it is more critical that the evidence be preserved.
C. The IS auditor has been requested to perform an investigation to capture evidence which may be used for legal purposes, and therefore, maintaining the integrity of the evidence should be the foremost goal. Improperly handled computer evidence is subject to being ruled inadmissible in a court of law.
D. While it is also important to assess all relevant evidence, it is more important to maintain the chain of custody, which ensures the integrity of evidence.

A5-13 A new business application has been designed in a large, complex organization and the business owner has requested that the various reports be viewed on a "need to know" basis. Which of the following access control methods would be the **BEST** method to achieve this requirement?

A. Mandatory
B. Role-based
C. Discretionary
D. Single sign-on

B is the correct answer.

Justification:

A. An access control system based on mandatory access control would be expensive, and difficult to implement and maintain in a large complex organization.
B. Role-based access control limits access according to job roles and responsibilities and would be the best method to allow only authorized users to view reports on a need-to-know basis.
C. Discretionary access control (DAC) is where the owner of the resources decides who should have access to that resource. Most access control systems are an implementation of DAC. This answer is not specific enough for this scenario.
D. Single sign-on is an access control technology used to manage access to multiple systems, networks and applications. This answer is not specific enough for this question.

A5-14 Which of the following is the **BEST** control to prevent the deletion of audit logs by unauthorized individuals in an organization?

A. Actions performed on log files should be tracked in a separate log.
B. Write access to audit logs should be disabled.
C. Only select personnel should have rights to view or delete audit logs.
D. Backups of audit logs should be performed periodically.

C is the correct answer.

Justification:
A. Having additional copies of log file activity would not prevent the original log files from being deleted.
B. For servers and applications to operate correctly, write access cannot be disabled.
C. Granting access to audit logs to only system administrators and security administrators would reduce the possibility of these files being deleted.
D. Frequent backups of audit logs would not prevent the logs from being deleted.

A5-15 A company is implementing a Dynamic Host Configuration Protocol. Given that the following conditions exist, which represents the **GREATEST** concern?

A. Most employees use laptops.
B. A packet filtering firewall is used.
C. The IP address space is smaller than the number of PCs.
D. Access to a network port is not restricted.

D is the correct answer.

Justification:
A. Dynamic Host Configuration Protocol provides convenience (an advantage) to the laptop users.
B. The existence of a firewall can be a security measure and would not normally be of concern.
C. A limited number of IP addresses can be addressed through network address translation or by increasing the number of IP addresses assigned to a particular subnet.
D. Given physical access to a port, anyone can connect to the internal network. This would allow individuals to connect that were not authorized to be on the corporate network.

A5-16 Which of the following is an effective preventive control to ensure that a database administrator (DBA) complies with the custodianship of the enterprise's data?

A. Exception reports
B. Segregation of duties
C. Review of access logs and activities
D. Management supervision

B is the correct answer.

Justification:
A. Exception reports are detective controls used to indicate when the activities of the database administrator (DBA) were performed without authorization.
B. Adequate segregation of duties (SoD) is a preventative control that can restrict the activities of the DBA to those that have been authorized by the data owners. SoD can restrict what a DBA can do by requiring more than one person to participate to complete a task.
C. Reviews of access logs are used to detect the activities performed by the DBA.
D. Management supervision of DBA activities is used to detect which DBA activities were not authorized.

A5-17 An employee has received a digital photo frame as an gift and has connected it to his/her work PC to transfer digital photos. The **PRIMARY** risk that this scenario introduces is that:

A. the photo frame storage media could be used to steal corporate data.
B. the drivers for the photo frame may be incompatible and crash the user's PC.
C. the employee may bring inappropriate photographs into the office.
D. the photo frame could be infected with malware.

D is the correct answer.

Justification:

A. Although any storage device could be used to steal data, the damage caused by malware could be widespread and severe for the enterprise, which is the more significant risk.
B. Although device drivers may be incompatible and crash the user's PC, the damage caused by malware could be widespread and severe for the enterprise.
C. Although inappropriate content could result, the damage caused by malware could be widespread and severe for the enterprise.
D. Any storage device can be a vehicle for infecting other computers with malware. There are several examples where it has been discovered that some devices are infected in the factory during the manufacturing process and controls should exist to prohibit employees from connecting any storage media devices to their company-issued PCs.

A5-18 An organization discovers that the computer of the chief financial officer has been infected with malware that includes a keystroke logger and a rootkit. The **FIRST** action to take would be to:

A. Contact the appropriate law enforcement authorities to begin an investigation.
B. Immediately ensure that no additional data are compromised.
C. Disconnect the PC from the network.
D. Update the antivirus signature on the pc to ensure that the malware or virus is detected and removed.

C is the correct answer.

Justification:

A. Although contacting law enforcement may be needed, the first step would be to halt data flow by disconnecting the computer from the network.
B. The first step is to disconnect the computer from the network thus ensuring that no additional data are compromised. and then, using proper forensic techniques, capture the information stored in temporary files, network connection information, programs loaded into memory and other information on the machine.
C. The most important task is to prevent further data compromise and preserve evidence by disconnecting the computer from the network.
D. Preserve the machine in a forensically sound condition and do not make any changes to it except to disconnect it from the network. Otherwise evidence would be destroyed by powering off the PC or updating the software on the PC. Information stored in temporary files, network connection information, programs loaded into memory, and other information may be lost.

A5-19 The IS auditor is reviewing findings from a prior IS audit of a hospital. One finding indicates that the organization was using email to communicate sensitive patient issues. The IT manager indicates that to address this finding, the organization has implemented digital signatures for all email users. What should the IS auditor's response be?

A. Digital signatures are not adequate to protect confidentiality.
B. Digital signatures are adequate to protect confidentiality.
C. The IS auditor should gather more information about the specific implementation.
D. The IS auditor should recommend implementation of digital watermarking for secure email.

A is the correct answer.

Justification:

A. Digital signatures are designed to provide authentication and nonrepudiation for email and other transmissions but are not adequate for confidentiality. This implementation is not adequate to address the prior-year's finding.
B. Digital signatures do not encrypt message contents, which means that an attacker who intercepts a message can read the message because the data are in plaintext.
C. Although gathering additional information is always a good step before drawing a conclusion on a finding, in this case the implemented solution simply does not provide confidentiality.
D. Digital watermarking is used to protect intellectual property rights for documents rather than to protect the confidentiality of email.

A5-20 Which of the following line media would provide the **BEST** security for a telecommunication network?

A. Broadband network digital transmission
B. Baseband network
C. Dialup
D. Dedicated lines

D is the correct answer.

Justification:

A. The secure use of broadband communications is subject to whether the network is shared with other users, the data are encrypted and the risk of network interruption.
B. A baseband network is one that is usually shared with many other users and requires encryption of traffic but still may allow some traffic analysis by an attacker.
C. A dial-up line is fairly secure because it is a private connection, but it is too slow to be considered for most commercial applications today.
D. Dedicated lines are set apart for a particular user or organization. Because there is no sharing of lines or intermediate entry points, the risk of interception or disruption of telecommunications messages is lower.

A5-21 To ensure that an organization is complying with privacy requirements, an IS auditor should **FIRST** review:

A. the IT infrastructure.
B. organizational policies, standards and procedures.
C. legal and regulatory requirements.
D. adherence to organizational policies, standards and procedures.

C is the correct answer.

Justification:

A. To comply with requirements, the IS auditor must first know what the requirements are. They can vary from one jurisdiction to another. The IT infrastructure is related to the implementation of the requirements.
B. The policies of the organization are subject to the legal requirements and should be checked for compliance after the legal requirements are reviewed.
C. To ensure that the organization is complying with privacy issues, an IS auditor should address legal and regulatory requirements first. To comply with legal and regulatory requirements, organizations need to adopt the appropriate infrastructure. After understanding the legal and regulatory requirements, an IS auditor should evaluate organizational policies, standards and procedures to determine whether they adequately address the privacy requirements, and then review the adherence to these specific policies, standards and procedures.
D. Checking for compliance is only done after the IS auditor is assured that the policies, standards and procedures are aligned with the legal requirements.

A5-22 A human resources company offers wireless Internet access to its guests, after authenticating with a generic user ID and password. The generic ID and password are requested from the reception desk. Which of the following controls **BEST** addresses the situation?

A. The password for the wireless network is changed on a weekly basis.
B. A stateful inspection firewall is used between the public wireless and company networks.
C. The public wireless network is physically segregated from the company network.
D. An intrusion detection system is deployed within the wireless network.

C is the correct answer.

Justification:

A. Changing the password for the wireless network does not secure against unauthorized access to the company network, especially because a guest could gain access to the wireless local area network at any time prior to the weekly password change interval.
B. A stateful inspection firewall will screen all packets from the wireless network into the company network; however, the configuration of the firewall would need to be audited and firewall compromises, although unlikely, are possible.
C. Keeping the wireless network physically separate from the company network is the best way to secure the company network from intrusion.
D. An intrusion detection system will detect intrusions but will not prevent unauthorized individuals from accessing the network.

A5-23 When reviewing the implementation of a local area network, an IS auditor should **FIRST** review the:

A. node list.
B. acceptance test report.
C. network diagram.
D. users list.

C is the correct answer.

Justification:

A. Verification of nodes from the node list would follow the review of the network diagram.
B. The review of the acceptance test report would follow the verification of nodes from the node list.
C. To properly review a local area network implementation, an IS auditor should first verify the network diagram to identify risk or single points of failure.
D. The users list would be reviewed after the acceptance test report.

A5-24 An IS auditor discovers that the configuration settings for password controls are more stringent for business users than for IT developers. Which of the following is the **BEST** action for the IS auditor to take?

A. Determine whether this is a policy violation and document it.
B. Document the observation as an exception.
C. Recommend that all password configuration settings be identical.
D. Recommend that logs of IT developer access are reviewed periodically.

A is the correct answer.

Justification:

A. If the policy documents the purpose and approval for different procedures, then an IS auditor only needs to document observations and tests as to whether the procedures are followed.
B. This condition would not be considered an exception if procedures are followed according to approved policies.
C. There may be valid reasons for these settings to be different; therefore, the auditor would not normally recommend changes before researching company policies and procedures.
D. While reviewing logs may be a good compensating control, the more important course of action would be to determine if policies are being followed.

A5-25 An organization is developing a new web-based application to process orders from customers. Which of the following security measures should be taken to protect this application from hackers?

A. Ensure that ports 80 and 443 are blocked at the firewall.
B. Inspect file and access permissions on all servers to ensure that all files have read-only access.
C. Perform a web application security review.
D. Make sure that only the IP addresses of existing customers are allowed through the firewall.

C is the correct answer.

Justification:

A. Port 80 must be open for a web application to work and port 443 for a Secured Hypertext Transmission Protocol to operate.
B. For customer orders to be placed, some data must be saved to the server. No customer orders could be placed on a read-only server.
C. Performing a web application security review is a necessary effort that would uncover security vulnerabilities that could be exploited by hackers.
D. Restricting IP addresses might be appropriate for some types of web applications but is not the best solution because a new customer could not place an order until the firewall rules were changed to allow the customer to connect.

A5-26 Which of the following types of penetration tests simulates a real attack and is used to test incident handling and response capability of the target?

A. Blind testing
B. Targeted testing
C. Double-blind testing
D. External testing

C is the correct answer.

Justification:

A. Blind testing is also known as black-box testing. This refers to a test where the penetration tester is not given any information and is forced to rely on publicly available information. This test simulates a real attack, except that the target organization is aware of the test being conducted.
B. Targeted testing is also known as white-box testing. This refers to a test where the penetration tester is provided with information and the target organization is also aware of the testing activities. In some cases, the tester is also provided with a limited-privilege account to be used as a starting point.
C. Double-blind testing is also known as zero-knowledge testing. This refers to a test where the penetration tester is not given any information and the target organization is not given any warning—both parties are "blind" to the test. This is the best scenario for testing response capability because the target will react as if the attack were real.
D. External testing refers to a test where an external penetration tester launches attacks on the target's network perimeter from outside the target network (typically from the Internet).

A5-27 An organization has requested that an IS auditor provide a recommendation to enhance the security and reliability of its Voice-over Internet Protocol (VoIP) system and data traffic. Which of the following would meet this objective?

A. VoIP infrastructure needs to be segregated using virtual local area networks.
B. Buffers need to be introduced at the VoIP endpoints.
C. Ensure that end-to-end encryption is enabled in the VoIP system.
D. Ensure that emergency backup power is available for all parts of the VoIP infrastructure.

A is the correct answer.

Justification:

A. Segregating the Voice-over Internet Protocol (VoIP) traffic using virtual local area networks (VLANs) would best protect the VoIP infrastructure from network-based attacks, potential eavesdropping and network traffic issues (which would help to ensure uptime).
B. The use of packet buffers at VoIP endpoints is a method to maintain call quality, not a security method.
C. Encryption is used when VoIP calls use the Internet (not the local LAN) for transport because the assumption is that the physical security of the building as well as the Ethernet switch and VLAN security is adequate.
D. The design of the network and the proper implementation of VLANs are more critical than ensuring that all devices are protected by emergency power.

A5-28 During a review of intrusion detection logs, an IS auditor notices traffic coming from the Internet, which appears to originate from the internal IP address of the company payroll server. Which of the following malicious activities would **MOST** likely cause this type of result?

A. A denial-of-service attack
B. Spoofing
C. Port scanning
D. A man-in-the-middle attack

B is the correct answer.

Justification:

A. A denial-of-service attack is designed to limit the availability of a resource and is characterized by a high number of requests that require response from the resource (usually a web site). The target spends so many resources responding to the attack requests that legitimate requests are not serviced. These attacks are most commonly launched from networks of compromised computers (botnets) and may involve attacks from multiple computers at once.

B. Spoofing is a form of impersonation where one computer tries to take on the identity of another computer. When an attack originates from the external network but uses an internal network address, the attacker is most likely trying to bypass firewalls and other network security controls by impersonating (or spoofing) the payroll server's internal network address. By impersonating the payroll server, the attacker may be able to access sensitive internal resources.

C. Port scanning is a reconnaissance technique that is designed to gather information about a target before a more active attack. Port scanning might be used to determine the internal address of the payroll server but would not normally create a log entry that indicated external traffic from an internal server address.

D. A man-in-the-middle attack is a form of active eavesdropping where the attacker intercepts a computerized conversation between two parties and then allows the conversation to continue by relaying the appropriate data to both parties, while simultaneously monitoring the same data passing through the attacker's conduit. This type of attack would not register as an attack originating from the payroll server, but instead it might be designed to hijack an authorized connection between a workstation and the payroll server.

A5-29 An IS auditor is reviewing an organization's information security policy, which requires encryption of all data placed on universal serial bus (USB) drives. The policy also requires that a specific encryption algorithm be used. Which of the following algorithms would provide the greatest assurance that data placed on USB drives is protected from unauthorized disclosure?

A. Data Encryption Standard
B. Message digest 5
C. Advanced Encryption Standard
D. Secure Shell

C is the correct answer.

Justification:

A. Data Encryption Standard (DES) is susceptible to brute force attacks and has been broken publicly; therefore, it does not provide assurance that data encrypted using DES will be protected from unauthorized disclosure.

B. Message digest 5 (MD5) is an algorithm used to generate a one-way hash of data (a fixed- length value) to test and verify data integrity. MD5 does not encrypt data but puts data through a mathematical process that cannot be reversed. As a result, MD5 could not be used to encrypt data on a universal serial bus (USB) drive.

C. Advanced Encryption Standard (AES) provides the strongest encryption of all of the choices listed and would provide the greatest assurance that data are protected. Recovering data encrypted with AES is considered computationally infeasible and so AES is the best choice for encrypting sensitive data.

D. Secure Shell (SSH) is a protocol that is used to establish a secure, encrypted, command-line shell session, typically for remote logon. Although SSH encrypts data transmitted during a session, SSH cannot encrypt data at rest, including data on USB drives. As a result, SSH is not appropriate for this scenario.

A5-30 During an IS audit of a global organization, the IS auditor discovers that the organization uses Voice-over Internet Protocol over the Internet as the sole means of voice connectivity among all offices. Which of the following presents the **MOST** significant risk for the organization's VoIP infrastructure?

A. Network equipment failure
B. Distributed denial-of-service attack
C. Premium-rate fraud (toll fraud)
D. Social engineering attack

B is the correct answer.

Justification:

A. The use of Voice-over Internet Protocol does not introduce any unique risk with respect to equipment failure, and redundancy can be used to address network failure.

B. A distributed denial-of-service (DDoS) attack would potentially disrupt the organization's ability to communicate among its offices and have the highest impact. In a traditional voice network, a DDoS attack would only affect the data network, not voice communications.

C. Toll fraud occurs when someone compromises the phone system and makes unauthorized long-distance calls. While toll fraud may cost the business money, the more severe risk would be the disruption of service.

D. Social engineering, which involves gathering sensitive information to launch an attack, can be exercised over any kind of telephony.

A5-31 Which of the following is the **MOST** effective control for restricting access to unauthorized Internet sites in an organization?

A. Routing outbound Internet traffic through a content-filtering proxy server
B. Routing inbound Internet traffic through a reverse proxy server
C. Implementing a firewall with appropriate access rules
D. Deploying client software utilities that block inappropriate content

A is the correct answer.

Justification:

A. A content-filtering proxy server will effectively monitor user access to Internet sites and block access to unauthorized web sites.
B. When a client web browser makes a request to an Internet site, those requests are outbound from the corporate network. A reverse proxy server is used to allow secure remote connection to a corporate site, not to control employee web access.
C. A firewall exists to block unauthorized inbound and outbound network traffic. Some firewalls can be used to block or allow access to certain sites, but the term firewall is generic—there are many types of firewalls, and this is not the best answer.
D. While client software utilities do exist to block inappropriate content, installing and maintaining additional software on a large number of PCs is less effective than controlling the access from a single, centralized proxy server.

A5-32 An internal audit function is reviewing an internally developed common gateway interface script for a web application. The IS auditor discovers that the script was not reviewed and tested by the quality control function. Which of the following types of risk is of **GREATEST** concern?

A. System unavailability
B. Exposure to malware
C. Unauthorized access
D. System integrity

C is the correct answer.

Justification:

A. While untested common gateway interfaces (CGIs) can cause the end-user web application to be compromised, this is not likely to make the system unavailable to other users.
B. Untested CGI scripts do not inherently lead to malware exposures.
C. Untested CGIs can have security weaknesses that allow unauthorized access to private systems because CGIs are typically executed on publicly available Internet servers.
D. While untested CGIs can cause the end-user web application to be compromised, this is not likely to significantly impact system integrity.

A5-33 An IS auditor is conducting a postimplementation review of an enterprise's network. Which of the following findings would be of **MOST** concern?

A. Wireless mobile devices are not password-protected.
B. Default passwords are not changed when installing network devices.
C. An outbound web proxy does not exist.
D. All communication links do not use encryption.

B is the correct answer.

Justification:

A. While mobile devices that are not password-protected would be a risk, it would not be as significant as unsecured network devices.
B. The most significant risk in this case would be if the factory default passwords are not changed on critical network equipment. This could allow anyone to change the configurations of network equipment.
C. The use of a web proxy is a good practice but may not be required depending on the enterprise.
D. Encryption is a good control for data security but is not appropriate to use for all communication links due to cost and complexity.

A5-34 An IS auditor is reviewing a third-party agreement for a new cloud-based accounting service provider. Which of the following considerations is the **MOST** important with regard to the privacy of the accounting data?

A. Data retention, backup and recovery
B. Return or destruction of information
C. Network and intrusion detection
D. A patch management process

B is the correct answer.

Justification:

A. Data retention, backup and recovery are important controls; however, they do not guarantee data privacy.
B. When reviewing a third-party agreement, the most important consideration with regard to the privacy of the data is the clause concerning the return or secure destruction of information at the end of the contract.
C. Network and intrusion detection are helpful when securing the data, but on their own, they do not guarantee data privacy stored at a third-party provider.
D. A patch management process helps secure servers and may prohibit unauthorized disclosure of data; however, it does not affect the privacy of the data.

A5-35 Which of the following is the **MOST** effective control when granting temporary access to vendors?

A. Vendor access corresponds to the service level agreement.
B User accounts are created with expiration dates and are based on services provided.
C. Administrator access is provided for a limited period.
D. User IDs are deleted when the work is completed.

B is the correct answer.

Justification:

A. The service level agreement may have a provision for providing access, but this is not a control; it would merely define the need for access.
B. The most effective control is to ensure that the granting of temporary access is based on services to be provided and that there is an expiration date (automated is best) associated with each unique ID. The use of an identity management system enforces temporary and permanent access for users, at the same time ensuring proper accounting of their activities.
C. Vendors may require administrator access for a limited period during the time of service. However, it is important to ensure that the level of access granted is set according to least privilege and that access during this period is monitored.
D. Deleting these user IDs after the work is completed is necessary, but if not automated, the deletion could be overlooked. The access should only be granted at the level of work required.

A5-36 During a logical access controls review, an IS auditor observes that user accounts are shared. The **GREATEST** risk resulting from this situation is that:

A. an unauthorized user may use the ID to gain access.
B. user access management is time consuming.
C. passwords are easily guessed.
D. user accountability may not be established.

D is the correct answer.

Justification:

A. The ability of unauthorized users to use a shared ID is more likely than of an individual ID—but the misuse of another person's ID is always a risk.
B. Using shared IDs would not pose an increased risk due to work effort required for managing access.
C. Shared user IDs do not necessarily have easily guessed passwords.
D. The use of a user ID by more than one individual precludes knowing who, in fact, used that ID to access a system; therefore, it is impossible to hold anyone accountable.

A5-37 An IS auditor is assessing a biometric system used to protect physical access to a data center containing regulated data. Which of the following observations is the **GREATEST** concern to the auditor?

A. Administrative access to the biometric scanners or the access control system is permitted over a virtual private network.
B. Biometric scanners are not installed in restricted areas.
C. Data transmitted between the biometric scanners and the access control system do not use a securely encrypted tunnel.
D. Biometric system risk analysis was last conducted three years ago.

C is the correct answer.

Justification:

A. Generally, virtual private network software provides a secure tunnel so that remote administration functions can be performed. This is not a concern.
B. Biometric scanners are best located in restricted areas to prevent tampering, but video surveillance is an acceptable mitigating control. The greatest concern is lack of a securely encrypted tunnel between the scanners and the access control system.
C. Data transmitted between the biometric scanners and the access control system should use a securely encrypted tunnel to protect the confidentially of the biometric data.
D. The biometric risk analysis should be reperformed periodically, but an analysis performed three years ago is not necessarily a cause for concern.

A5-38 When auditing a role-based access control system, the IS auditor noticed that some IT security employees have system administrator privileges on some servers, which allows them to modify or delete transaction logs. Which would be the **BEST** recommendation that the IS auditor should make?

A. Ensure that these employees are adequately supervised.
B. Ensure that backups of the transaction logs are retained.
C. Implement controls to detect the changes.
D. Write transaction logs in real time to Write Once and Read Many drives.

D is the correct answer.

Justification:

A. IT security employees cannot be supervised in the traditional sense unless the supervisor were to monitor each keystroke entered on a workstation, which is obviously not a realistic option.
B. Retaining backups of the transaction logs does not prevent the files from unauthorized modification prior to backup.
C. The log files themselves are the main evidence that an unauthorized change was made, which is a sufficient detective control. Protecting the log files from modification requires preventive controls such as securely writing the logs.
D. Allowing IT security employees access to transaction logs is often unavoidable because having system administrator privileges is required for them to do their job. The best control in this case, to avoid unauthorized modifications of transaction logs, is to write the transaction logs to WORM drive media in real time. It is important to note that simply backing up the transaction logs to tape is not adequate because data could be modified prior (typically at night) to the daily backup job execution.

A5-39 During an IS audit of a bank, the IS auditor is assessing whether the enterprise properly manages staff member access to the operating system. The IS auditor should determine whether the enterprise performs:

A. periodic review of user activity logs.
B. verification of user authorization at the field level.
C. review of data communication access activity logs.
D. periodic review of changing data files.

A is the correct answer.

Justification:

A. General operating system access control functions include logging user activities, events, etc. Reviewing these logs may identify users performing activities that should not have been permitted.
B. Verification of user authorization at the field level is a database- and/or an application-level access control function and not applicable to an operating system.
C. Review of data communication access activity logs is a network control feature.
D. Periodic review of changing data files is related to a change control process.

A5-40 An IS auditor performing an audit of the newly installed Voice-over Internet Protocol system was inspecting the wiring closets on each floor of a building. What would be the **GREATEST** concern?

A. The local area network (LAN) switches are not connected to uninterruptible power supply units.
B. Network cabling is disorganized and not properly labeled.
C. The telephones are using the same cable used for LAN connections.
D. The wiring closet also contains power lines and breaker panels.

A is the correct answer.

Justification:

A. Voice-over Internet Protocol (VoIP) telephone systems use standard network cabling and typically each telephone gets power over the network cable (power over Ethernet) from the wiring closet where the network switch is installed. If the local area network switches do not have backup power, the phones will lose power if there is a utility interruption and potentially not be able to make emergency calls.
B. While improper cabling can create reliability issues, the more critical issue in this case would be the lack of power protection.
C. An advantage of VoIP telephone systems is that they use the same cable types and even network switches as standard PC network connections. Therefore, this would not be a concern.
D. As long as the power and telephone equipment are separated, this would not be a significant risk.

A5-41 When reviewing an organization's logical access security to its remote systems, which of the following would be of **GREATEST** concern to an IS auditor?

A. Passwords are shared.
B. Unencrypted passwords are used.
C. Redundant logon IDs exist.
D. Third-party users possess administrator access.

B is the correct answer.

Justification:

A. The passwords should not be shared, but this is less important than ensuring that the password files are encrypted.
B. When evaluating the technical aspects of logical security, unencrypted passwords represent the greatest risk because it would be assumed that remote access would be over an untrusted network where passwords could be discovered.
C. Checking for the redundancy of logon IDs is essential but is less important than ensuring that the passwords are encrypted.
D. There may be business requirements such as the use of contractors that requires them to have system access, so this may not be a concern.

A5-42 During an IS risk assessment of a health care organization regarding protected health care information (PHI), an IS auditor interviews IS management. Which of the following findings from the interviews would be of **MOST** concern to the IS auditor?

A. The organization does not encrypt all of its outgoing email messages.
B. Staff have to type "[PHI]" in the subject field of email messages to be encrypted.
C. An individual's computer screen saver function is disabled.
D. Server configuration requires the user to change the password annually.

B is the correct answer.

Justification:

A. Encrypting all outgoing email is expensive and is not common business practice.
B. There will always be human-error risk that staff members forget to type certain words in the subject field. The organization should have automated encryption set up for outgoing email for employees working with protected health care information (PHI) to protect sensitive information.
C. Disabling the screen saver function increases the risk that sensitive data can be exposed to other employees; however, the risk is not as great as exposing the data to unauthorized individuals outside the organization.
D. While changing the password annually is a concern, the risk is not as great as exposing the data to unauthorized individuals outside the organization.

A5-43 Which of the following is the responsibility of information asset owners?

A. Implementation of information security within applications
B. Assignment of criticality levels to data
C. Implementation of access rules to data and programs
D. Provision of physical and logical security for data

B is the correct answer.

Justification:

A. Implementation of information security within an application is the responsibility of the data custodians based on the requirements set by the data owner.
B. It is the responsibility of owners to define the criticality (and sensitivity) levels of information assets.
C. Implementation of access rules is a responsibility of data custodians based on the requirements set by the data owner.
D. Provision of physical and logical security for data is the responsibility of the security administrator.

A5-44 An IS auditor reviewing a network log discovers that an employee ran elevated commands on their PC by invoking the task scheduler to launch restricted applications. This is an example what type of attack?

A. A race condition
B. A privilege escalation
C. A buffer overflow
D. An impersonation

B is the correct answer.

Justification:

A. A race condition exploit involves the timing of two events and an action that causes one event to happen later than expected. The scenario given is not an example of a race condition exploit.
B. A privilege escalation is a type of attack where higher-level system authority is obtained by various methods. In this example, the task scheduler service runs with administrator permissions, and a security flaw allows programs launched by the scheduler to run at the same permission level.
C. Buffer overflows involve applications of actions that take advantage of a defect in the way an application or system uses memory. By overloading the memory storage mechanism, the system will perform in unexpected ways. The scenario given is not an example of a buffer overflow exploit.
D. Impersonation attacks involve an error in the identification of a privileged user. The scenario given is not an example of this exploit.

A5-45 An IS auditor is reviewing an organization to ensure that evidence related to a data breach case is preserved. Which of the following choices would be of **MOST** concern to the IS auditor?

A. End users are not aware of incident reporting procedures.
B. Log servers are not on a separate network.
C. Backups are not performed consistently.
D. There is no chain of custody policy.

D is the correct answer.

Justification:

A. End users should be made aware of incident reporting procedures, but this is not likely to affect data integrity related to the breach. The IS auditor would be more concerned that the organization's policy exists and provides for proper evidence handling.
B. Having log servers segregated on a separate network might be a good idea because ensuring the integrity of log server data is important. However, it is more critical to ensure that the chain of custody policy is in place.
C. While not having valid backups would be a concern, the more important concern would be a lack of a chain of custody policy. Data breach evidence is not normally retrieved from backups.
D. Organizations should have a policy in place that directs employees to follow certain procedures when collecting evidence that may be used in a court of law. Chain of custody involves documentation of how digital evidence is acquired, processed, handled, stored and protected, and who handled the evidence and why. If there is no policy in place, it is unlikely that employees will ensure that the chain of custody is maintained during any data breach investigation.

A5-46 An IS auditor is reviewing access controls for a manufacturing organization. During the review, the IS auditor discovers that data owners have the ability to change access controls for a low-risk application. The **BEST** course of action for the IS auditor is to:

A. recommend that mandatory access control be implemented.
B. report this as a finding to upper management
C. report this to the data owners to determine whether it is an exception.
D. not report this issue because discretionary access controls are in place.

D is the correct answer.

Justification:

A. Recommending mandatory access control is not correct because it is more appropriate for data owners to have discretionary access controls (DAC) in a low-risk application.
B. The use of DAC may not be an exception and, until confirmed, should not be reported as an issue.
C. While an IS auditor may consult with data owners regarding whether this access is allowed normally, the IS auditor should not rely on the auditee to determine whether this is an issue.
D. DAC allows data owners to modify access, which is a normal procedure and is a characteristic of DAC.

A5-47 Electromagnetic emissions from a terminal represent a risk because they:

A. could damage or erase nearby storage media.
B. can disrupt processor functions.
C. could have adverse health effects on personnel.
D. can be detected and displayed.

D is the correct answer.

Justification:
A. While a strong magnetic field can erase certain storage media, normally terminals are designed to limit these emissions; therefore, this is not normally a concern.
B. Electromagnetic emissions should not cause disruption of central processing units.
C. Most electromagnetic emissions are low level and do not pose a significant health risk.
D. Emissions can be detected by sophisticated equipment and displayed, thus giving unauthorized persons access to data. TEMPEST is a term referring to the investigation and study of compromising emanations of unintentional intelligence-bearing signals that, if intercepted and analyzed, may reveal their contents.

A5-48 Security administration procedures require read-only access to:

A. access control tables.
B. security log files.
C. logging options.
D. user profiles.

B is the correct answer.

Justification:
A. Security administration procedures require write access to access control tables to manage and update the privileges according to authorized business requirements.
B. Security administration procedures require read-only access to security log files to ensure that, once generated, the logs are not modified. Logs provide evidence and track suspicious transactions and activities.
C. Logging options require write access to allow the administrator to update the way the transactions and user activities are monitored, captured, stored, processed and reported.
D. The security administrator is often responsible for user-facing issues such as managing user roles, profiles and settings. This requires the administrator to have more than read-only access.

A5-49 With the help of a security officer, granting access to data is the responsibility of:

A. data owners.
B. programmers.
C. system analysts.
D. librarians.

A is the correct answer.

Justification:

A. Data owners are responsible for the access to and use of data. Written authorization for users to gain access to computerized information should be provided by the data owners. Security administration with the owners' approval sets up access rules stipulating which users or group of users are authorized to access data or files and the level of authorized access (e.g., read or update).

B. Programmers will develop the access control software that will regulate the ways that users can access the data (update, read, delete, etc.), but the programmers do not have responsibility for determining who gets access to data.

C. Systems analysts work with the owners and programmers to design access controls according to the rules set by the owners.

D. The librarians enforce the access control procedures they have been given but do not determine who gets access.

A5-50 The **FIRST** step in data classification is to:

A. establish ownership.
B. perform a criticality analysis.
C. define access rules.
D. create a data dictionary.

A is the correct answer.

Justification:

A. Data classification is necessary to define access rules based on a need-to-do and need-to- know basis. The data owner is responsible for defining the access rules; therefore, establishing ownership is the first step in data classification.

B. A criticality analysis is required to determine the appropriate levels of protection of data, according to the data classification.

C. Access rules are set up dependent on the data classification.

D. Input for a data dictionary is prepared from the results of the data classification process.

A5-51 During the review of a biometrics system operation, an IS auditor should **FIRST** review the stage of:

A. enrollment.
B. identification.
C. verification.
D. storage.

A is the correct answer.

Justification:

A. The users of a biometric device must first be enrolled in the device.
B. The device captures a physical or behavioral image of the human, identifies the unique features and uses an algorithm to convert them into a string of numbers stored as a template to be used in the matching processes.
C. A user applying for access will be verified against the stored enrolled value.
D. The biometric stores sensitive personal information, so the storage must be secure.

A5-52 A hacker could obtain passwords without the use of computer tools or programs through the technique of:

A. social engineering.
B. sniffers.
C. back doors.
D. Trojan horses.

A is the correct answer.

Justification:

A. Social engineering is based on the divulgence of private information through dialogues, interviews, inquiries, etc., in which a user may be indiscreet regarding their or someone else's personal data.
B. A sniffer is a computer tool to monitor the traffic in networks.
C. Back doors are computer programs left by hackers to exploit vulnerabilities.
D. Trojan horses are computer programs that pretend to supplant a real program; thus, the functionality of the program is not authorized and is usually malicious in nature.

A5-53 The reliability of an application system's audit trail may be questionable if:

A. user IDs are recorded in the audit trail.
B. the security administrator has read-only rights to the audit file.
C. date and time stamps are recorded when an action occurs.
D. users can amend audit trail records when correcting system errors.

D is the correct answer.

Justification:

A. An audit trail must record the identity of the person or process involved in the logged activity to establish accountability.
B. Restricting the administrator to read-only access will protect the audit file from alteration.
C. Data and time stamps should be recorded in the logs to enable the reconstruction and correlation of events on multiple systems.
D. An audit trail is not effective if the details in it can be amended.

A5-54 While conducting an audit, an IS auditor detects the presence of a virus. What should be the IS auditor's **NEXT** step?

A. Observe the response mechanism.
B. Clear the virus from the network.
C. Inform appropriate personnel immediately.
D. Ensure deletion of the virus.

C is the correct answer.

Justification:
A. Observing the response mechanism should be done after informing appropriate personnel. This will enable an IS auditor to examine the actual workability and effectiveness of the response system.
B. The IS auditor is neither authorized nor capable in most cases of removing the virus from the network.
C. The first thing an IS auditor should do after detecting the virus is to alert the organization to its presence, then wait for their response.
D. An IS auditor should not make changes to the system being audited; ensuring the deletion of the virus is a management responsibility.

A5-55 The implementation of access controls **FIRST** requires:

A. a classification of IS resources.
B. the labeling of IS resources.
C. the creation of an access control list.
D. an inventory of IS resources.

D is the correct answer.

Justification:
A. The first step in implementing access controls is an inventory of IS resources, which is the basis for classification.
B. Labeling resources cannot be done without first determining the resources' classifications.
C. The access control list would not be done without a meaningful classification of resources.
D. The first step in implementing access controls is an inventory of IS resources, which is the basis for establishing ownership and classification.

A5-56 Which of the following is an example of the defense in-depth security principle?

A. Using two firewalls to consecutively check the incoming network traffic
B. Using a firewall as well as logical access controls on the hosts to control incoming network traffic
C. Lack of physical signs on the outside of a computer center building
D. Using two firewalls in parallel to check different types of incoming traffic

B is the correct answer.

Justification:
A. Use of two firewalls would not represent an effective defense in-depth strategy because the same attack could circumvent both devices. By using two different products, the probability of both products having the same vulnerabilities is diminished.
B. Defense in-depth means using different security mechanisms that back each other up. When network traffic passes the firewall unintentionally, the logical access controls form a second line of defense.
C. Having no physical signs on the outside of a computer center building is a single security measure known as security by obscurity.
D. Using two firewalls in parallel to check different types of incoming traffic provides redundancy but is only a single security mechanism and, therefore, no different than having a single firewall checking all traffic.

A5-57 Which of the following would be the **BEST** access control procedure?

A. The data owner formally authorizes access and an administrator implements the user authorization tables.
B. Authorized staff implements the user authorization tables and the data owner approves them.
C. The data owner and an IS manager jointly create and update the user authorization tables.
D. The data owner creates and updates the user authorization tables.

A is the correct answer.

Justification:

A. The data owner holds the privilege and responsibility for formally establishing the access rights. An IS administrator should then implement or update user authorization tables at the direction of the owner.
B. The owner sets the rules and conditions for access. It is best to obtain approval before implementing the tables.
C. The data owner may consult with the IS manager to set out access control rules, but the responsibility for appropriate access remains with the data owner. The IT department should set up the access control tables at the direction of the owner.
D. The data owner would not usually manage updates to the authorization tables.

A5-58 Which of the following would **MOST** effectively reduce social engineering incidents?

A. Security awareness training
B. Increased physical security measures
C. Email monitoring policy
D. Intrusion detection systems

A is the correct answer.

Justification:

A. Social engineering exploits human nature and weaknesses to obtain information and access privileges. By increasing employee awareness of security issues, it is possible to reduce the number of successful social engineering incidents.
B. In most cases, social engineering incidents do not require the physical presence of the intruder. Therefore, increased physical security measures would not prevent the incident.
C. An email monitoring policy informs users that all email in the organization is subject to monitoring; it does not protect the users from potential security incidents and intruders.
D. Intrusion detection systems are used to detect irregular or abnormal traffic patterns.

A5-59 An information security policy stating that "the display of passwords must be masked or suppressed" addresses which of the following attack methods?

A. Piggybacking
B. Dumpster diving
C. Shoulder surfing
D. Impersonation

C is the correct answer.

Justification:

A. Piggybacking refers to unauthorized persons following, either physically or virtually, authorized persons into restricted areas. Masking the display of passwords would not prevent someone from tailgating an authorized person.
B. This policy only refers to "the display of passwords," not dumpster diving (looking through an organization's trash for valuable information).
C. If a password is displayed on a monitor, any person or camera nearby could look over the shoulder of the user to obtain the password.
D. Impersonation refers to someone acting as an employee in an attempt to retrieve desired information.

A5-60 To ensure compliance with a security policy requiring that passwords be a combination of letters and numbers, an IS auditor should recommend that:

A. the company policy be changed.
B. passwords are periodically changed.
C. an automated password management tool be used.
D. security awareness training is delivered.

C is the correct answer.

Justification:

A. The policy is appropriate and does not require change. Changing the policy would not ensure compliance.
B. Having a requirement to periodically change passwords is good practice and should be in the password policy.
C. The use of an automated password management tool is a preventive control measure. The software would prevent repetition (semantic) and would enforce syntactic rules, thus making the passwords robust. It would also provide a method for ensuring frequent changes and would prevent the same user from reusing his/her old password for a designated period of time.
D. Security awareness training would not enforce compliance.

A5-61 An IS auditor reviewing digital rights management applications should expect to find an extensive use for which of the following technologies?

A. Digitalized signatures
B. Hashing
C. Parsing
D. Steganography

D is the correct answer.

Justification:

A. Digitalized signatures are the scans of a signature (not the same as a digital signature) and not related to digital rights management.
B. Hashing creates a message hash or digest, which is used to ensure the integrity of the message; it is usually considered a part of cryptography.
C. Parsing is the process of splitting up a continuous stream of characters for analytical purposes and is widely applied in the design of programming languages or in data entry editing.
D. Steganography is a technique for concealing the existence of messages or information within another message. An increasingly important steganographical technique is digital watermarking, which hides data within data (e.g., by encoding rights information in a picture or music file without altering the picture or music's perceivable aesthetic qualities).

A5-62 The information security policy that states "each individual must have his/her badge read at every controlled door" addresses which of the following attack methods?

A. Piggybacking
B. Shoulder surfing
C. Dumpster diving
D. Impersonation

A is the correct answer.

Justification:

A. Piggybacking refers to unauthorized persons following authorized persons, either physically or virtually, into restricted areas. This policy addresses the polite behavior problem of holding doors open for a stranger. If every employee must have their badge read at every controlled door, no unauthorized person could enter the sensitive area.
B. Shoulder surfing (looking over the shoulder of a person to view sensitive information on a screen or desk) would not be prevented by the implementation of this policy.
C. Dumpster diving, looking through an organization's trash for valuable information, could be done outside the company's physical perimeter; therefore, this policy would not address this attack method.
D. Impersonation refers to a social engineer acting as an employee, trying to retrieve the desired information. Some forms of social engineering attacks could join an impersonation attack and piggybacking, but this information security policy does not address the impersonation attack.

A5-63 Which of the following presents an inherent risk with no distinct identifiable preventive controls?

A. Piggybacking
B. Viruses
C. Data diddling
D. Unauthorized application shutdown

C is the correct answer.

Justification:

A. Piggybacking is the act of following an authorized person through a secured door and can be prevented by the use of deadman doors. Logical piggybacking is an attempt to gain access through someone who has the rights (e.g., electronically attaching to an authorized telecommunication link to possibly intercept transmissions). This could be prevented by encrypting the message.

B. Viruses are malicious program code inserted into another executable code that can self-replicate and spread from computer to computer via sharing of computer disks, transfer of logic over telecommunication lines or direct contact with an infected machine. Antivirus software can be used to protect the computer against viruses.

C. Data diddling involves changing data before they are entered into the computer. It is one of the most common abuses because it requires limited technical knowledge and occurs before computer security can protect the data. There are only compensating controls for data diddling.

D. The shutdown of an application can be initiated through terminals or microcomputers connected directly (online) or indirectly (dial-up line) to the computer. Only individuals knowing the high-level logon ID and password can initiate the shutdown process, which is effective if there are proper access controls.

A5-64 The **MOST** important difference between hashing and encryption is that hashing:

A. is irreversible.
B. output is the same length as the original message.
C. is concerned with integrity and security.
D. is the same at the sending and receiving end.

A is the correct answer.

Justification:

A. Hashing works one way—by applying a hashing algorithm to a message, a message hash/digest is created. If the same hashing algorithm is applied to the message digest, it will not result in the original message. As such, hashing is irreversible, while encryption is reversible. This is the basic difference between hashing and encryption.

B. Hashing creates a fixed-length output that is usually smaller than the original message, and encryption creates an output that is usually the same length as the original message.

C. Hashing is used to verify the integrity of the message and does not address security. The same hashing algorithm is used at the sending and receiving ends to generate and verify the message hash/digest.

D. Encryption may use different keys or a reverse process at the sending and receiving ends to encrypt and decrypt.

A5-65 Which of the following cryptography options would increase overhead/cost?

A. The encryption is symmetric rather than asymmetric.
B. A long asymmetric encryption key is used.
C. The hash is encrypted rather than the message.
D. A secret key is used.

B is the correct answer.

Justification:
A. An asymmetric algorithm requires more processing time than symmetric algorithms.
B. Computer processing time is increased for longer asymmetric encryption keys, and the increase may be disproportionate. For example, one benchmark showed that doubling the length of an RSA key from 512 bits to 1,024 bits caused the decrypt time to increase nearly six-fold.
C. A hash is usually shorter than the original message; therefore, a smaller overhead is required if the hash is encrypted rather than the message.
D. Use of a secret key, as a symmetric encryption key, is generally small and used for the purpose of encrypting user data.

A5-66 The **MOST** important factor in planning a black box penetration test is:

A. the documentation of the planned testing procedure.
B. a realistic evaluation of the environment architecture to determine scope.
C. knowledge by the management staff of the client organization.
D. scheduling and deciding on the timed length of the test.

C is the correct answer.

Justification:
A. A penetration test should be carefully planned and executed, but the most important factor is proper approvals.
B. In a black box penetration test, the environment is not known to the testing organization.
C. Black box penetration testing assumes no prior knowledge of the infrastructure to be tested. Testers simulate an attack from someone who is unfamiliar with the system. It is important to have management knowledge of the proceedings so that if the test is identified by the monitoring systems, the legality of the actions can be determined quickly.
D. A test must be scheduled so as to minimize the risk of affecting critical operations; however, this is part of working with the management of the organization.

A5-67 An organization allows for the use of universal serial bus drives to transfer operational data between offices. Which of the following is the **GREATEST** risk associated with the use of these devices?

A. Files are not backed up
B. Theft of the devices
C. Use of the devices for personal purposes
D. Introduction of malware into the network

B is the correct answer.

Justification:
A. While this is a risk, theft of an unencrypted device is a greater risk.
B. Because universal serial bus (USB) drives tend to be small, they are susceptible to theft or loss. This represents the greatest risk to the organization.
C. Use of USB drives for personal purposes is a violation of company policy; however, this is not the greatest risk.
D. Good general IT controls will include the scanning of USB drives for malware once they are inserted in a computer. The risk of malware in an otherwise robust environment is not as great as the risk of loss or theft.

A5-68 When performing a computer forensic investigation, in regard to the evidence gathered, an IS auditor should be **MOST** concerned with:

A. analysis.
B. evaluation.
C. preservation.
D. disclosure.

C is the correct answer.

Justification:
A. Analysis is important but not the primary concern related to evidence in a forensic investigation.
B. Evaluation is important but not the primary concern related to evidence in a forensic investigation.
C. Preservation and documentation of evidence for review by law enforcement and judicial authorities are of primary concern when investigating. Failure to properly preserve the evidence could jeopardize the admissibility of the evidence in legal proceedings.
D. Disclosure is important but not of primary concern to the IS auditor in a forensic investigation.

A5-69 A certificate authority (CA) can delegate the processes of:

A. revocation and suspension of a subscriber's certificate.
B. generation and distribution of the CA public key.
C. establishing a link between the requesting entity and its public key.
D. issuing and distributing subscriber certificates.

C is the correct answer.

Justification:
A. Revocation and suspension of the subscriber certificate are functions of the subscriber certificate life cycle management, which the certificate authority (CA) must perform.
B. Generation and distribution of the CA public key is a part of the CA key life cycle management process and, as such, cannot be delegated.
C. Establishing a link between the requesting entity and its public key is a function of a registration authority. This may or may not be performed by a CA; therefore, this function can be delegated.
D. Issuance and distribution of the subscriber certificate are functions of the subscriber certificate life cycle management, which the CA must perform.

A5-70 Which of the following results in a denial-of-service attack?

A. Brute force attack
B. Ping of death
C. Leapfrog attack
D. Negative acknowledgment attack

B is the correct answer.

Justification:
A. A brute force attack is typically a text attack that exhausts all possible key combinations used against encryption keys or passwords.
B. The use of Ping with a packet size higher than 65 KB and no fragmentation flag on will cause a denial of service.
C. A leapfrog attack, the act of telneting through one or more hosts to preclude a trace, makes use of user ID and password information obtained illicitly from one host to compromise another host.
D. A negative acknowledgment is a penetration technique that capitalizes on a potential weakness in an operating system that does not handle asynchronous interrupts properly, leaving the system in an unprotected state during such interrupts.

A5-71 Which of the following is an advantage of elliptic curve encryption over RSA encryption?

A. Computation speed
B. Ability to support digital signatures
C. Simpler key distribution
D. Message integrity controls

A is the correct answer.

Justification:
A. The main advantage of elliptic curve encryption (ECC) over RSA encryption is its computation speed. This is due in part to the use of much smaller keys in the ECC algorithm than in RSA.
B. Both encryption methods support digital signatures.
C. Both encryption methods are used for public key encryption and distribution.
D. Both ECC and RSA offer message integrity controls.

A5-72 Which of the following would be the **BEST** overall control for an Internet business looking for confidentiality, reliability and integrity of data?

A. Secure Sockets Layer
B. Intrusion detection system
C. Public key infrastructure
D. Virtual private network

A is the correct answer.

Justification:
A. Secure Sockets Layer (SSL) is used for many ecommerce applications to set up a secure channel for communications providing confidentiality through a combination of public and symmetric key encryption and integrity through hash message authentication code.
B. An intrusion detection system will log network activity but is not used for protecting traffic over the Internet.
C. Public key infrastructure is used in conjunction with SSL or for securing communications such as ecommerce and email.
D. A virtual private network (VPN) is a generic term for a communications tunnel that can provide confidentiality, integrity and authentication (reliability). A VPN can operate at different levels of the Open Systems Interconnection stack and may not always be used in conjunction with encryption. SSL can be called a type of VPN.

A5-73 Which of the following preventive controls **BEST** helps secure a web application?

A. Password masking
B. Developer training
C. Use of encryption
D. Vulnerability testing

B is the correct answer.

Justification:
A. Password masking is a necessary preventive control but is not the best way to secure an application.
B. Of the given choices, teaching developers to write secure code is the best way to secure a web application.
C. Encryption will protect data but is not sufficient to secure an application because other flaws in coding could compromise the application and data. Ensuring that applications are designed in a secure way is the best way to secure an application. This is accomplished by ensuring that developers are adequately educated on secure coding practices.
D. Vulnerability testing can help to ensure the security of web applications; however, the best preventive control is developer education because building secure applications from the start is more effective.

A5-74 Which of the following antivirus software implementation strategies would be the **MOST** effective in an interconnected corporate network?

A. Server-based antivirus software
B. Enterprise-based antivirus software
C. Workstation-based antivirus software
D. Perimeter-based antivirus software

B is the correct answer.

Justification:
A. An effective antivirus solution must be a combination of server-, network- and perimeter-based scanning and protection.
B. An important means of controlling the spread of viruses is to deploy an enterprisewide antivirus solution that will monitor and analyze traffic at many points. This provides a layered defense model that is more likely to detect malware regardless of how it comes into the organization—through a universal serial bus (USB) or portable storage, a network, an infected download or malicious web application.
C. Only checking for a virus on workstations would not be adequate because malware can infect many network devices or servers as well.
D. Because malware can enter an organization through many different methods, only checking for malware at the perimeter is not enough to protect the organization.

A5-75 Which of the following would be of **MOST** concern to an IS auditor reviewing a virtual private network implementation? Computers on the network that are located:

A. on the enterprise's internal network.
B. at the backup site.
C. in employees' homes.
D. at the enterprise's remote offices.

C is the correct answer.

Justification:

A. On an enterprise's internal network, there should be security policies and controls in place to detect and halt an outside attack that uses an internal machine as a staging platform.
B. Computers at the backup site are subject to the corporate security policy and, therefore, are not high-risk computers.
C. One risk of a virtual private network implementation is the chance of allowing high-risk computers onto the enterprise's network. All machines that are allowed onto the virtual network should be subject to the same security policy. Home computers are least subject to the corporate security policies and, therefore, are high-risk computers. Once a computer is hacked and "owned," any network that trusts that computer is at risk. Implementation and adherence to corporate security policy is easier when all computers on the network are on the enterprise's campus.
D. Computers on the network that are at the enterprise's remote offices, perhaps with different IS and security employees who have different ideas about security, are riskier than computers in the main office or backup site, but obviously less risky than home computers.

A5-76 The **PRIMARY** reason for using digital signatures is to ensure data:

A. confidentiality.
B. integrity.
C. availability.
D. correctness.

B is the correct answer.

Justification:

A. A digital signature does not, in itself, address message confidentiality.
B. Digital signatures provide integrity because the digital signature of a signed message (file, mail, document, etc.) changes every time a single bit of the document changes; thus, a signed document cannot be altered. A digital signature provides for message integrity, nonrepudiation and proof of origin.
C. Availability is not related to digital signatures.
D. In general, correctness is not related to digital signatures. A digital signature guarantee data integrity, however cannot ensure correctness of signed data.

A5-77 Which of the following is an example of a passive cybersecurity attack?

A. Traffic analysis
B. Masquerading
C. Denial-of-service
D. Email spoofing

A is the correct answer.

Justification:

A. Cybersecurity threats/vulnerabilities are divided into passive and active attacks. A passive attack is one that monitors or captures network traffic but does not in any way modify, insert or delete the traffic. Examples of passive attacks include network analysis, eavesdropping and traffic analysis.

B. Because masquerading alters the data by modifying the origin, it is an active attack.

C. Because a denial-of–service attack floods the network with traffic or sends malformed packets over the network, it is an active attack.

D. Because email spoofing alters the email header, it is an active attack.

A5-78 An IS auditor is reviewing security incident management procedures for the company. Which of the following choices is the **MOST** important consideration?

A. Chain of custody of electronic evidence
B. System breach notification procedures
C. Escalation procedures to external agencies
D. Procedures to recover lost data

A is the correct answer.

Justification:

A. The preservation of evidence is the most important consideration in regard to security incident management. If data and evidence are not collected properly, valuable information could be lost and would not be admissible in a court of law should the company decide to pursue litigation.

B. System breach notification is an important aspect and, in many cases, may even be required by laws and regulations; however, the security incident may not be a breach and the notification procedure might not apply.

C. Escalation procedures to external agencies such as the local police or special agencies dealing in cybercrime are important. However, without proper chain of custody procedures, vital evidence may be lost and would not be admissible in a court of law should the company decide to pursue litigation.

D. While having procedures in place to recover lost data is important, it is critical to ensure that evidence is protected to ensure follow-up and investigation.

A5-79 An accuracy measure for a biometric system is:

A. system response time.
B. registration time.
C. input file size.
D. false-acceptance rate.

D is the correct answer.

Justification:

A. An important consideration in the implementation of biometrics is the time required to process a user. If the system is too slow then it will impact productivity and lead to frustration. However, this is not an accuracy measure.

B. The registration time is a measure of the effort taken to enroll a user in the system. This is not an accuracy measure.

C. The file size to retain biometric information varies depending on the type of biometric solution selected. This is not an accuracy measure.

D. Three main accuracy measures are used for a biometric solution: false-rejection rate (FRR), cross-error rate (CER) and false-acceptance rate (FAR). FRR is a measure of how often valid individuals are rejected. FAR is a measure of how often invalid individuals are accepted. CER is a measure of when the false-rejection rate equals the false-acceptance rate.

A5-80 An IS auditor evaluating logical access controls should **FIRST**:

A. document the controls applied to the potential access paths to the system.
B. test controls over the access paths to determine if they are functional.
C. evaluate the security environment in relation to written policies and practices.
D. obtain an understanding of the security risk to information processing.

D is the correct answer.

Justification:

A. Documentation and evaluation is the second step in assessing the adequacy, efficiency and effectiveness of the controls and is based on the risk to the system that necessitates the controls.

B. The third step is to test the access paths—to determine if the controls are functioning.

C. It is only after the risk is determined and the controls documented that the IS auditor can evaluate the security environment to assess its adequacy through review of the written policies, observation of practices and comparison of them to appropriate security good practices.

D. When evaluating logical access controls, an IS auditor should first obtain an understanding of the security risk facing information processing by reviewing relevant documentation, by inquiries, and conducting a risk assessment. This is necessary so that the IS auditor can ensure the controls are adequate to address risk.

A5-81 Which of the following is the **MOST** secure way to remove data from obsolete magnetic tapes during a disposal?

A. Overwriting the tapes
B. Initializing the tape labels
C. Degaussing the tapes
D. Erasing the tapes

C is the correct answer.

Justification:

A. Overwriting the tapes is a good practice, but if the tapes have contained sensitive information then it is necessary to degauss them.
B. Initializing the tape labels would not remove the data on the tape and could lead to compromise of the data on the tape.
C. The best way to handle obsolete magnetic tapes is to degauss them. Degaussing is the application of a coercive magnetic force to the tape media. This action leaves a very low residue of magnetic induction, essentially erasing the data completely from the tapes.
D. Erasing the tapes will make the data unreadable except for sophisticated attacks; therefore, tapes containing sensitive data should be degaussed.

A5-82 The review of router access control lists should be conducted during:

A. an environmental review.
B. a network security review.
C. a business continuity review.
D. a data integrity review.

B is the correct answer.

Justification:

A. Environmental reviews examine physical security such as power and physical access. They do not require a review of the router access control lists.
B. Network security reviews include reviewing router access control lists, port scanning, internal and external connections to the system, etc.
C. Business continuity reviews ensure the business continuity plan is up to date, adequate to protect the organization and tested, and do not require a review of the router access control lists.
D. Data integrity reviews validate data accuracy and protect from improper alterations, but do not require a review of the router access control lists.

A5-83 Which of the following components is responsible for the collection of data in an intrusion detection system?

A. Analyzer
B. Administration console
C. User interface
D. Sensor

D is the correct answer.

Justification:
A. Analyzers receive input from sensors and determine the presence of and type of intrusive activity.
B. An administration console is the management interface component of an intrusion detection system (IDS).
C. A user interface allows the administrators to interact with the IDS.
D. Sensors are responsible for collecting data. Sensors may be attached to a network, server or other location and may gather data from many points for later analysis.

A5-84 Which of the following is the **MOST** significant function of a corporate public key infrastructure and certificate authority employing X.509 digital certificates?

A. It provides the public/private key set for the encryption and signature services used by email and file space.
B. It binds a digital certificate and its public key to an individual subscriber's identity.
C. It provides the authoritative source for employee identity and personal details.
D. It provides the authoritative authentication source for object access.

B is the correct answer.

Justification:
A. While some email applications depend on public key infrastructure (PKI)-issued certificates for nonrepudiation, the purpose of PKI is to provide authentication of the individual and link an individual with their private key. The certificate authority (CA) does not ordinarily create the user's private key.
B. PKI is primarily used to gain assurance that protected data or services originated from a legitimate source. The process to ensure the validity of the subscriber identity by linking to the digital certificate/public key is strict and rigorous.
C. Personal details are not stored in or provided by components in the PKI.
D. Authentication services within operating systems and applications may be built on PKI-issued certificates, but PKI does not provide authentication services for object access.

A5-85 A digital signature contains a message digest to:

A. show if the message has been altered after transmission.
B. define the encryption algorithm.
C. confirm the identity of the originator.
D. enable message transmission in a digital format.

A is the correct answer.

Justification:
A. The message digest is calculated and included in a digital signature to prove that the message has not been altered. The message digest sent with the message should have the same value as the recalculation of the digest of the received message.
B. The message digest does not define the algorithm; it is there to ensure integrity.
C. The message digest does not confirm the identity of the user; it is there to ensure integrity.
D. The message digest does not enable the transmission in digital format; it is there to ensure integrity.

A5-86 Which of the following manages the digital certificate life cycle to ensure adequate security and controls exist in digital signature applications related to ecommerce?

A. Registration authority
B. Certificate authority
C. Certification revocation list
D. Certification practice statement

B is the correct answer.

Justification:
A. A registration authority is an optional entity that is responsible for the administrative tasks associated with registering the end entity that is the subject of the certificate issued by the certificate authority (CA).
B. The CA maintains a directory of digital certificates for the reference of those receiving them. It manages the certificate life cycle, including certificate directory maintenance and certificate revocation list (CRL) maintenance and publication.
C. A CRL is an instrument for checking the continued validity of the certificates for which the CA has responsibility. A certificate that is put on a CRL can no longer be trusted.
D. A certification practice statement is a detailed set of rules governing the certificate authority's operations.

A5-87 A Transmission Control Protocol/Internet Protocol (TCP/IP)-based environment is exposed to the Internet. Which of the following **BEST** ensures that complete encryption and authentication protocols exist for protecting information while transmitted?

A. Work is completed in tunnel mode with IP security.
B. A digital signature with RSA has been implemented.
C. Digital certificates with RSA are being used.
D. Work is being completed in TCP services.

A is the correct answer.

Justification:
A. Tunnel mode with Internet Protocol (IP) security provides encryption and authentication of the complete IP package. To accomplish this, the authentication header and encapsulating security payload services can be nested. This is known as IP Security.
B. A digital signature with RSA provides authentication and integrity but not confidentiality.
C. Digital certificates with RSA provide authentication and integrity but do not provide encryption.
D. Transmission Control Protocol services do not provide encryption and authentication.

A5-88 Digital signatures require the:

A. signer to have a public key and the receiver to have a private key.
B. signer to have a private key and the receiver to have a public key.
C. signer and receiver to have a public key.
D. signer and receiver to have a private key.

B is the correct answer.

Justification:

A. If a sender encrypts a message with a public key, it will provide confidential transmission to the receiver with the private key.
B. Digital signatures are intended to verify to a recipient the integrity of the data and the identity of the sender. The digital signature standard is based on the sender encrypting a digest of the message with their private key and the receiver validating the message with the public key.
C. Asymmetric key cryptography always works with key pairs. Therefore, a message encrypted with a public key could only be opened with a private key.
D. If both the sender and receiver have a private key there would be no way to validate the digital signature.

A5-89 The feature of a digital signature that ensures the sender cannot later deny generating and sending the message is called:

A. data integrity.
B. authentication.
C. nonrepudiation.
D. replay protection.

C is the correct answer.

Justification:

A. Data integrity refers to changes in the plaintext message that would result in the recipient failing to compute the same message hash.
B. Because only the claimed sender has the private key used to create the digital signature, authentication ensures that the message has been sent by the claimed sender.
C. Integrity, authentication, nonrepudiation and replay protection are all features of a digital signature. Nonrepudiation ensures that the claimed sender cannot later deny generating and sending the message.
D. Replay protection is a method that a recipient can use to check that the message was not intercepted and re-sent (replayed).

A5-90 During the collection of forensic evidence, which of the following actions would **MOST** likely result in the destruction or corruption of evidence on a compromised system?

A. Dumping the memory content to a file
B. Generating disk images of the compromised system
C. Rebooting the system
D. Removing the system from the network

C is the correct answer.

Justification:

A. Copying the memory contents is a normal forensics procedure where possible. Done carefully, it will not corrupt the evidence.
B. Proper forensics procedures require creating two copies of the images of the system for analysis. Hash values ensure that the copies are accurate.
C. Rebooting the system may result in a change in the system state and the loss of files and important evidence stored in memory.
D. When investigating a system, it is recommended to disconnect it from the network to minimize external infection or access.

A5-91 An IS auditor is reviewing Secure Sockets Layer enabled web sites for the company. Which of the following choices would be the **HIGHEST** risk?

A. Expired digital certificates
B. Self-signed digital certificates
C. Using the same digital certificate for multiple web sites
D. Using 56-bit digital certificates

B is the correct answer.

Justification:

A. An expired certificate leads to blocked access to the web site leading to unwanted downtime. However, there is no loss of data. Therefore, the comparative risk is lower.
B. Self-signed digital certificates are not signed by a certificate authority (CA) and can be created by anyone. Thus, they can be used by attackers to impersonate a web site, which may lead to data theft or perpetrate a man-in-the-middle attack.
C. Using the same digital certificate is not a significant risk. Wildcard digital certificates may be used for multiple subdomain web sites.
D. 56-bit digital certificates may be needed to connect with older versions of operating systems (OSs) or browsers. While they have a lower strength than 128-bit or 256-bit digital certificates, the comparative risk of a self-signed certificate is higher.

A5-92 Which of the following controls would **BEST** detect intrusion?

A. User IDs and user privileges are granted through authorized procedures.
B. Automatic logoff is used when a workstation is inactive for a particular period of time.
C. Automatic logoff of the system occurs after a specified number of unsuccessful attempts.
D. Unsuccessful logon attempts are monitored by the security administrator.

D is the correct answer.

Justification:

A. User IDs and the granting of user privileges define a policy. This is a type of administrative or managerial control that may prevent intrusion but would not detect it.
B. Automatic logoff is a method of preventing access through unattended or inactive terminals but is not a detective control.
C. Unsuccessful attempts to log on are a method for preventing intrusion, not detecting it.
D. Intrusion is detected by the active monitoring and review of unsuccessful logon attempts.

A5-93 Which of the following is the **BEST** control over a guest wireless ID that is given to vendor staff?

A. Assignment of a renewable user ID which expires daily
B. A write-once log to monitor the vendor's activities on the system
C. Use of a user ID format similar to that used by employees
D. Ensuring that wireless network encryption is configured properly

A is the correct answer.

Justification:

A. A renewable user ID which expires daily would be a good control because it would ensure that wireless access will automatically terminate daily and cannot be used without authorization.
B. While it is recommended to monitor vendor activities while vendor staff are on the system, this is a detective control and thus is not as strong as a preventive control.
C. The user ID format does not change the overall security of the wireless connection.
D. Controls related to the encryption of the wireless network are important; however, the access to that network is a more critical issue.

A5-94 An IS auditor performing a telecommunication access control review should be concerned **PRIMARILY** with the:

A. maintenance of access logs of usage of various system resources.
B. authorization and authentication of the user prior to granting access to system resources.
C. adequate protection of stored data on servers by encryption or other means.
D. accountability system and the ability to identify any terminal accessing system resources.

B is the correct answer.

Justification:

A. The maintenance of access logs of usage of system resources is a detective control. A preventive control should be used first.
B. The authorization and authentication of users before granting them access to system resources (networks, servers, applications, etc.) is the most significant aspect in a telecommunication access control review because it is a preventive control. Weak controls at this level can affect all other aspects of security.
C. The adequate protection of data being stored on servers by encryption or other means is a method of protecting stored information and is not a network access issue.
D. The accountability system and the ability to identify any terminal accessing system resources deal with controlling access through the identification of a terminal or device attempting to connect to the network. This is called node authentication and is not as good as authenticating the user sitting at that node.

A5-95 An IS auditor suspects an incident is occurring while an audit is being performed on a financial system. What should the IS auditor do **FIRST**?

A. Request that the system be shut down to preserve evidence.
B. Report the incident to management.
C. Ask for immediate suspension of the suspect accounts.
D. Investigate the source and nature of the incident.

B is the correct answer.

Justification:

A. The IS auditor should follow the incident response process of the organization. The auditor is not authorized to shut the system down.
B. Reporting the suspected incident to management will help initiate the incident response process, which is the most appropriate action. Management is responsible for making decisions regarding the appropriate response. It is not the IS auditor's role to respond to incidents during an audit.
C. The IS auditor is not authorized to lead the investigation or to suspend user accounts. The auditor should report the incident to management.
D. Management is responsible to set up and follow an incident management plan; that is not the responsibility of the IS auditor.

A5-96 When using public key encryption to secure data being transmitted across a network:

A. both the key used to encrypt and decrypt the data are public.
B. the key used to encrypt is private, but the key used to decrypt the data is public.
C. the key used to encrypt is public, but the key used to decrypt the data is private.
D. both the key used to encrypt and decrypt the data are private.

C is the correct answer.

Justification:

A. The public and private keys always work as a pair—if a public key is used to encrypt a message, the corresponding private key MUST be used to decrypt the message.
B. If the message is encrypted with a private key, that will provide proof of origin but not message security or confidentiality.
C. Public key encryption, also known as asymmetric key cryptography, uses a public key to encrypt the message and a private key to decrypt it.
D. Using two private keys would not be possible with asymmetric encryption.

A5-97 The technique used to ensure security in virtual private networks is called:

A. data encapsulation.
B. data wrapping.
C. data transformation.
D. data hashing.

A is the correct answer.

Justification:

A. Encapsulation, or tunneling, is a technique used to encrypt the traffic payload so that it can be securely transmitted over an insecure network.
B. Wrapping is used where the original packet is wrapped in another packet but is not directly related to security.
C. To transform or change the state of the communication would not be used for security.
D. Hashing is used in virtual private networks to ensure message integrity.

A5-98 During an audit of a telecommunications system, an IS auditor finds that the risk of intercepting data transmitted to and from remote sites is very high. The **MOST** effective control for reducing this exposure is:

A. encryption.
B. callback modems.
C. message authentication.
D. dedicated leased lines.

A is the correct answer.

Justification:
A. Encryption of data is the most secure method of protecting confidential data from exposure.
B. A callback system is used to ensure that a user is only logging in from a known location. It is not effective to protect the transmitted data from interception.
C. Message authentication is used to prove message integrity and source but not confidentiality.
D. It is more difficult to intercept traffic traversing a dedicated leased line than it is to intercept data on a shared network, but the only way to really protect the confidentiality of data is to encrypt it.

A5-99 An Internet-based attack using password sniffing can:

A. enable one party to act as if they are another party.
B. cause modification to the contents of certain transactions.
C. be used to gain access to systems containing proprietary information.
D. result in major problems with billing systems and transaction processing agreements.

C is the correct answer.

Justification:
A. Spoofing attacks can be used to enable one party to act as if they are another party.
B. Data modification attacks can be used to modify the contents of certain transactions.
C. Password sniffing attacks can be used to gain access to systems on which proprietary information is stored.
D. Repudiation of transactions can cause major problems with billing systems and transaction processing agreements.

A5-100 Which of the following controls would be the **MOST** comprehensive in a remote access network with multiple and diverse subsystems?

A. Proxy server
B. Firewall installation
C. Demilitarized zone
D. Virtual private network

D is the correct answer.

Justification:
A. A proxy server is a type of firewall installation used as an intermediary to filter and control traffic between internal and external parties.
B. While firewall installations are the primary line of defense, they would need to have encryption and a virtual private network (VPN) to secure remote access traffic.
C. A demilitarized zone (DMZ) is an isolated network used to permit outsiders to access certain corporate information in a semi-trusted environment. The DMZ may host a web server or other external facing services. Traffic to a DMZ is not usually encrypted unless it is terminating on a VPN located in the DMZ.
D. The best way to secure remote access is through the use of encrypted VPNs. This would allow remote users a secure connection to the main systems.

A5-101 During an audit of an enterprise that is dedicated to ecommerce, the IS manager states that digital signatures are used when receiving communications from customers. To substantiate this, an IS auditor must prove that which of the following is used?

A. A biometric, digitalized and encrypted parameter with the customer's public key
B. A hash of the data that is transmitted and encrypted with the customer's private key
C. A hash of the data that is transmitted and encrypted with the customer's public key
D. The customer's scanned signature encrypted with the customer's public key

B is the correct answer.

Justification:
A. Biometrics are not used in digital signatures or public key encryption.
B. The calculation of a hash, or digest, of the data that are transmitted, and its encryption require the private key of the client (sender) and is called a signature of the message, or digital signature. The receiver hashes the received message and compares the hash they compute with the received hash, after the digital signature has been decrypted with the sender's public key. If the hash values are the same, the conclusion would be that there is integrity in the data that have arrived, and the origin is authenticated. The concept of encrypting the hash with the private key of the originator provides nonrepudiation because it can only be decrypted with their public key, and the private key would not be known to the recipient. Simply put, in a key-pair situation, anything that can be decrypted by a sender's public key must have been encrypted with their private key, so they must have been the sender (i.e., nonrepudiation).
C. It would not be correct to encrypt the hash with the customer's public key because then the recipient would need access to the customer's private key to decrypt the digital signature.
D. A scan of the customer's signature would be known as a digitized signature, not a digital signature, and would be of little or no value in this scenario.

A5-102 When planning an audit of a network setup, an IS auditor should give highest priority to obtaining which of the following network documentation?

A. Wiring and schematic diagram
B. Users' lists and responsibilities
C. Application lists and their details
D. Backup and recovery procedures

A is the correct answer.

Justification:
A. The wiring and schematic diagram of the network is necessary to carry out a network audit. The IS auditor needs to know what equipment, configuration and addressing is used on the network to perform an audit of the network setup.
B. When performing an audit of network setup, the users' lists would not be of value.
C. Application lists are not required to audit network configuration.
D. Backup and recovery procedures are important but not as important as knowing the network layout.

A5-103 Which of the following should an IS auditor be **MOST** concerned about in a financial application?

A. Programmers have access to source code in user acceptance testing environment.
B. Secondary controls are documented for identified role conflicts.
C. The information security officer does not authorize all application changes.
D. Programmers have access to the production database.

D is the correct answer.

Justification:
A. Programmers who have access to application source code are not of concern to the IS auditor because programmers need access to source code to do their jobs. User acceptance testing (UAT) environment is a separate from production environment, changes cannot be moved into production environment without prior authorization.
B. When segregation of duties conflicts are identified, secondary controls should be in place to mitigate risk. While the IS auditor reviews secondary controls, in this case the greater concern is programmers having access to the production database.
C. The information security officer is not likely to authorize all application changes; therefore, this is not a concern for an IS auditor.
D. Programmers who have access to the production database are considered to be a segregation of duties conflict.

A5-104 Which of the following is the **MAIN** reason an organization should have an incident response plan? The plan helps to:

A. ensure prompt communication of adverse events to relevant management.
B. contain costs related to maintaining disaster recovery plan capabilities.
C. ensure that customers are promptly notified of issues such as security breaches.
D. minimize the duration and impact of system outages and security incidents.

D is the correct answer.

Justification:
A. Incident response plans generally deal with a wide range of possible issues. While it is important to have a proper communication of adverse event within the organization, the primary objective is to reduce impact of incidents.
B. An effective incident response plan could minimize damage to the organization, which minimizes costs, but the main purpose of the incident response plan is to minimize damage. Possible damage could include nonfinancial metrics, such as damage to a company's reputation.
C. While an incident response plan includes elements such as when and how to contact customers about a significant incident, the primary purpose of the plan is to minimize the impact.
D. An incident response plan helps minimize the impact of an incident because it provides a controlled response to incidents. The phases of the plan include planning, detection, evaluation, containment, eradication, escalation, response, recovery, reporting, postincident review and a review of lessons learned.

A5-105 Email message authenticity and confidentiality is **BEST** achieved by signing the message using the:

A. sender's private key and encrypting the message using the receiver's public key.
B. sender's public key and encrypting the message using the receiver's private key.
C. receiver's private key and encrypting the message using the sender's public key.
D. receiver's public key and encrypting the message using the sender's private key.

A is the correct answer.

Justification:

A. By signing the message with the sender's private key, the receiver can verify its authenticity using the sender's public key. Encrypting with the receiver's public key provides confidentiality.
B. Signing can only occur using the sender's private key.
C. The sender would not have access to the receiver's private key.
D. By encrypting the message with the receiver's public key, only the receiver can decrypt the message using their own private key. The receiver's private key is confidential and, therefore, unknown to the sender. Messages encrypted using the sender's private key can be read by anyone with the sender's public key.

A5-106 An organization is considering connecting a critical PC-based system to the Internet. Which of the following would provide the **BEST** protection against hacking?

A. An application-level gateway
B. A remote access server
C. A proxy server
D. Port scanning

A is the correct answer.

Justification:

A. An application-level gateway is the best way to protect against hacking because it can be configured with detailed rules that describe the type of user or connection that is or is not permitted. It analyzes, in detail, each package—not only in layers one through four of the Open System Interconnection model, but also layers five through seven, which means that it reviews the commands of each higher-level protocol (Hypertext Transmission Protocol, File Transfer Protocol, Simple Network Management Protocol, etc.).
B. For a remote access server, there is a device (server) that asks for a username and password before entering the network. This is good when accessing private networks, but it can be mapped or scanned from the Internet, creating security exposure.
C. Proxy servers can provide excellent protection, but depending on the type of proxy, they may not be able to examine traffic as effectively as an application gateway. For proxy servers to work, an individual is needed who really knows how to do this, and applications can use different ports for the different sections of the program.
D. Port scanning is used to detect vulnerabilities or open ports on a network, but not when trying to control what comes from the Internet, or when all the ports available need to be controlled. For example, the port for Ping (echo request) could be blocked and the IP addresses would be available for the application and browsing but would not respond to Ping.

A5-107 Which of the following is the **MOST** secure and economical method for connecting a private network over the Internet in a small- to medium-sized organization?

A. Virtual private network
B. Dedicated line
C. Leased line
D. Integrated services digital network

A is the correct answer.

Justification:
A. The most secure method is a virtual private network, using encryption, authentication and tunneling to allow data to travel securely from a private network to the Internet.
B. A dedicated line is quite expensive and only needed when there are specific confidentiality and availability needs.
C. A leased line is an expensive but private option, but rarely a good option today.
D. Integrated services digital network is not encrypted and would need additional security to be a valid option.

A5-108 The potential for unauthorized system access by way of terminals or workstations within an organization's facility is increased when:

A. connecting points are available in the facility to connect laptops to the network.
B. users take precautions to keep their passwords confidential.
C. terminals with password protection are located in insecure locations.
D. terminals are located within the facility in small clusters under the supervision of an administrator.

A is the correct answer.

Justification:
A. Any person with wrongful intentions can connect a laptop to the network. The insecure connecting points make unauthorized access possible if the individual has knowledge of a valid user ID and password. The other choices are controls for preventing unauthorized network access.
B. If system passwords are not readily available for intruders to use, they must guess, introducing an additional factor and requires time.
C. System passwords provide protection against unauthorized use of terminals located in insecure locations.
D. Supervision is a very effective control when used to monitor access to a small operating unit or production resources.

A5-109 Which of the following functions is performed by a virtual private network?

A. Hiding information from sniffers on the net
B. Enforcing security policies
C. Detecting misuse or mistakes
D. Regulating access

A is the correct answer.

Justification:
A. A virtual private network (VPN) hides information from sniffers on the Internet using tunneling. It works based on encapsulation and encryption of sensitive traffic.
B. A VPN does support security policies related to secure communications, but its primary purpose is to protect data in transit.
C. A VPN does not check the content of packets, so it cannot detect misuse or mistakes.
D. A VPN is not used to regulate access. A user may have to log in to use a VPN, but that is not the purpose of the VPN.

A5-110 Applying a digital signature to data traveling in a network provides:

A. confidentiality and integrity.
B. security and nonrepudiation.
C. integrity and nonrepudiation.
D. confidentiality and nonrepudiation.

C is the correct answer.

Justification:
A. A digital signature does not encrypt the message, so it cannot provide confidentiality.
B. A digital signature does not encrypt the message, so it cannot provide security.
C. A digital signature is created by signing a hash of a message with the private key of the sender. This provides for the integrity (through the hash) and the proof of origin (nonrepudiation) of the message.
D. A digital signature does not provide confidentiality.

A5-111 Which of the following would an IS auditor consider a weakness when performing an audit of an organization that uses a public key infrastructure with digital certificates for its business-to-consumer transactions via the Internet?

A. Customers are widely dispersed geographically, but the certificate authorities (CAs) are not.
B. Customers can make their transactions from any computer or mobile device.
C. The CA has several data processing subcenters to administer certificates.
D. The organization is the owner of the CA.

D is the correct answer.

Justification:
A. It is common to use a single certificate authority (CA). They do not need to be geographically dispersed.
B. The use of public key infrastructure and certificates allows flexible secure communications from many devices.
C. The CA will often have redundancy and failover capabilities to alternate data centers.
D. If the CA belongs to the same organization, this would pose a risk. The management of a CA must be based on trusted and secure procedures. If the organization has not set in place the controls to manage the registration, distribution and revocation of certificates this could lead to a compromise of the certificates and loss of trust.

A5-112 Which of the following is the **MOST** reliable method to ensure identity of sender for messages transferred across Internet?

A. Digital signatures
B. Asymmetric cryptography
C. Digital certificates
D. Message authentication code

C is the correct answer.

Justification:
A. Digital signatures are used for both authentication and integrity, but the identity of the sender would still be confirmed by the digital certificate.
B. Asymmetric cryptography, such as public key infrastructure, appears to authenticate the sender but is vulnerable to a man-in-the-middle attack.
C. Digital certificates are issued by a trusted third party. The message sender attaches the certificate and the recipient can verify authenticity with the certificate repository.
D. Message authentication code is used for message integrity verification.

A5-113 Which of the following is the **BEST** way for an IS auditor to determine the effectiveness of a security awareness and training program?

A. Review the security training program.
B. Ask the security administrator.
C. Interview a sample of employees.
D. Review the security reminders to employees.

C is the correct answer.

Justification:

A. A security training program may be well designed, but the results of the program will be determined by employee awareness.
B. Asking the security administrator would not show the effectiveness of a security awareness and training program because such a program should target more than just the administrator.
C. Interviewing a sample of employees is the best way to determine the effectiveness of a security awareness and training program because overall awareness must be determined, and effective security is dependent on people. Reviewing the security training program would not be the ultimate indicator of the effectiveness of the awareness training.
D. Reviewing the security reminders to the employees is not the best way to find out the effectiveness of the training awareness because sending reminders may result in little actual awareness.

A5-114 A laptop computer belonging to a company database administrator (DBA) and containing a file of production database passwords has been stolen. What should the organization do **FIRST**?

A. Send a report to the IS audit department.
B. Change the name of the DBA account.
C. Suspend the DBA account.
D. Change the database password.

D is the correct answer.

Justification:

A. While the IS audit department should be notified, this should not be the first action.
B. Changing the database administrator (DBA) account name could impact production database servers and thus would not be a good idea.
C. Suspending the DBA account could impact the production database servers and may not be effective if there is more than one DBA account sharing the same database password. The thief may guess the account names of the other DBAs.
D. The password should be changed immediately because there is no way to know whether it has been compromised.

A5-115 If inadequate, which of the following would be the **MOST** likely contributor to a denial-of-service attack?

A. Router configuration and rules
B. Design of the internal network
C. Updates to the router system software
D. Audit testing and review techniques

A is the correct answer.

Justification:

A. Improper router configuration and rules could lead to an exposure to denial-of-service (DoS) attacks.
B. An inefficient design of the internal network may also lead to a DoS but this is not as high a risk as router misconfiguration errors.
C. Updates to router software has led to a DoS in the past, but this is a subset of router configuration and rules.
D. Audit testing and review techniques can cause a DoS if tests disable systems or applications, but this is not the most likely risk.

A5-116 The Secure Sockets Layer protocol ensures the confidentiality of a message by using:

A. symmetric encryption.
B. message authentication codes.
C. hash function.
D. digital signature certificates.

A is the correct answer.

Justification:
A. Secure Sockets Layer (SSL) uses a symmetric key for message encryption.
B. A message authentication code is used for ensuring data integrity.
C. Hash function is used for generating a message digest which can provide message integrity; it is not used for message encryption.
D. Digital signature certificates are used by SSL for server authentication.

A5-117 The **PRIMARY** goal of a web site certificate is:

A. authentication of the web site that will be surfed.
B. authentication of the user who surfs through that site.
C. preventing surfing of the web site by hackers.
D. the same purpose as that of a digital certificate.

A is the correct answer.

Justification:
A. Authenticating the site to be surfed is the primary goal of a web certificate.
B. Authentication of a user is achieved through passwords and not by a web site certificate.
C. The site certificate does not prevent hacking, nor does it authenticate a person.
D. Web site certificates may serve the same purpose as a digital certificate, but the goal of certificates is authentication.

A5-118 An IS auditor performing detailed network assessments and access control reviews should **FIRST**:

A. determine the points of entry into the network.
B. evaluate users' access authorization.
C. assess users' identification and authorization.
D. evaluate the domain-controlling server configuration.

A is the correct answer.

Justification:
A. In performing detailed network assessments and access control reviews, an IS auditor should first determine the points of entry to the system and review the points of entry, accordingly, for appropriate controls.
B. Evaluation of user access authorization is an implementation issue for appropriate controls for the points of entry.
C. Assessment of user identification and authorization are implementation issues for appropriate controls for the points of entry.
D. Evaluation of the domain-controlling server configuration is not the first area to be reviewed. It will be reviewed once the network entry points have been identified.

A5-119 The **MOST** serious challenge in the operation of an intrusion detection system is:

A. filtering false-positives alerts.
B. learning vendor-specific protocols.
C. updating detection signatures.
D. blocking eligible connections.

A is the correct answer.

Justification:

A. **Because of the configuration and the way intrusion detection system (IDS) technology operates, the main problem in operating IDSs is the recognition (detection) of events that are not really security incidents—false positives, the equivalent of a false alarm. An IS auditor needs to be aware of this and should check for implementation of related controls (such as IDS tuning) and incident handling procedures (such as the screening process) to know if an event is a security incident or a false positive.**
B. It might be necessary to learn vendor-specific protocols or commands for interacting with IDS, however most vendors provide relevant documentation and trainings which could be quickly mastered by qualified IT personnel.
C. It is necessary to regularly update detection signatures, however majority of modern IDSs systems has built-in modules providing automated and secure updates.
D. Blocking suspicious connections is a characteristic of Intrusion Prevention Systems, which are different type of network security systems.

AS5-120 An IS auditor performing an audit has determined that developers have been granted administrative access to the virtual machine management console to manage their own servers used for software development and testing. Which of the following choices would be of **MOST** concern for the IS auditor?

A. Developers have the ability to create or de-provision servers.
B. Developers could gain elevated access to production servers.
C. Developers can affect the performance of production servers with their applications.
D. Developers could install unapproved applications to any servers.

A is the correct answer.

Justification:

A. **Virtualization offers the ability to create or destroy virtual machines (VMs) through the administrative interface with administrative access. While a developer would be unlikely to de-provision a production server, the administrative console would grant him/her the ability to do this, which would be a significant risk.**
B. When properly configured, the administrative console of a virtual server host does not allow an individual to bypass the authentication of the guest operating system (OS) to access the server. In this case, while the developers could potentially start, stop or even de-provision a production VM, they could not gain elevated access to the OS of the guest through the administrative interface.
C. While there could be instances where a software development team might use resource-intensive applications that could cause performance issues for the virtual host, the greater risk would be the ability to de-provision VMs.
D. When properly configured, the administrative console of a virtual server host does not allow an individual to bypass the authentication of the guest OS to access the server; therefore, the concern that unauthorized software could be installed is not valid.

A5-121 Which of the following findings would be of **GREATEST** concern to an IS auditor during a review of logical access to an application?

A. Some developers have update access to production data.
B. The file storing the application ID password is in cleartext in the production code.
C. The change control team has knowledge of the application ID password.
D. The application does not enforce the use of strong passwords.

B is the correct answer.

Justification:

A. Developers might need limited update access to production data to perform their jobs and this access, when approved and reviewed by management, is acceptable even though it does pose a risk.
B. Compromise of the application ID password can result in untraceable, unauthorized changes to production data; storing the password in cleartext poses the greatest risk. While the production code may be protected from update access, it is viewable by development teams.
C. Knowledge of the application ID password by the change control team does not pose a great concern if adequate separation of duties exists between change control and development activities. There may be occasions when the application ID needs to be used by change control in the production environment.
D. While the lack of a strong password policy and configuration can result in compromised accounts, the risk is lower than if the application ID password is compromised because the application ID password does not allow for traceability.

A5-122 The management of an organization has decided to establish a security awareness program. Which of the following would **MOST** likely be a part of the program?

A. Using an intrusion detection system to report incidents
B. Mandating the use of passwords to access all software
C. Installing an efficient user log system to track the actions of each user
D. Training provided on a regular basis to all current and new employees

D is the correct answer.

Justification:

A. Using an intrusion detection system to report incidents that occur is an implementation of a security program and is not effective in establishing a security awareness program.
B. Mandating the use of passwords is a policy decision, not an awareness issue.
C. Installing an efficient user log system is not a part of an awareness program.
D. Regular training is an important part of a security awareness program.

A5-123 A company determined that its web site was compromised, and a rootkit was installed on the server hosting the application. Which of the following choices would have **MOST** likely prevented the incident?

A. A host-based intrusion prevention system
B. A network-based intrusion detection system
C. A firewall
D. Operating system patching

A is the correct answer.

Justification:

A. A host-based intrusion prevention system (IPS) prevents unauthorized changes to the host. If a malware attack attempted to install a rootkit, the IPS would refuse to permit the installation without the consent of an administrator.

B. A network-based intrusion detection system (IDS) relies on attack signatures based on known exploits and attack patterns. If the IDS is not kept up to date with the latest signatures, or the attacker is able to create or gain access to an exploit unknown to the IDS, it will go undetected. A web server exploit performed through the web application itself, such as a Structured Query Language injection attack, would not appear to be an attack to the network-based IDS.

C. A firewall by itself does not protect a web server because the ports required for users to access the web server must be open in the firewall. Web server attacks are typically performed over the same ports that are open for normal web traffic. Therefore, a firewall does not protect the web server.

D. Operating system (OS) patching will make exploitation of the server more difficult for the attacker and less likely. However, attacks on the web application and server OS may succeed based on issues unrelated to any unpatched server vulnerabilities, and the host-based IPS should detect any attempts to change files on the server, regardless of how access was obtained.

A5-124 The role of the certificate authority (CA) as a third party is to:

A. provide secured communication and networking services based on certificates.
B. host a repository of certificates with the corresponding public and secret keys issued by that CA.
C. act as a trusted intermediary between two communication partners.
D. confirm the identity of the entity owning a certificate issued by that CA.

D is the correct answer.

Justification:

A. Providing a communication infrastructure is not a certificate authority (CA) activity.

B. The secret keys belonging to the certificates would not be archived at the CA.

C. The CA can contribute to authenticating the communicating partners to each other, but the CA is not involved in the communication stream itself.

D. The primary activity of a CA is to issue certificates. The primary role of the CA is to check the identity of the entity owning a certificate and to confirm the integrity of any certificate it issued.

A5-125 Which of the following types of penetration tests effectively evaluates the incident handling and response capability of the system administrator?

A. Targeted testing
B. Internal testing
C. Double-blind testing
D. External testing

C is the correct answer.

Justification:

A. In targeted testing, penetration testers are provided with information related to target and network design and the target's IT team is aware of the testing activities.

B. Internal testing refers to attacks and control circumvention attempts on the target from within the perimeter. The system administrator is typically aware of the testing activities.

C. In double-blind testing, the penetration tester has little or limited knowledge about the target system, and personnel at the target site have not been informed that a test is being performed. Because the administrator and security staff at the target are not aware of the test, it can effectively evaluate the incident handling and response capability of the system administrator.

D. External testing is a generic term that refers to attacks and control circumvention attempts on the target from outside the target system. The system administrator may or may not be aware of the testing activities, so this is not the correct answer. (Note: Rather than concentrating on specific terms, CISA candidates should understand the differences between various types of penetration testing.)

Question **A5-126** refers to the following diagram.

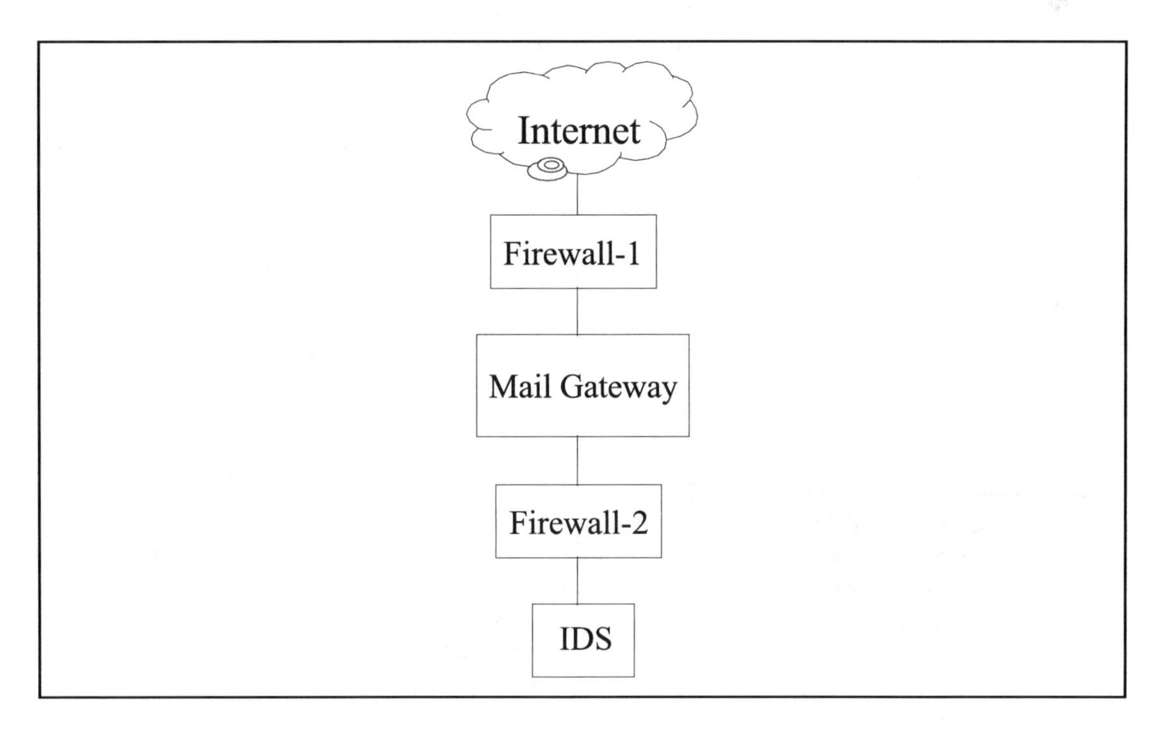

A5-126 Email traffic from the Internet is routed via firewall-1 to the mail gateway. Mail is routed from the mail gateway, via firewall-2, to the mail recipients in the internal network. Other traffic is not allowed. For example, the firewalls do not allow direct traffic from the Internet to the internal network. The intrusion detection system (IDS) detects traffic for the internal network that did not originate from the mail gateway. The **FIRST** action triggered by the IDS should be to:

A. alert the appropriate staff.
B. create an entry in the log.
C. close firewall-2.
D. close firewall-1.

B is the correct answer.

Justification:

A. The first action taken by an intrusion detection system (IDS) will be to create a log entry and then alert the appropriate staff.

B. Creating an entry in the log is the first step taken by a network IDS. The IDS may also be configured to send an alert to the administrator, send a note to the firewall and may even be configured to record the suspicious packet.

C. Traffic for the internal network that did not originate from the mail gateway is a sign that firewall-1 is not functioning properly. This may have been be caused by an attack from a hacker. After the IDS has logged the suspicious traffic, it may signal firewall-2 to close, thus preventing damage to the internal network. After closing firewall-2, the malfunctioning of firewall-1 can be investigated. The IDS should trigger the closing of firewall-2 either automatically or by manual intervention. Between the detection by the IDS and a response from the system administrator, valuable time can be lost, in which a hacker could also compromise firewall-2.

D. The IDS will usually only protect the internal network by closing firewall-2 and will not close the externally facing firewall-1.

A5-127 An organization has experienced a large amount of traffic being re-routed from its Voice-over Internet Protocol packet network. The organization believes it is a victim of eavesdropping. Which of the following could result in eavesdropping of VoIP traffic?

A. Corruption of the Address Resolution Protocol cache in Ethernet switches
B. Use of a default administrator password on the analog phone switch
C. Deploying virtual local area networks without enabling encryption
D. End users having access to software tools such as packet sniffer applications

A is the correct answer.

Justification:

A. On an Ethernet switch there is a data table known as the Address Resolution Protocol (ARP) cache, which stores mappings between media access control and IP addresses. During normal operations, Ethernet switches only allow directed traffic to flow between the ports involved in the conversation and no other ports can see that traffic. However, if the ARP cache is intentionally corrupted with an ARP poisoning attack, some Ethernet switches simply "flood" the directed traffic to all ports of the switch, which could allow an attacker to monitor traffic not normally visible to the port where the attacker was connected, and thereby eavesdrop on Voice-over Internet Protocol (VoIP) traffic.

B. VoIP systems do not use analog switches and inadequate administrator security controls would not be an issue.

C. VoIP data are not normally encrypted in a LAN environment because the controls regarding VLAN security are adequate.

D. Most software tools such as packet sniffers cannot make changes to LAN devices, such as the VLAN configuration of an Ethernet switch used for VoIP. Therefore, the use of software utilities of this type is not a risk.

Question **A5-128** refers to the following diagram.

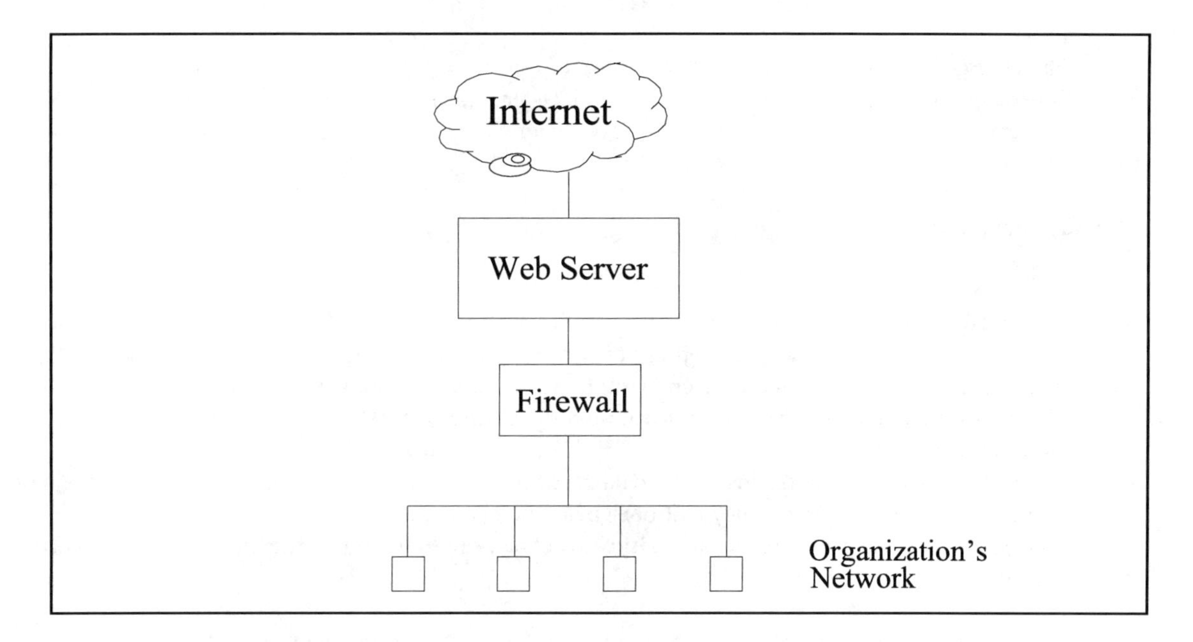

A5-128 To detect attack attempts that the firewall is unable to recognize, an IS auditor should recommend placing a network intrusion detection system between the:

A. Firewall and the organization's network.
B. Internet and the firewall.
C. Internet and the web server.
D. Web server and the firewall.

A is the correct answer.

Justification:

A. Attack attempts that could not be recognized by the firewall will be detected if a network-based intrusion detection system (IDS) is placed between the firewall and the organization's network.

B. A network-based IDS placed between the Internet and the firewall will detect attack attempts, whether they are or are not noticed by the firewall.

C. Placing an IDS outside of the web server will identify attacks directed at the web server but will not detect attacks missed by the firewall.

D. Placing the IDS after the web server would identify attacks that have made it past the web server but will not indicate whether the firewall would have been able to detect the attacks.

A5-129 An IS auditor is reviewing the physical security controls of a data center and notices several areas for concern. Which of the following areas is the **MOST** important?

A. The emergency power off button cover is missing.
B. Scheduled maintenance of the fire suppression system was not performed.
C. There are no security cameras inside the data center.
D. The emergency exit door is blocked.

D is the correct answer.

Justification:
A. The emergency power off button issue is a significant concern, but life safety is the highest priority.
B. The primary purpose of the fire suppression system is to protect the equipment and building. The lack of scheduled maintenance is a concern; however, this does not indicate that the system would not function as required. The more critical issue is the emergency exit because life safety is the highest priority.
C. The lack of security cameras inside the data center may be a significant concern; however, the more significant issue is the emergency exit door being blocked.
D. Life safety is always the highest priority; therefore, the blocking of the emergency exit is the most serious problem.

A5-130 Which of the following choices **BEST** helps information owners to properly classify data?

A. Understanding of technical controls that protect data
B. Training on organizational policies and standards
C. Use of an automated data leak prevention tool
D. Understanding which people need to access the data

B is the correct answer.

Justification:
A. While understanding how the data are protected is important, these controls might not be applied properly if the data classification schema is not well understood.
B. While implementing data classification, it is most essential that organizational policies and standards, including the data classification schema, are understood by the owner or custodian of the data so they can be properly classified.
C. While an automated data leak prevention (DLP) tool may enhance productivity, the users of the application would still need to understand what classification schema was in place.
D. In terms of protecting the data, the data requirements of end users are critical, but if the data owner does not understand what data classification schema is in place, it would be likely that inappropriate access to sensitive data might be granted by the data owner.

A5-131 While auditing an internally developed web-application, an IS auditor determines that all business users share a common access profile. Which of the following is the **MOST** relevant recommendation to prevent the risk on unauthorized data modification?

A. Enable detailed logging of user actions.
B. Customize user access profiles per job responsibility.
C. Enforce strong password policy for all accounts.
D. Implement regular access rights review.

B is the correct answer.

Justification:

A. Logging is a detective control and often a secondary recommendation in the event that technical issues or costs prohibit implementation of preventive controls.
B. The strongest control is a preventive control that is automated through the system. Developing additional access profiles would ensure that the system restricts users to privileges defined by their job responsibilities and that an audit trail exists for those user actions.
C. While a enforcing password policy is a type of preventive control, it is not as effective as removing excessive access rights from users who do not need it to perform their job duties.
D. Access right review will not help in this scenario, because all profiles have similar set of access rights.

A5-132 Which of the following is the **MOST** important security consideration to an organization that wants to move a business application to external cloud-service (PaaS) provided by a vendor?

A. Classification and categories of data process by the application.
B. Cost of hosting the application internally versus externally.
C. A reputation of a vendor on the market and feedbacks from clients.
D. Drop of application performance due to use of shared services.

A is the correct answer.

Justification:

A. Types of data and its sensitivity is a primary consideration, as there might be legal obligations related to data hosting and its level of protection (e.g. personal information, banking information, health information, etc.)
B. Cost is an important factor for an organization to consider during the move to cloud, however the highest risk is to violate data privacy laws.
C. A reputation of a vendor on the market is an important factor for an organization to consider during the move to cloud, however the highest risk is to violate data privacy laws.
D. Drop of application performance due to use of shared services is an important factor for an organization to consider during the move to cloud, however the highest risk is to violate data privacy laws.

A5-133 Which of the following is **BEST** suited for secure communications within a small group?

A. Key distribution center
B. Certificate authority
C. Web of trust
D. Kerberos Authentication System

C is the correct answer.

Justification:

A. A key distribution center is a part of a Kerberos implementation suitable for internal communication for a large group within an institution, and it will distribute symmetric keys for each session.
B. Certificate authority is a trusted third party that ensures the authenticity of the owner of the certificate. This is necessary for large groups and formal communication.
C. Web of trust is a key distribution method suitable for communication in a small group. It is used by tools such as pretty good privacy and distributes the public keys of users within a group.
D. A Kerberos Authentication System extends the function of a key distribution center by generating "tickets" to define the facilities on networked machines, which are accessible to each user.

A5-134 Which of the following is the **MOST** important action in recovering from a cyberattack?

A. Activating an incident response team
B. Hiring cyberforensic investigators
C. Executing a business continuity plan
D. Preserving evidence

A is the correct answer.

Justification:

A. Hopefully the incident response team and procedures were set up prior to the cyberattack. The first step is to activate the team, contain the incident and keep the business operational.
B. When a cyberattack is suspected, cyberforensic investigators should be used to set up alarms, catch intruders within the network, and track and trace them over the Internet. The use of cyberforensic experts is only done after the incident has been identified.
C. The most important objective in recovering from a cyberattack is to keep the business operational, but most attacks will not require the activation or use of the business continuity plan.
D. The primary objective for the business is to stay in business. In a noncriminal investigation this may even mean that some evidence is lost.

A5-135 What method might an IS auditor use to test wireless security at branch office locations?

A. War dialing
B. Social engineering
C. War driving
D. Password cracking

C is the correct answer.

Justification:

A. War dialing is a technique for gaining access to a computer or a network through the dialing of defined blocks of telephone numbers, with the hope of getting an answer from a modem.
B. Social engineering is a technique used to gather information that can assist an attacker in gaining logical or physical access to data or resources. Social engineering exploits human weaknesses.
C. War driving is a technique for locating and gaining access to wireless networks by driving or walking around a building with a wireless-equipped computer.
D. Password crackers are tools used to guess users' passwords by trying combinations and dictionary words. Once a wireless device has been identified, password crackers may be used to try to attack it.

A5-136 Which of the following intrusion detection systems will **MOST** likely generate false alarms resulting from normal network activity?

A. Statistical-based
B. Signature-based
C. Neural network
D. Host-based

A is the correct answer.

Justification:

A. A statistical-based intrusion detection system (IDS) relies on a definition of known and expected behavior of systems. Because normal network activity may, at times, include unexpected behavior (e.g., a sudden massive download by multiple users), these activities will be flagged as suspicious.
B. A signature-based IDS is limited to its predefined set of detection rules, just like a virus scanner. Signature-based systems traditionally have low levels of false positives but may be weak at detecting new attacks.
C. A neural network combines the statistical- and signature-based IDSs to create a hybrid and better system.
D. Host-based is another type of IDS, but it would not be used to monitor network activity.

A5-137 When auditing security for a data center, an IS auditor should look for the presence of a voltage regulator to ensure that the:

A. hardware is protected against power surges.
B. integrity is maintained if the main power is interrupted.
C. immediate power will be available if the main power is lost.
D. hardware is protected against long-term power fluctuations.

A is the correct answer.

Justification:

A. A voltage regulator protects against short-term power fluctuations.
B. A voltage regulator does not maintain the integrity if power is interrupted or lost.
C. An uninterruptible power supply (UPS) is used to provide constant power even if main power is lost.
D. A voltage regulator protects against short-term power fluctuations.

A5-138 In an organization where an IT security baseline has been defined an IS auditor should **FIRST** ensure:

A. implementation.
B. compliance.
C. documentation.
D. sufficiency.

D is the correct answer.

Justification:

A. The first step is to review the baseline to ensure that it is adequate or sufficient to meet the security requirements of the organization. Then the IS auditor will ensure that it is implemented and measure compliance.
B. Compliance cannot be measured until the baseline has been implemented, but the IS auditor must first ensure that the correct baseline is being implemented.
C. After the baseline has been defined, it must be documented, and the IS auditor will check that the baseline is appropriate before checking for implementation.
D. An IS auditor should first evaluate the definition of the minimum baseline level by ensuring the sufficiency of the control baseline to meet security requirements.

A5-139 Which of the following environmental controls is appropriate to protect computer equipment against short-term reductions in electrical power?

A. Power line conditioners
B. Surge protective devices
C. Alternative power supplies
D. Interruptible power supplies

A is the correct answer.

Justification:

A. **Power line conditioners are used to compensate for peaks and valleys in the power supply and reduce peaks in the power flow to what is needed by the machine. Any valleys are removed by power stored in the equipment.**
B. Surge protection devices protect against high-voltage bursts.
C. Alternative power supplies are intended for power failures that last for longer periods and are normally coupled with other devices such as an uninterruptible power supply to compensate for the power loss until the alternate power supply becomes available.
D. An interruptible power supply would cause the equipment to come down whenever there was a power failure.

A5-140 An IS auditor inspected a windowless room containing phone switching and networking equipment and documentation binders. The room was equipped with two handheld fire extinguishers—one filled with carbon dioxide (CO_2), the other filled with halon gas. Which of the following should be given the **HIGHEST** priority in the IS auditor's report?

A. The halon extinguisher should be removed because halon has a negative impact on the atmospheric ozone layer.
B. Both fire suppression systems present a risk of suffocation when used in a closed room.
C. The CO_2 extinguisher should be removed, because CO_2 is ineffective for suppressing fires involving solid combustibles (paper).
D. The documentation binders should be removed from the equipment room to reduce potential risk.

B is the correct answer.

Justification:

A. The Montreal Protocol allows existing halon installations to remain, although some countries may have laws that require its removal.
B. **Protecting people's lives should always be of highest priority in fire suppression activities. Carbon dioxide (CO_2) and halon both reduce the oxygen ratio in the atmosphere, which can induce serious personal hazards. In many countries, installing or refilling halon fire suppression systems is not allowed.**
C. CO_2 extinguishers can be used on most types of fires, and their use in a server room would be appropriate.
D. Although not of highest priority, removal of the documentation would probably reduce some of the risk.

A5-141 What is a risk associated with attempting to control physical access to sensitive areas such as computer rooms using card keys or locks?

A. Unauthorized individuals wait for controlled doors to open and walk in behind those authorized.
B. The contingency plan for the organization cannot effectively test controlled access practices.
C. Access cards, keys and pads can be easily duplicated allowing easy compromise of the control.
D. Removing access for those who are no longer authorized is complex.

A is the correct answer.

Justification:

A. Piggybacking or tailgating can compromise the physical access controls.
B. The testing of controlled access would be of minimal concern in a disaster recovery environment.
C. Duplicating access control cards or keys is technically challenging.
D. An access control system should have easily followed procedures for managing user access throughout the access life cycle.

A5-142 An organization with extremely high security requirements is evaluating the effectiveness of biometric systems. Which of the following performance indicators is **MOST** important?

A. False-acceptance rate
B. Equal-error rate
C. False-rejection rate
D. False-identification rate

A is the correct answer.

Justification:

A. False-acceptance rate (FAR) is the frequency of accepting an unauthorized person as authorized, thereby granting access when it should be denied. In an organization with high security requirements, limiting the number of false acceptances is more important that the impact on the false reject rate.
B. Equal-error rate (EER) (also called the crossover error rate) is the point where the FAR equals the false-rejection rate (FRR). This is the criteria used to measure the optimal accuracy of the biometric system, but in a highly secure environment, the FAR is more important that the EER.
C. FRR denies an authorized person access, but this is less important than the FAR because it is better to deny access to an authorized individual than to grant access to an unauthorized individual.
D. False-identification rate (FIR) is the probability that an authorized person is identified, but is assigned a false ID.

A5-143 Which of the following groups would create **MOST** concern to an IS auditor if they have full access to the production database?

A. Application developers
B. System administrators
C. Business users
D. Information security team

A is the correct answer.

Justification:

A. Application developers having the access to production environment bear the highest risk. Due to their focus on delivery of changes, they tend to bypass quality assurance controls installing deficient changes into production environment.
B. System administrators may require full production access to conduct their administration duties; however, they should be monitored for unauthorized activity.
C. Business users might not need a full access to database. Such set up might result in negatives scenarios (fraud), however developers having a direct access to production environment is a higher concern.
D. The data recovery team will need full access to make sure the complete database is recoverable.

A5-144 The **BEST** overall quantitative measure of the performance of biometric control devices is:

A. false-rejection rate.
B. false-acceptance rate.
C. equal-error rate.
D. estimated-error rate.

C is the correct answer.

Justification:

A. The false-rejection rate (FRR) only measures the number of times an authorized person is denied entry.
B. The false-acceptance rate (FAR) only measures the number of times an unauthorized person may be accepted as authorized.
C. A low equal-error rate (EER) is a combination of a low FRR and a low FAR. EER, expressed as a percentage, is a measure of the number of times that the FRR and FAR are equal. A low EER is the measure of the more effective biometrics control device.
D. The estimated-error rate is not a valid biometric term.

A5-145 Which of the following is the **MOST** effective control over visitor access to a data center?

A. Visitors are escorted.
B. Visitor badges are required.
C. Visitors sign in.
D. Visitors are spot-checked by operators.

A is the correct answer.

Justification:

A. Escorting visitors will provide the best assurance that visitors have permission to access defined areas within the data processing facility.
B. Requiring visitors to wear badges is a good practice, but not a reliable control.
C. Requiring that visitors sign in is good practice, but not a reliable control. After visitors are in the building, the sign-in process will not prevent them from accessing unauthorized areas.
D. Visitors should be accompanied at all times while they are on the premises, not only when they are in the data processing facility.

A5-146 In a public key infrastructure, a registration authority:

A. verifies information supplied by the subject requesting a certificate.
B. issues the certificate after the required attributes are verified and the keys are generated.
C. digitally signs a message to achieve nonrepudiation of the signed message.
D. registers signed messages to protect them from future repudiation.

A is the correct answer.

Justification:
A. **A registration authority is responsible for verifying information supplied by the subject requesting a certificate and verifies the requestor's right to request a certificate on behalf of themselves or their organization.**
B. Certification authorities, not registration authorities, actually issue certificates once verification of the information has been completed.
C. The sender who has control of his/her private key signs the message, not the registration authority.
D. Registering signed messages is not a task performed by registration authorities.

A5-147 Confidentiality of the data transmitted in a wireless local area network is **BEST** protected if the session is:

A. restricted to predefined media access control addresses.
B. encrypted using static keys.
C. encrypted using dynamic keys.
D. initiated from devices that have encrypted storage.

C is the correct answer.

Justification:
A. Limiting the number of devices that can access the network via media access control address filtering is an inefficient control and does not address the issue of encrypting the session.
B. Encryption with static keys—using the same key for a long period of time—carries a risk that the key would be compromised.
C. **When using dynamic keys, the encryption key is changed frequently, thus reducing the risk of the key being compromised and the message being decrypted.**
D. Encryption of the data on the connected device (laptop, smart phone, etc.) addresses the confidentiality of the data on the device, not the wireless session.

A5-148 Which of the following provides the **MOST** relevant information for proactively strengthening security settings?

A. Bastion host
B. Intrusion detection system
C. Honeypot
D. Intrusion prevention system

C is the correct answer.

Justification:
A. A bastion host is a hardened system used to host services. It does not provide information about an attack.
B. Intrusion detection systems are designed to detect and address an attack in progress and stop it as soon as possible.
C. **The design of a honeypot is such that it lures the hacker and provides clues as to the hacker's methods and strategies, and the resources required to address such attacks. A honeypot allows the attack to continue, so as to obtain information about the hacker's strategy and methods.**
D. Intrusion prevention systems are designed to detect and address an attack in progress and stop it as soon as possible.

A5-149 Over the long term, which of the following has the greatest potential to improve the security incident response process?

A. A walk-through review of incident response procedures
B. Simulation exercises performed by incident response team
C. Ongoing security training for users
D. Documenting responses to an incident

B is the correct answer.

Justification:
A. A walk-through is a good first step to evaluate the incident response plan, but the lessons learned from incidents will provide more meaningful long-term benefits.
B. Simulation exercises to find the gaps and shortcomings in the actual incident response processes will help improve the process over time.
C. Training the users and members of the incident response team will improve the effectiveness of the team but learning from the lessons of previous incidents will generate the greatest benefit.
D. Documenting all incidents is important to allow later analysis and review but is not as important as the results of the analysis.

A5-150 When reviewing an intrusion detection system, an IS auditor should be **MOST** concerned about which of the following?

A. High number of false-positive alarms
B. Low coverage of network traffic
C. Network performance downgrade
D. Default detection settings

B is the correct answer.

Justification:
A. Although the number of false-positives is a serious issue, the problem will be known and can be corrected.
B. The cybersecurity attacks might not be timely identified if only small portion of network traffic is analyzed.
C. Intrusion detection system might decrease an overall network performance, however it is a secondary risk in this case.
D. It is a good practice to customize IDS settings to specific network perimeter, however there is a higher likelihood to miss the attacks due to insufficient network coverage.

A5-151 Distributed denial-of-service attacks on Internet sites are typically evoked by hackers using which of the following?

A. Logic bombs
B. Phishing site
C. Spyware
D. Botnets

D is the correct answer.

Justification:
A. Logic bombs are programs designed to destroy or modify data at a specific event or time in the future.
B. Phishing is an attack, normally via email, pretending to be an authorized person or organization requesting information.
C. Spyware is a program that picks up information from PC drives by making copies of their contents.
D. A botnet is a number of Internet-connected devices, each of which is running one or more bots. Botnets can be used to perform distributed denial-of-service attack (DDoS attack), steal data, send spam, and allows the attacker to access the device and its connection.

A5-152 Validated digital signatures in an email software application will:

A. help detect spam.
B. provide confidentiality.
C. add to the workload of gateway servers.
D. significantly reduce available bandwidth.

A is the correct answer.

Justification:

A. Validated electronic signatures are based on qualified certificates that are created by a certificate authority, with the technical standards required to ensure the key can neither be forced nor reproduced in a reasonable time. Such certificates are only delivered through a registration authority after a proof of identity has been passed. Using strong signatures in email traffic, nonrepudiation can be assured, and a sender can be tracked. The recipient can configure his/her email server or client to automatically delete emails from specific senders.

B. For confidentiality issues, one must use encryption, not a signature.

C. Without any filters directly applied on mail gateway servers to block traffic without strong signatures, the workload will not increase. Using filters directly on a gateway server will result in an overhead less than antivirus software imposes.

D. Digital signatures are only a few bytes in size and will not slash bandwidth. Even if gateway servers were to check certificate revocation lists, there is little overhead.

A5-153 In transport mode, the use of the Encapsulating Security Payload protocol is advantageous over the authentication header protocol because it provides:

A. connectionless integrity.
B. data origin authentication.
C. antireplay service.
D. confidentiality.

D is the correct answer.

Justification:

A. Both forms of Internet Protocol security (IPSec), authentication header (AH) and encapsulating security payload (ESP), provide connectionless integrity.

B. Both AH and ESP authenticate data origin.

C. The time stamps used in IPSec will prevent replay attacks.

D. Only the ESP protocol provides confidentiality via encryption.

A5-154 IS management recently replaced its existing wired local area network with a wireless infrastructure to accommodate the increased use of mobile devices within the organization. This will increase the risk of which of the following attacks?

A. Port scanning
B. Back door
C. Man-in-the-middle
D. War driving

D is the correct answer.

Justification:
A. Port scanning will often target the external firewall of the organization. Use of wireless will not affect this.
B. A back door is an opening implanted into or left in software that enables an unauthorized entry into a system.
C. Man-in-the-middle attacks intercept a message and can read, replace or modify it.
D. A war driving attack uses a wireless Ethernet card, set in promiscuous mode, and a powerful antenna to penetrate wireless systems from outside.

A5-155 Which of the following is the **GREATEST** concern associated with the use of peer-to-peer computing?

A. Virus infection
B. Data leakage
C. Network performance issues
D. Unauthorized software usage

B is the correct answer.

Justification:
A. While peer-to-peer computing does increase the risk of virus infection, the risk of data leakage is more severe, especially if it contains proprietary data or intellectual property.
B. Peer-to-peer computing can share the contents of a user hard drive over the Internet. The risk that sensitive data could be shared with others is the greatest concern.
C. Peer-to-peer computing may use more network bandwidth and, therefore, may create performance issues. However, data leakage is a more severe risk.
D. Peer-to-peer computing may be used to download or share unauthorized software, which users could install on their PCs unless other controls prevent it. However, data leakage is a more severe risk.

A5-156 The IS management of a multinational company is considering upgrading its existing virtual private network to support Voice-over Internet Protocol communication via tunneling. Which of the following considerations should be **PRIMARILY** addressed?

A. Reliability and quality of service
B. Means of authentication
C. Privacy of voice transmissions
D. Confidentiality of data transmissions

A is the correct answer.

Justification:

A. Reliability and quality of service (QoS) are the primary considerations to be addressed. Voice communications require consistent levels of service, which may be provided through QoS and class of service controls.
B. The company currently has a virtual private network (VPN); authentication has been implemented by the VPN using tunneling.
C. Privacy of voice transmissions is provided by the VPN protocol.
D. The company currently has a VPN; confidentiality of both data and Voice-over Internet Protocol traffic has been implemented by the VPN using tunneling.

A5-157 Which of the following antispam filtering methods has the **LOWEST** possibility of false-positive alerts?

A. Rule-based
B. Check-sum based
C. Heuristic filtering
D. Statistic-based

B is the correct answer.

Justification:

A. Rule-based filtering will trigger false-positive alert each time a key word is met in the message.
B. The advantage of this type of filtering is that it lets ordinary users help identify spam, and not just administrators, thus vastly increasing the pool of spam fighters. The disadvantage is that spammers can insert unique invisible gibberish—known as hashbusters—into the middle of each of their messages, thus making each message unique and having a different checksum. This leads to an arms race between the developers of the checksum software and the developers of the spam-generating software.
C. A heuristic is a technique designed for solving a problem more quickly when classic methods are too slow, or for finding an approximate solution when classic methods fail to find any exact solution. This is achieved by trading optimality, completeness, accuracy, or precision for speed. In a way, it can be considered a shortcut.
D. Statistical filtering analyze frequency of each word within the message and then evaluating the message as a whole. Therefore, it can ignore a suspicious keyword if the entire message is within normal bounds, however prone to false-positive alerts.

A5-158 Which of the following public key infrastructure (PKI) elements describes procedure for disabling a compromised private key?

A. Certificate revocation list
B. Certification practice statement
C. Certificate policy
D. PKI disclosure statement

B is the correct answer.

Justification:
A. The certificate revocation list is a list of certificates that have been revoked before their scheduled expiration date.
B. The certification practice statement is the how-to document used in policy-based public key infrastructure (PKI).
C. The certificate policy sets the requirements that are subsequently implemented by the CPS.
D. The PKI disclosure statement covers critical items such as the warranties, limitations and obligations that legally bind each party.

A5-159 The use of residual biometric information to gain unauthorized access is an example of which of the following attacks?

A. Replay
B. Brute force
C. Cryptographic
D. Mimic

A is the correct answer.

Justification:
A. Residual biometric characteristics, such as fingerprints left on a biometric capture device, may be reused by an attacker to gain unauthorized access.
B. A brute force attack involves feeding the biometric capture device numerous different biometric samples.
C. A cryptographic attack targets the algorithm or the encrypted data.
D. In a mimic attack, the attacker reproduces characteristics similar to those of the enrolled user such as forging a signature or imitating a voice.

A5-160 An IS auditor is reviewing system access and discovers an excessive number of users with privileged access. The IS auditor discusses the situation with the system administrator, who states that some personnel in other departments need privileged access and management has approved the access. Which of the following would be the **BEST** course of action for the IS auditor?

A. Determine whether compensating controls are in place.
B. Document the issue in the audit report.
C. Recommend an update to the procedures.
D. Discuss the issue with senior management.

A is the correct answer.

Justification:
A. An excessive number of users with privileged access is not necessarily an issue if compensating controls are in place.
B. An IS auditor should gather additional information before presenting the situation in the report.
C. An update to procedures would not address a potential weakness in logical security and may not be feasible if individuals are required to have this access to perform their jobs.
D. The IS auditor should gather additional information before reporting the item to senior management.

A5-161 Two-factor authentication can be circumvented through which of the following attacks?

A. Denial-of-service
B. Man-in-the-middle
C. Key logging
D. Brute force

B is the correct answer.

Justification:

A. A denial-of-service attack does not have a relationship to authentication.
B. A man-in-the-middle attack is similar to piggybacking in that the attacker pretends to be the legitimate destination, and then merely retransmits whatever is sent by the authorized user along with additional transactions after authentication has been accepted. This is done in many instances of bank fraud.
C. Key logging could circumvent single-factor authentication but not two-factor authentication.
D. Brute force could circumvent single-factor authentication but not two-factor authentication.

A5-162 An organization can ensure that the recipients of emails from its employees can authenticate the identity of the sender by:

A. digitally signing all email messages.
B. encrypting all email messages.
C. compressing all email messages.
D. password protecting all email messages.

A is the correct answer.

Justification:

A. By digitally signing all email messages, the receiver will be able to validate the authenticity of the sender.
B. Encrypting all email messages would ensure that only the intended recipient will be able to open the message; however, it would not ensure the authenticity of the sender.
C. Compressing all email messages would reduce the size of the message but would not ensure authenticity.
D. Password protecting all email messages would ensure that only those who have the password would be able to open the message; however, it would not ensure authenticity of the sender.

SCENARIO

A scenario is a mini-case study that describes a situation or an organization and requires candidates to answer one or more questions based on the information provided. A scenario can focus on a specific domain or on several domains. The CISA exam will include scenarios.

QUESTIONS A5-163 THROUGH A5-164 REFER TO THE FOLLOWING INFORMATION:

Company XYZ has outsourced production support to service provider ABC located in another country. The ABC service provider personnel remotely connect to the corporate network of the XYZ outsourcing entity over the Internet.

A5-163 Which of the following would provide the **BEST** assurance that only authorized users of ABC connect over the Internet for production support to XYZ?

A. Single sign-on authentication
B. Password complexity requirements
C. Two-factor authentication
D. Internet Protocol address restrictions

C is the correct answer.

Justification:

A. Single sign-on authentication provides a single access point to system resources. It would not be best in this situation.
B. While password complexity requirements would help prevent unauthorized access, two-factor authentication is a more effective control for this scenario.
C. Two-factor authentication is the best method to provide a secure connection because it uses two factors, typically "what you have" (for example, a device to generate one-time-passwords), "what you are" (for example, biometric characteristics) or "what you know" (for example, a personal identification number or password). Using a password in and of itself without the use of one or more of the other factors mentioned is not the best for this scenario.
D. Internet Protocol addresses can always change or be spoofed and, therefore, are not the best form of authentication for the scenario mentioned.

SEE INFORMATION PRECEDING QUESTION A5-163

A5-164 Which of the following would **BEST** provide assurance that transmission of information is secure while the production support team at ABC is providing support to XYZ?

A. Secret key encryption
B. Dynamic Internet Protocol address and port
C. Hash functions
D. Virtual private network tunnel

D is the correct answer.

Justification:

A. Secret key encryption would require sharing of the same key at the source and destination and involve an additional step for encrypting and decrypting data at each end. This is not a feasible solution given the scenario.
B. Using a dynamic Internet Protocol address and port is not an effective control because an attacker could easily find the new address using the domain name system.
C. While the use of a cryptographic hash function may be helpful to validate the integrity of data files, in this case it would not be useful for a production support team connecting remotely.
D. As ABC and XYZ are communicating over the Internet, which is an untrusted network, establishing an encrypted virtual private network tunnel would best ensure that the transmission of information was secure.

A5-165 The **PRIMARY** purpose of installing data leak prevention software is to:

A. restrict user access to confidential files stored on servers.
B. detect attempts to destroy sensitive data in an internal network.
C. block external systems from accessing internal resources.
D. control confidential documents leaving the internal network.

D is the correct answer.

Justification:

A. Access privileges to confidential files stored on the server will be controlled through digital rights management (DRM) software.
B. Potential attacks to systems on the internal network would normally be controlled through an intrusion detection system (IDS) and intrusion prevention system (IPS) as well as by security controls of the systems themselves. Data leak prevention (DLP) systems focus on data leaving the enterprise.
C. Controlling what external systems can access internal resources is the function of a firewall rather than a DLP system.
D. A server running a DLP software application uses predefined criteria to check whether any confidential documents or data are leaving the internal network.

A5-166 Which of the following is a control that can be implemented to reduce risk of internal fraud if application programmers are allowed to move programs into the production environment in a small organization?

A. Post-implementation functional testing
B. Registration and review of changes
C. Validation of user requirements
D. User acceptance testing

B is the correct answer.

Justification:

A. Independent postimplementation testing would not be as effective because the system could be accepted by the end user without detecting the undocumented functionality.
B. An independent review of the changes to the program in production could identify potential unauthorized changes, versions or functionality that the programmer had put into production.
C. An independent review of user requirements would not be as effective because the system could meet user requirements and still include undocumented functionalities.
D. An independent review of user acceptance would not be as effective because the system could be accepted by the end users, and the undocumented functionalities could remain undetected.

A5-167 A characteristic of User Datagram Protocol in network communications is:

A. packets may arrive out of order.
B. increased communication latency.
C. incompatibility with packet broadcast.
D. error correction may slow down processing.

A is the correct answer.

Justification:

A. User Datagram Protocol (UDP) uses a simple transmission model without implicit handshaking routines for providing reliability, ordering or data integrity. Thus, UDP provides an unreliable service and datagrams may arrive out of order, appear duplicated or get dropped.
B. The advantage of UDP is that the lack of error checking allows for reduced latency. Time-sensitive applications, such as online video or audio, often use UDP because of the reduced latency of this protocol.
C. UDP is compatible with packet broadcast (sending to all on the local network) and multicasting (sending to all subscribers).
D. UDP assumes that error checking and correction is either not necessary or performed in the application, avoiding the overhead of such processing at the network interface level.

A5-168 Which of the following choices is the **MOST** effective control that should be implemented to ensure accountability for application users accessing sensitive data in the human resource management system (HRMS) and among interfacing applications to the HRMS?

A. Two-factor authentication
B. A digital certificate
C. Audit trails
D. Single sign-on authentication

C is the correct answer.

Justification:

A. Two-factor authentication would enhance security while logging into the human resource management system application; however, it will not establish accountability for actions taken subsequent to login.
B. A digital certificate will also enhance login security to conclusively authenticate users logging into the application. However, it will not establish accountability because user ID and transaction details will not be captured without an audit trail.
C. Audit trails capture which user, at what time, and date, along with other details, has performed the transaction and this helps in establishing accountability among application users.
D. Single sign-on authentication allows users to log in seamlessly to the application, thus easing the authentication process. However, this would also not establish accountability.

A5-169 An IS auditor reviewing wireless network security determines that the Dynamic Host Configuration Protocol is disabled at all wireless access points. This practice:

A. reduces the risk of unauthorized access to the network.
B. is not suitable for small networks.
C. automatically provides an IP address to anyone.
D. increases the risk associated with Wireless Encryption Protocol (WEP).

A is the correct answer.

Justification:

A. Dynamic Host Configuration Protocol (DHCP) automatically assigns IP addresses to anyone connecting to the network. With DHCP disabled, static IP addresses must be used, and this requires either administrator support or a higher level of technical skill to attach to the network and gain Internet access.
B. DHCP is suitable for networks of all sizes from home networks to large complex organizations.
C. DHCP does not provide IP addresses when disabled.
D. Disabling of the DHCP makes it more difficult to exploit the well-known weaknesses in Wireless Encryption Protocol.

A5-170 Which of the following is **MOST** indicative of the effectiveness of an information security awareness program?

A. Employees report more information regarding security incidents.
B. All employees have signed the information security policy.
C. Most employees have attended an awareness session.
D. Information security responsibilities have been included in job descriptions.

A is the correct answer.

Justification:

A. Although the promotion of security awareness is a preventive control, it can also be a detective measure because it encourages people to identify and report possible security violations. The reporting of incidents implies that employees are acting as a consequence of the awareness program.
B. The existence of evidence that all employees have signed the security policy does not ensure that security responsibilities have been understood and applied.
C. One of the objectives of the security awareness program is to inform the employees of what is expected of them and what their responsibilities are, but this knowledge does not ensure that employees will perform their activities in a secure manner.
D. The documentation of roles and responsibilities in job descriptions is not an indicator of the effectiveness of the awareness program.

A5-171 An organization stores and transmits sensitive customer information within a secure wired network. It has implemented an additional wireless local area network (WLAN) to support general-purpose staff computing needs. A few employees with WLAN access have legitimate business reasons for also accessing customer information. Which of the following represents the **BEST** control to ensure separation of the two networks?

A. Establish two physically separate networks.
B. Implement virtual local area network segmentation.
C. Install a dedicated router between the two networks.
D. Install a firewall between the networks.

D is the correct answer.

Justification:

A. While having two physically separate networks would ensure the security of customer data, it would make it impossible for authorized wireless users to access that data.
B. While a VLAN would provide separation of the two networks, it is possible, with sufficient knowledge, for an attacker to gain access to one VLAN from the other.
C. A dedicated router between the two networks would separate them; however, this would be less secure than a firewall.
D. In this case, a firewall could be used as a strong control to allow authorized users on the wireless network to access the wired network.

A5-172 From a control perspective, the **PRIMARY** objective of classifying information assets is to:

A. establish guidelines for the level of access controls that should be assigned.
B. ensure access controls are assigned to all information assets.
C. assist management and auditors in risk assessment.
D. identify which assets need to be insured against losses.

A is the correct answer.

Justification:

A. Information has varying degrees of sensitivity and criticality in meeting business objectives. By assigning classes or levels of sensitivity and criticality to information resources, management can establish guidelines for the level of access controls that should be assigned. End user management and the security administrator will use these classifications in their risk assessment process to assign a given class to each asset.
B. Not all information needs to be protected through access controls. Overprotecting data would be expensive.
C. The classification of information is usually based on the risk assessment, not the other way around.
D. Insuring assets is valid; however, this is not the primary objective of information classification.

A5-173 An IS auditor reviewing access controls for a client-server environment should **FIRST**:

A. evaluate the encryption technique.
B. identify the network access points.
C. review the identity management system.
D. review the application level access controls.

B is the correct answer.

Justification:

A. Evaluating encryption techniques would be performed at a later stage of the review.
B. A client-server environment typically contains several access points and uses distributed techniques, increasing the risk of unauthorized access to data and processing. To evaluate the security of the client server environment, all network access points should be identified.
C. Reviewing the identity management system would be performed at a later stage of the review.
D. Reviewing the application level access controls would be performed at a later stage of the review.

A5-174 To prevent Internet Protocol (IP) spoofing attacks, a firewall should be configured to drop a packet for which the sender of a packet:

A. specifies the route that a packet should take through the network (the source routing field is enabled).
B. puts multiple destination hosts (the destination field has a broadcast address).
C. indicates that the computer should immediately stop using the TCP connection (a reset flag is turned on).
D. allows use of dynamic routing instead of static routing (Open Shortest Path First protocol is enabled).

A is the correct answer.

Justification:

A. Internet Protocol (IP) spoofing takes advantage of the source-routing option in the IP. With this option enabled, an attacker can insert a spoofed source IP address. The packet will travel the network according to the information within the source-routing field, bypassing the logic in each router, including dynamic and static routing.
B. If a packet has a broadcast destination address, it is definitely suspicious and if allowed to pass will be sent to all addresses in the subnet. This is not related to IP spoofing.
C. Turning on the reset flag is part of the normal procedure to end a Transmission Control Protocol connection.
D. The use of dynamic or static routing will not represent a spoofing attack.

A5-175 An IS auditor is reviewing a manufacturing company and finds that mainframe users at a remote site connect to the mainframe at headquarters over the Internet via Telnet. Which of the following offers the **STRONGEST** security?

A. Use of a point-to-point leased line
B. Use of a firewall rule to allow only the Internet Protocol address of the remote site
C. Use of two-factor authentication
D. Use of a nonstandard port for Telnet

A is the correct answer.

Justification:

A. A leased line will effectively extend the local area network of the headquarters to the remote site, and the mainframe Telnet connection would travel over the private line, which would be less of a security risk when using an insecure protocol such as Telnet.

B. A firewall rule at the headquarters network to only allow Telnet connections from the Internet Protocol (IP) address assigned to the remote site would make the connection more secure than the current arrangement, but a dedicated leased line is the most secure option of those listed.

C. While two-factor authentication would enhance the login security, it would not secure the transmission channel against eavesdropping, and, therefore, a leased line would be a better option.

D. Attacks on network services start with the assumption that network services use the standard Transmission Control Protocol/IP port number assigned for the service, which is port 23 for Telnet. By reconfiguring the host and client, a different port can be used. Assigning a nonstandard port for services is a good general security practice because it makes it more difficult to determine what service is using the port; however, in this case, creating a leased-line connection to the remote site would be a better solution.

A5-176 There is a concern that the risk of unauthorized access may increase after implementing a single sign-on process. To prevent unauthorized access, the **MOST** important action is to:

A. monitor failed authentication attempts.
B. review log files regularly.
C. deactivate unused accounts promptly.
D. mandate a strong password policy.

D is the correct answer.

Justification:

A. Ensuring that all failed authentication attempts are monitored is a good practice but is not a preventive control.

B. Reviewing the log files can increase the probability of detecting unauthorized access but will not prevent unauthorized access.

C. Ensuring that all unused accounts are deactivated is important; however, unauthorized access may occur via a regularly used account.

D. Strong passwords are important in any environment but take on special importance in an SSO environment, where a user enters a password only one time and thereafter has general access throughout the environment. Of the options given, only a strong password policy offers broad preventative effects.

A5-177 An IS auditor reviewing the implementation of an intrusion detection system (IDS) should be **MOST** concerned if:

A. IDS sensors are placed outside of the firewall.
B. a behavior-based IDS is causing many false alarms.
C. a signature-based IDS is weak against new types of attacks.
D. the IDS is used to detect encrypted traffic.

B is the correct answer.

Justification:

A. An organization can place sensors outside of the firewall to detect attacks. These sensors are placed in highly sensitive areas and on extranets.
B. An excessive number of false alarms from a behavior-based intrusion detection system (IDS) indicates that additional tuning is needed. False positives cannot be eliminated entirely, but ignoring this warning sign may negate the value of the system by causing those responsible for monitoring its warnings to become convinced that anything reported is false.
C. Being weak against new types of attacks is expected from a signature-based IDS because it can only recognize attacks that have been previously identified.
D. An IDS cannot detect attacks within encrypted traffic, but there may be good reason to detect the presence of encrypted traffic, such as when a next-generation firewall is configured to terminate encrypted connections at the perimeter. In such cases, detecting encrypted packets flowing past the firewall could indicate improper configuration or even a compromise of the firewall itself.

A5-178 Which of the following **BEST** describes the role of a directory server in a public key infrastructure?

A. Encrypts the information transmitted over the network
B. Makes other users' certificates available to applications
C. Facilitates the implementation of a password policy
D. Stores certificate revocation lists

B is the correct answer.

Justification:

A. Encrypting the information transmitted over the network is a role performed by a security server.
B. A directory server makes other users' certificates available to applications.
C. Facilitating the implementation of a password policy is not relevant to public key infrastructure .
D. Storing certificate revocation lists is a role performed by a security server.

A5-179 An IS auditor is reviewing an organization's network operations center (NOC). Which of the following choices is of the **GREATEST** concern? The use of:

A. a wet pipe-based fire suppression system.
B. a rented rack space in the NOC.
C. a carbon dioxide-based fire suppression system.
D. an uninterrupted power supply with 10 minutes of backup power.

C is the correct answer.

Justification:

A. Wet pipe systems may damage computer equipment, but they are safe for humans and not as damaging as carbon dioxide (CO2) systems.
B. Rented rack space is not a concern as long as security controls are maintained. Most organizations rent server rack space.
C. CO2 systems should not be used in areas where people are present, because their function will cause suffocation in the event of a fire. Controls should consider personnel safety first.
D. Depending on the system, a few minutes might be all that is needed for a graceful shutdown. However, a CO2 system is dangerous for personnel.

A5-180 Inadequate programming and coding practices increase the risk of:

A. social engineering.
B. buffer overflow exploitation.
C. synchronize flood.
D. brute force attacks.

B is the correct answer.

Justification:

A. Social engineering attempts to gather sensitive information from people and primarily relies on human behavior. This is not a programming or coding problem.

B. Buffer overflow exploitation may occur when programs do not check the length of the data that are input into a program. An attacker can send data that exceed the length of a buffer and overwrite part of the program with arbitrary code, which will then be executed with the privileges of the program. The countermeasure is proper programming and good coding practices.

C. A synchronize (SYN) flood is a form of denial-of-service attack in which an attacker sends a succession of SYN requests to a target system. A SYN flood is not related to programming and coding practices.

D. Brute force attacks are used against passwords and are not related to programming and coding practices.

A5-181 During an access control review for a mainframe application, an IS auditor discovers user security groups without designated owners. The **PRIMARY** reason that this is a concern to the IS auditor is that, without ownership, there is no one with clear responsibility for:

A. updating group metadata.
B. reviewing existing user access.
C. approval of user access.
D. removing terminated users.

C is the correct answer.

Justification:

A. Updating data about the group is not a great concern when compared to unauthorized access.

B. While the periodic review of user accounts is a good practice, this is a detective control and not as robust as preventing unauthorized access to the group in the first place.

C. Without an owner to provide approval for user access to the group, unauthorized individuals could potentially gain access to any sensitive data within the rights of the group.

D. Revoking access to terminated users is a compensating control for the normal termination process and is also a detective control.

A5-182 An IS auditor discovers that uniform resource locators (URLs) for online control self-assessment questionnaires are sent using URL shortening services. The use of URL shortening services would **MOST** likely increase the risk of which of the following attacks?

A. Spoofing
B. Phishing
C. Buffer overflow
D. Denial of service

B is the correct answer.

Justification:

A. Spoofing applies to source addressing, while uniform resource locator (URL) shortening applies to destination addressing.
B. URL shortening services have been adopted by hackers to fool users and spread malware (i.e., phishing)
C. Buffer overflows are not generally associated with URL shortening.
D. Denial-of-service attacks are not affected by URL shortening services.

A5-183 When installing an intrusion detection system, which of the following is **MOST** important?

A. Properly locating it in the network architecture
B. Preventing denial-of-service attacks
C. Identifying messages that need to be quarantined
D. Minimizing the rejection errors

A is the correct answer.

Justification:

A. Proper location of an intrusion detection system (IDS) in the network is the most important decision during installation. A poorly located IDS could leave key areas of the network unprotected.
B. A network IDS will monitor network traffic and a host-based IDS will monitor activity on the host, but it has no capability of preventing a denial-of-service (DoS) attack.
C. Configuring an IDS can be a challenge because it may require the IDS to "learn" what normal activity is, but the most important part of the installation is to install it in the right places.
D. An IDS is only a monitoring device and does not reject traffic. Rejection errors would apply to a biometric device.

A5-184 Which of the following is the **BEST** criterion for evaluating the adequacy of an organization's security awareness program?

A. Senior management is aware of critical information assets and demonstrates an adequate concern for their protection.
B. Job descriptions contain clear statements of accountability for information security.
C. In accordance with the degree of risk and business impact, there is adequate funding for security efforts.
D. No actual incidents have occurred that have caused a loss or a public embarrassment.

B is the correct answer.

Justification:

A. Senior management's level of awareness and concern for information assets is a criterion for evaluating the importance that they attach to those assets and their protection, but it is not as meaningful as having job descriptions that require all staff to be responsible for information security.
B. The inclusion of security responsibilities in job descriptions is a key factor in demonstrating the maturity of the security program and helps ensure that staff and management are aware of their roles with respect to information security.
C. Funding is important but having funding does not ensure that the security program is effective or adequate.
D. The number of incidents that have occurred is a criterion for evaluating the adequacy of the risk management program, but it is not a criterion for evaluating a security program.

A5-185 Which of the following features of a public key infrastructure is **MOST** closely associated with proving that an online transaction was authorized by a specific customer?

A. Nonrepudiation
B. Encryption
C. Authentication
D. Integrity

A is the correct answer.

Justification:

A. Nonrepudiation, achieved through the use of digital signatures, prevents the senders from later denying that they generated and sent the message.
B. Encryption plays a role in creating digital signatures, which are used to provide nonrepudiation, but encryption is also used for other purposes, whereas nonrepudiation is entirely concerned with ensuring that specific actions can be traced to specific actors in a manner beyond reasonable doubt.
C. Authentication is necessary to establish the identification of all parties to a communication but does not play a central role in the scenario described.
D. Integrity ensures that transactions are accurate but does not provide the identification of the customer.

A5-186 After reviewing its business processes, a large organization is deploying a new web application based on a Voice-over Internet Protocol technology. Which of the following is the **MOST** appropriate approach for implementing access control that will facilitate security management of the VoIP web application?

A. Fine-grained access control
B. Role-based access control
C. Access control lists
D. Network/service access control

B is the correct answer.

Justification:
A. Fine-grained access control on Voice-over Internet Protocol (VoIP) web applications does not scale to enterprisewide systems because it is primarily based on individual user identities and their specific technical privileges.
B. Authorization in this case can best be addressed by role-based access control (RBAC) technology. RBAC controls access according to job roles or functions. RBAC is easy to manage and can enforce strong and efficient access controls in large-scale web environments including VoIP implementation.
C. Access control lists on VoIP web applications do not scale to enterprisewide systems because they are primarily based on individual user identities and their specific technical privileges.
D. Network/service addresses VoIP availability but does not address application-level access or authorization.

A5-187 During a logical access controls review, an IS auditor observes that user accounts are shared. The **GREATEST** risk resulting from this situation is that:

A. an unauthorized user may use the ID to gain access.
B. user access management is time consuming.
C. user accountability is not established.
D. passwords are easily guessed.

C is the correct answer.

Justification:
A. The risk of an unauthorized user accessing the system with a shared ID is no greater than an unauthorized user accessing the system with a unique user ID.
B. Access management would not be any different with shared IDs.
C. The use of a single user ID by more than one individual precludes knowing who, in fact, used that ID to access a system; therefore, it is more difficult to hold anyone accountable.
D. Shared user IDs do not necessarily have easily guessed passwords.

A5-188 To protect a Voice-over Internet Protocol infrastructure against a denial-of-service attack, it is **MOST** important to secure the:

A. access control servers.
B. session border controllers.
C. backbone gateways.
D. intrusion detection system.

B is the correct answer.

Justification:

A. Securing the access control server may prevent account alteration or lockout but is not the primary protection against denial-of-service (DoS) attacks.

B. Session border controllers enhance the security in the access network and in the core. In the access network, they hide a user's real address and provide a managed public address. This public address can be monitored, minimizing the opportunities for scanning and DoS attacks. Session border controllers permit access to clients behind firewalls while maintaining the firewall's effectiveness. In the core, session border controllers protect the users and the network. They hide network topology and users' real addresses. They can also monitor bandwidth and quality of service.

C. Backbone gateways are isolated and not readily accessible to hackers, so this is not a location of DoS attacks.

D. Intrusion detection systems monitor traffic, but do not protect against DoS attacks.

A5-189 In an online banking application, which of the following would **BEST** protect against identity theft?

A. Encryption of personal password
B. Restricting the user to a specific terminal
C. Two-factor authentication
D. Periodic review of access logs

C is the correct answer.

Justification:

A. A password alone is only single-factor authentication and could be guessed or broken.

B. Restricting the user to a specific terminal is not a practical alternative for an online application because the users may need to log in from multiple devices.

C. Two-factor authentication requires two independent methods for establishing identity and privileges. Factors include something you know such as a password; something you have such as a token; and something you are which is biometric. Requiring two of these factors makes identity theft more difficult.

D. Periodic review of access logs is a detective control and does not protect against identity theft.

A5-190 An IS auditor has found that employees are emailing sensitive company information to public web-based email domains. Which of the following is the **BEST** remediation option for the IS auditor to recommend?

A. Encrypted mail accounts
B. Training and awareness
C. Activity monitoring
D. Data loss prevention

D is the correct answer.

Justification:

A. Encrypted email accounts will secure the information being sent but will not prevent an employee from sending the information to an unauthorized person.
B. Training and awareness may influence employee behavior but are not effective as preventative controls when dealing with intentional exfiltration.
C. Activity monitoring is a detective control and will not prevent data from leaving the network.
D. Data loss prevention is an automated preventive tool that can block sensitive information from leaving the network, while at the same time logging the offenders. This is a better choice than relying on training and awareness because it works equally well when there is intent to steal data.

A5-191 Which of the following potentially blocks hacking attempts?

A. Intrusion detection system
B. Honeypot system
C. Intrusion prevention system
D. Network security scanner

C is the correct answer.

Justification:

A. An intrusion detection system is a detective control.
B. A honeypot solution captures intruder activity or traps the intruders when they attempt to explore a simulated target.
C. An intrusion prevention system is deployed as an inline device on a network or host that can detect and block hacking attempts.
D. A network security scanner identifies vulnerabilities but does not remediate them.

A5-192 A web server is attacked and compromised. Organizational policy states that incident response should balance containment of an attack with retaining freedom for later legal action against an attacker. Under the circumstances, which of the following should be performed **FIRST**?

A. Dump the volatile storage data to a disk.
B. Run the server in a fail-safe mode.
C. Disconnect the web server from the network.
D. Shut down the web server.

C is the correct answer.

Justification:

A. Dumping the volatile storage data to a disk may be used at the investigation stage but does not contain an attack in progress.
B. To run the server in a fail-safe mode, the server needs to be shut down.
C. The first action is to disconnect the web server from the network to secure the device for investigation, contain the damage and prevent more actions by the attacker.
D. Shutting down the server could potentially erase information that might be needed for a forensic investigation or to develop a strategy to prevent future similar attacks.

A5-193 What is the **BEST** approach to mitigate the risk of a phishing attack?

A. Intrusion detection
B. Security assessment
C. Strong authentication
D. User education

D is the correct answer.

Justification:

A. Intrusion detection systems (IDSs) will capture network or host traffic for analysis and may detect malicious activity but are not generally effective against phishing attacks.
B. Assessing security does not mitigate the risk. Phishing is based on social engineering and often distributed through email.
C. Phishing attacks can be mounted in various ways, often through email; strong two-factor authentication cannot mitigate most types of phishing attacks.
D. The best way to mitigate the risk of phishing is to educate users to take caution with suspicious Internet communications and not to trust them until verified. Users may require regular training to recognize suspicious web pages and email as the means and methods of threat actors evolve.

A5-194 A key IT systems developer has suddenly resigned from an enterprise. Which of the following will be the **MOST** important action?

A. Set up an exit interview with human resources.
B. Initiate the handover process to ensure continuity of the project.
C. Terminate the developer's logical access to IT resources.
D. Ensure that management signs off on the termination paperwork.

C is the correct answer.

Justification:

A. The interview with human resources (HR) is also an important process if it is conducted by the last date of employment, but it is of secondary importance compared to removing the developer's access to systems.
B. As long as the handover process to a designated employee is conducted by the last date of employment, there should be no problems.
C. To protect IT assets, terminating logical access to IT resources is the first and most important action to take after management has confirmed the employee's clear intention to leave the enterprise.
D. Ensuring that management signs off on termination paperwork is important, but not as critical as terminating access to the IT systems.

A5-195 Which of the following is a passive attack to a network?

A. Message modification
B. Masquerading
C. Denial-of-service
D. Traffic analysis

D is the correct answer.

Justification:
A. Message modification involves the capturing of a message and making unauthorized changes or deletions, changing the sequence or delaying transmission of captured messages. An attack that modifies the data would be an active attack.
B. Masquerading is an active attack in which the intruder presents an identity other than the original identity.
C. Denial-of-service occurs when a computer connected to the Internet is flooded with data and/or requests that must be processed. This is an active attack.
D. Traffic analysis allows a watching threat actor to determine the nature of the flow of traffic between defined hosts, which may allow the threat actor to guess the type of communication taking place without taking an active role.

A5-196 The **MOST** likely explanation for a successful social engineering attack is:

A. computer error.
B. judgment error.
C. expertise.
D. technology.

B is the correct answer.

Justification:
A. Social engineering focuses on human behavior.
B. Social engineering is fundamentally about obtaining from someone a level of trust that is not warranted.
C. Generally, social engineering attacks do not require significant expertise; often, the attacker is not proficient in information technology or systems.
D. Technology may facilitate social engineering, but it is fundamentally about obtaining human trust.

A5-197 A company is planning to install a network-based intrusion detection system to protect the web site that it hosts. Where should the device be installed?

A. On the local network
B. Outside the firewall
C. In the demilitarized zone
D. On the server that hosts the web site

C is the correct answer.

Justification:

A. While an intrusion detection system (IDS) can be installed on the local network to ensure that systems are not subject to internal attacks, a company's public web server would not normally be installed on the local network, but rather in the demilitarized zone (DMZ).
B. It is not unusual to place a network IDS outside of the firewall just to watch the traffic that is reaching the firewall, but this would not be used to specifically protect the web application.
C. Network-based IDSs detect attack attempts by monitoring network traffic. A public web server is typically placed on the protected network segment known as the demilitarized zone (DMZ). An IDS installed in the DMZ detects and reports on malicious activity originating from the Internet as well as the internal network, thus allowing the administrator to act.
D. A host-based IDS would be installed on the web server, but a network-based IDS would not.

A5-198 An IS auditor is evaluating a virtual machine (VM)-based architecture used for all programming and testing environments. The production architecture is a three-tier physical architecture. What is the **MOST** important IT control to test to ensure availability and confidentiality of the web application in production?

A. Server configuration has been hardened appropriately.
B. Allocated physical resources are available.
C. System administrators are trained to use the VM architecture.
D. The VM server is included in the disaster recovery plan.

A is the correct answer.

Justification:

A. The most important control to test in this configuration is the server configuration hardening. It is important to patch known vulnerabilities and to disable all non-required functions before production, especially when production architecture is different from development and testing architecture.
B. The greatest risk is associated with the difference between the testing and production environments. Ensuring that physical resources are available is a relatively low risk and easily addressed.
C. Virtual machines (VMs) are often used for optimizing programming and testing infrastructure. In this scenario, the development environment (VM architecture) is different from the production infrastructure (physical three-tier). Because the VMs are not related to the web application in production, there is no real requirement for the system administrators to be familiar with a virtual environment.
D. Because the VMs are only used in a development environment and not in production, it may not be necessary to include VMs in the disaster recovery plan.

A5-199 In what capacity would an IS auditor **MOST** likely see a hash function applied?

A. Authentication
B. Identification
C. Authorization
D. Encryption

A is the correct answer.

Justification:

A. **The purpose of a hash function is to produce a "fingerprint" of data that can be used to ensure integrity and authentication. A hash of a password also provides for authentication of a user or process attempting to access resources.**
B. Hash functions are not used for identification. They are used to validate the authenticity of the identity.
C. Hash functions are not typically used to provide authorization. Authorization is provided after the authentication has been established.
D. Hash functions do not encrypt data.

A5-200 The **BEST** filter rule for protecting a network from being used as an amplifier in a denial-of-service attack is to deny all:

A. outgoing traffic with source addresses external to the network.
B. incoming traffic with discernible spoofed IP source addresses.
C. incoming traffic that includes options set in the Internet Protocol.
D. incoming traffic whose destination address belongs to critical hosts.

A is the correct answer.

Justification:

A. **Outgoing traffic with an Internet Protocol (IP) source address different than the internal IP range in the network is invalid. In most of the cases, it signals a denial-of-service attack originated by an internal user or by a previously compromised internal machine; in both cases, applying this filter will stop the infected machine from participating in the attack.**
B. Denying incoming traffic will not prevent an internal machine from participating in an attack on an outside target.
C. Incoming traffic will have the IP options set according to the type of traffic. This is a normal condition.
D. Denying incoming traffic to internal hosts will prevent legitimate traffic.

A5-201 The purpose of a mantrap controlling access to a computer facility is **PRIMARILY** to:

A. prevent piggybacking.
B. prevent toxic gases from entering the data center.
C. starve a fire of oxygen.
D. prevent rapid movement in or out of the facility.

A is the correct answer.

Justification:

A. **The intended purpose of a mantrap controlling access to a computer facility is primarily to prevent piggybacking.**
B. Preventing toxic gases from entering the data center could be accomplished with a single self-closing door.
C. Starving a fire of oxygen could be accomplished with a single self-closing fire door.
D. A rapid exit may be necessary in some circumstances (e.g., a fire).

A5-202 Which of the following should be a concern for an IS auditor reviewing an organization's cloud computing strategy which is based on a software as a service (SaaS) model with an external provider?

A. Workstation upgrades must be performed.
B. Long-term software acquisition costs are higher.
C. Contract with the provider does not include onsite technical support.
D. Incident handling procedures with the provider are not well defined.

D is the correct answer.

Justification:
A. Unless organization workstations are obsolete, upgrading should not be an issue with a software as a service (SaaS) model because most applications running as SaaS use common technologies that allow a user to run the software on different devices.
B. The reduction of software acquisition costs is one of the benefits of SaaS.
C. A SaaS provider does not normally have onsite support for the organization.
D. A SaaS provider does not normally have onsite support for the organization. Therefore, incident handling procedures between the organization and its provider are critical for the detection, communication and resolution of incidents, including effective lines of communication and escalation processes.

A5-203 A company has decided to implement an electronic signature scheme based on a public key infrastructure. The user's private key will be stored on the computer's hard drive and protected by a password. The **MOST** significant risk of this approach is:

A. use of the user's electronic signature by another person if the password is compromised.
B. forgery by using another user's private key to sign a message with an electronic signature.
C. impersonation of a user by substitution of the user's public key with another person's public key.
D. forgery by substitution of another person's private key on the computer.

A is the correct answer.

Justification:
A. The user's digital signature is only protected by a password. Compromise of the password would enable access to the signature. This is the most significant risk.
B. Creating a digital signature with another user's private key would indicate that the message came from a different person, and therefore, the true user's credentials would not be forged.
C. Impersonation of a public key would require the modification of the certificate issued by the certificate authority. This is very difficult and least likely.
D. The substitution of another person's private key would not work because the digital signature would be validated with the original user's public key.

A5-204 Which of the following would be **BEST** prevented by a raised floor in the computer machine room?

A. Damage of wires around computers and servers
B. A power failure from static electricity
C. Shocks from earthquakes
D. Water flood damage

A is the correct answer.

Justification:

A. **The primary reason for having a raised floor is to enable ventilation systems, power cables and data cables to be installed underneath the floor. This eliminates the safety and damage risk posed when cables are placed in a spaghetti-like fashion on an open floor.**
B. Static electricity should be avoided in the machine room; therefore, measures such as specially manufactured carpet or shoes would be more appropriate for static prevention than a raised floor.
C. Raised floors do not address shocks from earthquakes. To address earthquakes, anti-seismic architecture would be required to establish a quake-resistant structural framework.
D. Computer equipment needs to be protected against water. However, a raised floor would not prevent damage to the machines in the event of overhead water pipe leakage.

A5-205 A business application system accesses a corporate database using a single ID and password embedded in a program. Which of the following would provide efficient access control over the organization's data?

A. Introduce a secondary authentication method such as card swipe.
B. Apply role-based permissions within the application system.
C. Have users input the ID and password for each database transaction.
D. Set an expiration period for the database password embedded in the program.

B is the correct answer.

Justification:

A. The issue is user permissions, not authentication; therefore, adding a stronger authentication does not improve the situation.
B. **This is a normal process to allow the application to communicate with the database. Therefore, the best control is to control access to the application and procedures to ensure that access to data is granted based on a user's role.**
C. Having a user input the ID and password for access would provide a better control because a database log would identify the initiator of the activity. However, this may not be efficient because each transaction would require a separate authentication process.
D. It is a good practice to set an expiration date for a password. However, this might not be practical for an ID automatically logged in from the program. Often, this type of password is set not to expire.

A5-206 An IS auditor selects a server for a penetration test that will be carried out by a technical specialist. Which of the following is **MOST** important?

A. The tools used to conduct the test
B. Certifications held by the IS auditor
C. Permission from the data owner of the server
D. An intrusion detection system is enabled

C is the correct answer.

Justification:
A. The choice of tools is important to ensure a valid test and prevent system failure; however, the permission of the owner is most important.
B. Whether the IS auditor holds certifications is not relevant to the effectiveness of the test.
C. The data owner should be informed of the risk associated with a penetration test, the timing of the test, what types of tests are to be conducted and other relevant details.
D. An intrusion detection system is not required for a penetration test.

A5-207 The **GREATEST** benefit of having well-defined data classification policies and procedures is:

A. a more accurate inventory of information assets.
B. a decreased cost of controls.
C. a reduced risk of inappropriate system access.
D. an improved regulatory compliance.

B is the correct answer.

Justification:
A. A more accurate inventory of information assets is a benefit but would not be the greatest benefit of the choices listed.
B. An important benefit of a well-defined data classification process would be to lower the cost of protecting data by ensuring that the appropriate controls are applied with respect to the sensitivity of the data. Without a proper classification framework, some security controls may be greater and, therefore, costlier than is required based on the data classification.
C. Classifying the data may assist in reducing the risk of inappropriate system access, but that would not be the greatest benefit.
D. Improved regulatory compliance would be a benefit; however, achieving a cost reduction would be a greater benefit.

A5-208 Which of the following criteria are **MOST** needed to ensure that log information is admissible in court? Ensure that data have been:

A. independently time stamped.
B. recorded by multiple logging systems.
C. encrypted by the most secure algorithm.
D. verified to ensure log integrity.

D is the correct answer.

Justification:

A. Independent time stamps are a key requirement in logging. This is one method of ensuring log integrity; however, this does not prevent information from being modified.
B. Having multiple logging resources may work to ensure redundancy; however, increased redundancy may not effectively add value to the credibility of log information.
C. The strength of the encryption algorithm may improve data confidentiality; however, this does not necessarily prevent data from being modified.
D. It is important to assure that log information existed at a certain point of time and it has not been altered. Therefore, evidential credibility of log information is enhanced when there is proof that no one has tampered with this information, something typically accomplished by maintaining a documented chain of custody.

A5-209 Which of the following is the **MOST** reliable form of single factor personal identification?

A. Smart card
B. Password
C. Photo identification
D. Iris scan

D is the correct answer.

Justification:

A. There is no guarantee that a smart card is being used by the correct person because it can be shared, stolen, or lost and found.
B. Passwords can be shared and, if written down, carry the risk of discovery.
C. Photo IDs can be forged or falsified.
D. Because no two irises are alike, identification and verification can be done with confidence.

A5-210 Which of the following controls would be **MOST** effective in reducing the risk of loss due to fraudulent online payment requests?

A. Transaction monitoring
B. Protecting web sessions using Secure Sockets Layer
C. Enforcing password complexity for authentication
D. Inputting validation checks on web forms

A is the correct answer.

Justification:

A. An electronic payment system could be the target of fraudulent activities. An unauthorized user could potentially enter false transactions. By monitoring transactions, the payment processor could identify potentially fraudulent transactions based on the typical usage patterns, monetary amounts, physical location of purchases, and other data that are part of the transaction process.

B. Using Secure Sockets Layer would help to ensure the secure transmission of data to and from the user's web browser and help to ensure that the end user has reached the correct web site, but this would not prevent fraudulent transactions.

C. Online transactions are not necessarily protected by passwords; for example, credit card transactions are not necessarily protected. The use of strong authentication would help to protect users of the system from fraud by attackers guessing passwords, but transaction monitoring would be the better control.

D. Inputting validation checks on web forms is important to ensure that attackers do not compromise the web site, but transaction monitoring would be the best control.

A5-211 Users are issued security tokens to be used in combination with a personal identification number (PIN) to access the corporate virtual private network. Regarding the PIN, what is the **MOST** important rule to be included in a security policy?

A. Users should not leave tokens where they could be stolen.
B. Users must never keep the token in the same bag as their laptop computer.
C. Users should select a PIN that is completely random, with no repeating digits.
D. Users should never write down their PIN.

D is the correct answer.

Justification:

A. Access to the token is of no value without the personal identification number (PIN); one cannot work without the other.

B. Access to the token is of no value without the PIN; one cannot work without the other.

C. The PIN does not need to be random as long as it is secret.

D. If a user writes their PIN on a slip of paper, an individual with the token, the slip of paper, and the computer could access the corporate network. A token and the PIN is a two-factor authentication method.

A5-212 A firewall is being deployed at a new location. Which of the following is the **MOST** important factor in ensuring a successful deployment?

A. Reviewing logs frequently
B. Testing and validating the rules
C. Training a local administrator at the new location
D. Sharing firewall administrative duties

B is the correct answer.

Justification:
A. A regular review of log files would not start until the deployment has been completed.
B. A mistake in the rule set can render a firewall ineffective or insecure. Therefore, testing and validating the rules is the most important factor in ensuring a successful deployment.
C. Training a local administrator may not be necessary if the firewalls are managed from a central location.
D. Having multiple administrators is a good idea, but not the most important for successful deployment.

A5-213 A data center has a badge-entry system. Which of the following is **MOST** important to protect the computing assets in the center?

A. Badge readers are installed in locations where tampering would be noticed.
B. The computer that controls the badge system is backed up frequently.
C. A process for promptly deactivating lost or stolen badges is followed.
D. All badge entry attempts are logged, whether or not they succeed.

C is the correct answer.

Justification:
A. Tampering with a badge reader cannot open the door, so this is irrelevant.
B. The configuration of the system does not change frequently; therefore, frequent backup is not necessary.
C. The biggest risk is from unauthorized individuals who can enter the data center, whether they are employees or not. Thus, having and following a process of deactivating lost or stolen badges is important.
D. Logging the entry attempts is important, but not as important as ensuring that a lost or stolen badge is disabled as quickly as possible.

A5-214 What is the **MOST** prevalent security risk when an organization implements remote virtual private network (VPN) access to its network?

A. Malicious code could be spread across the network.
B. The VPN logon could be spoofed.
C. Traffic could be sniffed and decrypted.
D. The VPN gateway could be compromised.

A is the correct answer.

Justification:
A. Virtual private network (VPN) is a mature technology; VPN devices are hard to break. However, when remote access is enabled, malicious code in a remote client could spread to the organization's network. One problem is when the VPN terminates inside the network and the encrypted VPN traffic goes through the firewall. This means that the firewall cannot adequately examine the traffic.
B. A secure VPN solution would use two-factor authentication to prevent spoofing.
C. Sniffing encrypted traffic does not generally provide an attack vector for its unauthorized decryption.
D. A misconfigured or poorly implemented VPN gateway could be subject to attack, but if it is located in a secure subnet, then the risk is reduced.

A5-215 The use of digital signatures:

A. requires the use of a one-time password generator.
B. provides encryption to a message.
C. validates the source of a message.
D. ensures message confidentiality.

C is the correct answer.

Justification:
A. A one-time password generator is not a requirement for using digital signatures.
B. A digital signature provides for integrity and proof of origin for a message but does not address confidentiality.
C. The use of a digital signature verifies the identity of the sender.
D. A digital signature does not ensure message confidentiality.

A5-216 The **FIRST** step in a successful attack to a system is:

A. gathering information.
B. gaining access.
C. denying services.
D. evading detection.

A is the correct answer.

Justification:
A. Successful attacks start by gathering information about the target system. This is done in advance so that the attacker gets to know the target systems and the potential vulnerabilities that can be exploited in the attack.
B. Once attackers have discovered potential vulnerabilities through information gathering, they will usually attempt to gain access.
C. An attacker will usually launch a denial of service as one of the last steps in the attack.
D. When attackers have gained access and possibly infected the victim with a rootkit, they will delete audit logs and take other steps to hide their tracks.

A5-217 Which of the following methods **BEST** mitigates the risk of disclosing confidential information through the use of social networking sites?

A. Providing security awareness training
B. Requiring a signed acceptable use policy
C. Monitoring the use of social media
D. Blocking access to social media

A is the correct answer.

Justification:
A. Providing security awareness training is the best method to mitigate the risk of disclosing confidential information on social networking sites. It is important to remember that users may access these services through other means such as mobile phones and home computers; therefore, awareness training is most critical.
B. Requiring a signed acceptable use policy can be a good control. However, if users are not aware of the risk, then this policy may not be effective.
C. Monitoring the use of social media through the use of a proxy server that tracks the web sites users visit is not an effective control because users may access these services through other means such as mobile phones and home computers.
D. Blocking the use of social media through network controls is not an effective control because users may access these services through other means such as mobile phones and home computers.

A5-218 An IS auditor finds that conference rooms have active network ports. Which of the following would prevent this discovery from causing concern?

A. The corporate network is using an intrusion prevention system.
B. This part of the network is isolated from the corporate network.
C. A single sign-on has been implemented in the corporate network.
D. Antivirus software is in place to protect the corporate network.

B is the correct answer.

Justification:

A. An intrusion prevention system may stop an attack, but it would be far better to restrict the ability of machines in the conference rooms from being able to access the corporate network altogether.
B. If the conference rooms have access to the corporate network, unauthorized users may be able to connect to the corporate network; therefore, both networks should be isolated either via a firewall or by being physically separated.
C. A single sign-on solution is used for access control but would not still leave a risk when unauthorized people have physical access to the corporate network.
D. Antivirus software would reduce the impact of possible viruses; however, unauthorized users would still be able to access the corporate network, which is the biggest risk.

A5-219 When conducting a penetration test of an IT system, an organization should be **MOST** concerned with:

A. the confidentiality of the report.
B. finding all weaknesses on the system.
C. restoring systems to the original state.
D. logging changes made to production systems.

C is the correct answer.

Justification:

A. A penetration test report is a sensitive document because it lists the vulnerabilities of the target system. However, the main requirement for the penetration test team is to restore the system to its original condition.
B. Finding all possible weaknesses is not possible in complex information systems.
C. After the test is completed, the systems must be restored to their original state. In performing the test, changes may have been made to firewall rules, user IDs created, or false files uploaded. These must all be cleaned up before the test is completed.
D. All changes made should be recorded, but the most important concern is to ensure that the changes are reversed at the end of the test.

A5-220 An IS auditor is reviewing a new web-based order entry system the week before it goes live. The IS auditor has identified that the application, as designed, may be missing several critical controls regarding how the system stores customer credit card information. The IS auditor should **FIRST**:

A. determine whether system developers have proper training on adequate security measures.
B. determine whether system administrators have disabled security controls for any reason.
C. verify that security requirements have been properly specified in the project plan.
D. validate whether security controls are based on requirements which are no longer valid.

C is the correct answer.

Justification:

A. While it is important for programmers to understand security, it is more important that the security requirements were properly stated in the project plan.
B. System administrators may have made changes to the controls, but it is assumed that the auditor is reviewing the system as designed a week prior to implementation so the administrators have not yet configured the system.
C. If there are significant security issues identified by an IS auditor, the first question is whether the security requirements were correct in the project plan. Depending on whether the requirements were included in the plan would affect the recommendations the auditor would make.
D. It is possible that security requirements will change over time based on new threats or vulnerabilities, but if critical controls are missing, this points toward a faulty design that was based on incomplete requirements.

A5-221 When protecting an organization's IT systems, which of the following is normally the next line of defense after the network firewall has been compromised?

A. Personal firewall
B. Antivirus programs
C. Intrusion detection system
D. Virtual local area network configuration

C is the correct answer.

Justification:

A. Personal firewalls would be later in the defensive strategy, being located on the endpoints.
B. Antivirus programs would be installed on endpoints as well as on the network, but the next layer of defense after a firewall is an intrusion detection system (IDS)/intrusion protection system.
C. An IDS would be the next line of defense after the firewall. It would detect anomalies in the network/server activity and try to detect the perpetrator.
D. Virtual local area network configurations are not intended to compensate for a compromise of the firewall. They are an architectural good practice.

A5-222 Which of the following is the **BEST** control to mitigate the risk of pharming attacks to an Internet banking application?

A. User registration and password policies
B. User security awareness
C. Use of intrusion detection/intrusion prevention systems
D. Domain name system server security hardening

D is the correct answer.

Justification:

A. User registration and password policies cannot mitigate pharming attacks because they do not prevent manipulation of domain name system (DNS) records.
B. User security awareness cannot mitigate pharming attacks because it does not prevent manipulation of DNS records.
C. The use of intrusion detection/intrusion prevention systems cannot mitigate pharming attacks because they do not prevent manipulation of DNS records.
D. The pharming attack redirects the traffic to an unauthorized web site by exploiting vulnerabilities of the DNS server. To avoid this kind of attack, it is necessary to eliminate any known vulnerability that could allow DNS poisoning. Older versions of DNS software are vulnerable to this kind of attack and should be patched.

A5-223 Which of the following would **MOST** effectively enhance the security of a challenge-response based authentication system?

A. Selecting a more robust algorithm to generate challenge strings
B. Implementing measures to prevent session hijacking attacks
C. Increasing the frequency of associated password changes
D. Increasing the length of authentication strings

B is the correct answer.

Justification:

A. Selecting a more robust algorithm will enhance the security; however, this may not be as important in terms of risk mitigation when compared to man-in-the-middle attacks.
B. Challenge response-based authentication is prone to session hijacking or man-in-the-middle attacks. Security management should be aware of this and engage in risk assessment and control design such as periodic authentication when they employ this technology.
C. Frequently changing passwords is a good security practice; however, the exposures lurking in communication pathways may pose a greater risk.
D. Increasing the length of authentication strings will not prevent man-in-the-middle or session hijacking attacks.

A5-224 When transmitting a payment instruction, which of the following will help verify that the instruction was not duplicated?

A. Using a cryptographic hashing algorithm
B. Enciphering the message digest
C. Calculating a checksum of the transaction
D. Using a sequence number and time stamp

D is the correct answer.

Justification:

A. Use of a cryptographic hashing algorithm against the entire message helps achieve data integrity but will not prevent duplicate processing.
B. Enciphering the message digest using the sender's private key, which signs the sender's digital signature to the document, helps in authenticating the source and integrity of the transaction but will not prevent duplicate processing.
C. A checksum can be used for data integrity but not to prevent duplicate transactions.
D. When transmitting data, a sequence number and/or time stamp built into the message to make it unique can be checked by the recipient to ensure that the message was not intercepted and replayed. This is known as replay protection and could be used to verify that a payment instruction was not duplicated.

A5-225 In wireless communication, which of the following controls allows the receiving device to verify that the received communications have not been altered in transit?

A. Device authentication and data origin authentication
B. Wireless intrusion detection and intrusion prevention systems
C. The use of cryptographic hashes
D. Packet headers and trailers

C is the correct answer.

Justification:

A. Device authentication and data origin authentication allow wireless endpoints to authenticate each other to prevent man-in-the-middle attacks and masquerading.
B. Wireless intrusion detection and intrusion prevention systems have the ability to detect misconfigured devices and rogue devices and detect and possibly stop certain types of attacks.
C. Calculating cryptographic hashes for wireless communications allows the receiving device to verify that the received communications have not been altered in transit. This prevents masquerading and message modification attacks.
D. Packet headers and trailers alone do not ensure that the content has not been altered because an attacker could alter both the data and the trailer.

A5-226 An organization is planning to replace its wired networks with wireless networks. Which of the following would **BEST** secure the wireless network from unauthorized access?

A. Implement Wired Equivalent Privacy.
B. Permit access to only authorized media access control addresses.
C. Disable open broadcast of service set identifiers.
D. Implement Wi-Fi Protected Access 2.

D is the correct answer.

Justification:

A. Wired Equivalent Privacy can be cracked within minutes. WEP uses a static key that has to be communicated to all authorized users, thus management is difficult. Also, there is a greater vulnerability if the static key is not changed at regular intervals.
B. The practice of allowing access based on media access control is not a solution because MAC addresses can be spoofed by attackers to gain access to the network.
C. Disabling open broadcast of service set identifiers is not an effective access control because many tools can detect a wireless access point that is not broadcasting.
D. Wi-Fi Protected Access (WPA) 2 implements most of the requirements of the IEEE 802.11i standard. The Advanced Encryption Standard used in WPA2 provides better security. Also, WPA2 supports both the Extensible Authentication Protocol and the pre-shared secret key authentication model.

A5-227 An IS auditor is reviewing a software-based firewall configuration. Which of the following represents the **GREATEST** vulnerability?

A. An implicit deny rule as the last rule in the rule base.
B. Installation on an operating system configured with default settings.
C. Rules permitting or denying access to systems or networks.
D. Configuration as a virtual private network endpoint.

B is the correct answer.

Justification:

A. Configuring a firewall with an implicit deny rule is common practice.
B. Default settings of most equipment—including operating systems—are often published and provide an intruder with predictable configuration information, which allows easier system compromise. To mitigate this risk, firewall software should be installed on a system using a hardened operating system that has limited functionality, providing only the services necessary to support the firewall software.
C. A firewall configuration should have rules allowing or denying access according to policy.
D. A firewall is often set up as the endpoint for a virtual private network.

A5-228 The **GREATEST** risk from an improperly implemented intrusion prevention system is:

A. too many alerts for system administrators to verify.
B. decreased network performance due to additional traffic.
C. blocking of critical systems or services due to false triggers.
D. reliance on specialized expertise within the IT organization.

C is the correct answer.

Justification:
A. A number of false positives may cause excessive administrator workload, but this is a relatively minor risk.
B. The intrusion prevention system will not generate any traffic that would impact network performance.
C. An IPS prevents a connection or service based on how it is programmed to react to specific incidents. If the IPS is triggered based on incorrectly defined or nonstandard behavior, it may block the service or connection of a critical internal system.
D. Configuring an IPS can take months of learning what is and what is not acceptable behavior, but this does not require specialized expertise.

A5-229 When reviewing a digital certificate verification process, which of the following findings represents the **MOST** significant risk?

A. There is no registration authority for reporting key compromises.
B. The certificate revocation list is not current.
C. Digital certificates contain a public key that is used to encrypt messages and verify digital signatures.
D. Subscribers report key compromises to the certificate authority.

B is the correct answer.

Justification:
A. The certificate authority (CA) can assume the responsibility if there is no registration authority.
B. If the certificate revocation list is not current, there could be a digital certificate that is not revoked that could be used for unauthorized or fraudulent activities.
C. Digital certificates contain a public key that is used to encrypt messages and verify digital signatures; therefore, this is not a risk.
D. Subscribers reporting key compromises to the CA is not a risk because reporting this to the CA enables the CA to take appropriate action.

A5-230 When using a digital signature, the message digest is computed by the:

A. sender only.
B. receiver only.
C. sender and receiver both.
D. certificate authority.

C is the correct answer.

Justification:
A. The message digest must be computed by the sender and the receiver to ensure message integrity.
B. The receiver will compute a digest of the received message to verify integrity of the received message.
C. A digital signature is an electronic identification of a person or entity. It is created by using asymmetric encryption. To verify integrity of data, the sender uses a cryptographic hashing algorithm against the entire message to create a message digest to be sent along with the message. Upon receipt of the message, the receiver will recompute the hash using the same algorithm.
D. The certificate authority (CA) issues certificates that link the public key with its owner. The CA does not compute digests of the messages to be communicated between the sender and receiver.

A5-231 Which of the following would effectively verify the originator of a transaction?

A. Using a secret password between the originator and the receiver
B. Encrypting the transaction with the receiver's public key
C. Using a portable document format to encapsulate transaction content
D. Digitally signing the transaction with the source's private key

D is the correct answer.

Justification:
A. Because they are a "shared secret" between the user and the system itself, passwords are considered a weaker means of authentication.
B. Encrypting the transaction with the recipient's public key will provide confidentiality for the information but will not verify the source.
C. Using a portable document format will protect the integrity of the content but not necessarily authorship.
D. A digital signature is an electronic identification of a person, created by using a public key algorithm, to verify the identity of the source of a transaction and the integrity of its content to a recipient.

A5-232 An organization has established a guest network for visitor access. Which of the following should be of **GREATEST** concern to an IS auditor?

A. A login screen is not displayed for guest users.
B. The guest network is not segregated from the production network.
C. Guest users who are logged in are not isolated from each other.
D. A single factor authentication technique is used to grant access.

B is the correct answer.

Justification:
A. Using a web captive portal, which displays a login screen in the user's web browser, is a good practice to authenticate guests. However, if the guest network is not segregated from the production network, users could introduce malware and potentially gain inappropriate access to systems and information.
B. The implication of this is that guests have access to the organization's network. Allowing untrusted users to connect to the organization's network could introduce malware and potentially allow these individuals inappropriate access to systems and information.
C. There are certain platforms in which it is allowable for guests to interact with one another. Also, guests could be warned to use only secured systems and a policy covering interaction among guests could be created.
D. Although a multifactor authentication technique is preferred, a single-factor authentication method should be adequate if properly implemented.

A5-233 Which of the following provides the **GREATEST** assurance for database password encryption?

A. Secure hash algorithm-256
B. Advanced encryption standard
C. Secure Shell
D. Triple data encryption standard

B is the correct answer.

Justification:
A. Hashing functions are often used to protect passwords, but hashing is not encryption.
B. The use of advanced encryption standard (AES) is a secure encryption algorithm that is appropriate for encrypting passwords.
C. Secure Shell may encrypt passwords that are being transmitted but does not encrypt data at rest.
D. Triple Data Encryption Standard is a valid encryption method; however, AES is a stronger and more recent encryption algorithm.

A5-234 The reason a certification and accreditation process is performed on critical systems is to ensure that:

A. Security compliance has been technically evaluated
B. Data have been encrypted and are ready to be stored
C. The systems have been tested to run on different platforms
D. The systems have followed the phases of a waterfall model

A is the correct answer.

Justification:

A. Certified and accredited systems are systems that have had their security compliance technically evaluated for running in a specific environment and configuration.
B. Certification tests security functionality, including encryption where that is required, but that is not the primary objective of the certification and accreditation (C&A) process.
C. Certified systems are evaluated to run in a specific environment.
D. A waterfall model is a software development methodology and not a reason for performing a C&A process.

A5-235 A perpetrator looking to gain access to and gather information about encrypted data being transmitted over a network would **MOST** likely use:

A. eavesdropping.
B. spoofing.
C. traffic analysis.
D. masquerading.

C is the correct answer.

Justification:

A. In eavesdropping, which is a passive attack, the intruder gathers the information flowing through the network with the intent of acquiring message contents for personal analysis or for third parties. Encrypted traffic is generally protected against eavesdropping
B. Spoofing is an active attack. In spoofing, a user receives an email that appears to have originated from one source when it actually was sent from another source.
C. In traffic analysis, which is a passive attack, an intruder determines the nature of the traffic flow between defined hosts and through an analysis of session length, frequency and message length, the intruder is able to guess the type of communication taking place. This typically is used when messages are encrypted, and eavesdropping would not yield any meaningful results.
D. In masquerading, the intruder presents an identity other than the original identity. This is an active attack.

A5-236 A hotel has placed a PC in the lobby to provide guests with Internet access. Which of the following presents the **GREATEST** risk for identity theft?

A. Web browser cookies are not automatically deleted.
B. The computer is improperly configured.
C. System updates have not been applied on the computer.
D. Session time out is not activated.

D is the correct answer.

Justification:
A. If web browser cookies are not automatically deleted, it might be possible to determine the web sites that a user has accessed. However, if sessions do not time out, it is easier for identity theft to occur.
B. If the PC is not configured properly and does not have antivirus software installed, there could be a risk of virus or malware infection. This could cause identity theft. However, if sessions do not time out, it is easier for identity theft to occur.
C. If system updates have not been applied, there could be a greater risk of virus or malware infection. This could cause identity theft. However, if sessions do not time out, it is easier for identity theft to occur.
D. If an authenticated session is inactive and unattended, it can be hijacked and used for illegal purposes. It might then be difficult to establish the intruder because a legitimate session was used.

A5-237 The **MOST** effective biometric control system is the one with:

A. the highest equal-error rate.
B. the lowest equal-error rate.
C. a false-rejection rate equal to the false-acceptance rate.
D. a false-rejection rate equal to the failure-to-enroll rate.

B is the correct answer.

Justification:
A. The biometric that has the highest equal-error rate (EER) is the most ineffective.
B. The EER of a biometric system denotes the percent at which the false-acceptance rate (FAR) is equal to the false-rejection rate (FRR). The biometric that has the lowest EER is the most effective.
C. For any biometric, there will be a measure at which the FRR will be equal to the FAR. This is the EER.
D. Failure-to-enroll rate (FER) is an aggregate measure of FRR.

A5-238 Which of the following is a form of two-factor user authentication?

A. A smart card and personal identification number
B. A unique User ID and complex, non-dictionary password
C. An iris scan and a fingerprint scan
D. A magnetic-strip card and a proximity badge

A is the correct answer.

Justification:
A. A smart card is something that a user has, while a personal identification number paired with the card is something the user knows. This is an example of two-factor authentication.
B. Both an ID and a password are something the user knows, so this pairing provides single-factor user authentication regardless of complexity.
C. Both an iris scan and a fingerprint scan are something the user is, so this pairing is not a basis for two-factor user authentication.
D. Both a magnetic card and a proximity badge are examples of something a user has, so these are not adequate for two-factor authentication.

A5-239 An IS auditor is reviewing the physical security measures of an organization. Regarding the access card system, the IS auditor should be **MOST** concerned that:

A. Non-personalized access cards are given to the cleaning staff, who use a sign-in sheet but show no proof of identity.
B. access cards are not labeled with the organization's name and address to facilitate easy return of a lost card.
C. card issuance and rights administration for the cards are done by different departments, causing unnecessary lead time for new cards.
D. the computer system used for programming the cards can only be replaced after three weeks in the event of a system failure.

A is the correct answer.

Justification:

A. Physical security is meant to control who is entering a secured area, so identification of all individuals is of utmost importance. It is not adequate to trust unknown external people by allowing them to write down their alleged name without proof (e.g., identity card, driver's license).
B. Having the name and address of the organization on the card may be a concern because a malicious finder could use a lost or stolen card to enter the organization's premises.
C. Separating card issuance from technical rights management is a method to ensure the proper segregation of duties so that no single person can produce a functioning card for a restricted area within the organization's premises. The long lead time is an inconvenience but not a serious audit risk.
D. System failure of the card programming device would normally not mean that the readers do not function anymore. It simply means that no new cards can be issued, so this option is minor compared to the threat of improper identification.

A5-240 When reviewing the procedures for the disposal of computers, which of the following should be the **GREATEST** concern for the IS auditor?

A. Hard disks are overwritten several times at the sector level but are not reformatted before leaving the organization.
B. All files and folders on hard disks are separately deleted, and the hard disks are formatted before leaving the organization.
C. Hard disks are rendered unreadable by hole-punching through the platters at specific positions before leaving the organization.
D. The transport of hard disks is escorted by internal security staff to a nearby metal recycling company, where the hard disks are registered and then shredded.

B is the correct answer.

Justification:

A. Overwriting a hard disk at the sector level would completely erase data, directories, indices and master file tables. Reformatting is not necessary because all contents are destroyed. Overwriting several times makes useless some forensic measures, which are able to reconstruct former contents of newly overwritten sectors by analyzing special magnetic features of the platter's surface.
B. Deleting and formatting only marks the sectors that contained files as being free. Publicly available tools are sufficient for someone to reconstruct data from hard drives prepared this way.
C. While hole-punching does not delete file contents, the hard disk cannot be used anymore, especially when head parking zones and track zero information are impacted. Reconstructing data would be extremely expensive because all analysis must be performed under a clean room atmosphere and is only possible within a short time frame or until the surface is corroded.
D. Data reconstruction from shredded hard disks is virtually impossible, especially when the scrap is mixed with other metal parts. If the transport can be secured and the destruction be proved as described in the option, this is a valid method of disposal.

A5-241 A new business application requires deviation from the standard configuration of the operating system (OS). What activity should the IS auditor recommend to the security manager as a **FIRST** response?

A. Initial rejection of the request because it is against the security policy
B. Approval of the exception to policy to meet business needs
C. Assessment of the risk and identification of compensating controls
D. Revision of the OS baseline configuration

C is the correct answer.

Justification:

A. The security policy may be waived with management approval to meet business requirements; it is not up to the security manager to refuse the deviation.
B. The security manager may make a case for deviation from the policy, but this should be based on a risk assessment and compensating controls. The deviation itself should be approved in accordance with a defined exception handling process.
C. Before approving any exception, the security manager should first check for compensating controls and assess the possible risk due to deviation.
D. Updating or revising the baseline configuration is not associated with requests for deviations.

A5-242 An organization has created a policy that defines the types of web sites that users are forbidden to access. What is the **MOST** effective technology to enforce this policy?

A. Stateful inspection firewall
B. Web content filter
C. Web cache server
D. Proxy server

B is the correct answer.

Justification:

A. A stateful inspection firewall is of little help in filtering web traffic because it does not review the content of the web site, nor does it take into consideration the site's classification.
B. A web content filter accepts or denies web communications according to the configured rules. To help the administrator properly configure the tool, organizations and vendors have made available uniform resource locator blacklists and classifications for millions of web sites.
C. A web cache server is designed to improve the speed of retrieving the most common or recently visited web pages.
D. A proxy server is incorrect because a proxy server services the request of its clients by forwarding requests to other servers. Many people incorrectly use proxy server as a synonym of web proxy server even though not all web proxy servers have content filtering capabilities.

A5-243 Which of the following specifically addresses how to detect cyberattacks against an organization's IT systems and how to recover from an attack?

A. An incident response plan
B. An IT contingency plan
C. A business continuity plan
D. A continuity of operations plan

A is the correct answer.

Justification:

A. The incident response plan (IRP) determines the information security responses to incidents such as cyberattacks on systems and/or networks. This plan establishes procedures to enable security personnel to identify, mitigate and recover from malicious computer incidents such as unauthorized access to a system or data, denial-of-service or unauthorized changes to system hardware or software.

B. The IT contingency plan addresses IT system disruptions and establishes procedures for recovering from a major application or general support system failure. The contingency plan deals with ways to recover from an unexpected failure, but it does not address the identification or prevention of cyberattacks.

C. The business continuity plan (BCP) addresses business processes and provides procedures for sustaining essential business operations while recovering from a significant disruption. While a cyberattack could be severe enough to require use of the BCP, the IRP would be used to determine which actions should be taken—both to stop the attack as well as to resume normal operations after the attack.

D. The continuity of operations plan addresses the subset of an organization's missions that are deemed most critical and contains procedures to sustain these functions at an alternate site for a short time period.

A5-244 The cryptographic hash sum of a message is recalculated by the receiver. This is to ensure:

A. the confidentiality of the message.
B. nonrepudiation by the sender.
C. the authenticity of the message.
D. the integrity of data transmitted by the sender.

D is the correct answer.

Justification:

A. A hash function ensures integrity of a message; encrypting with a secret key provides confidentiality.

B. Signing the message with the private key of the sender ensures nonrepudiation and authenticity.

C. Authenticity of the message is provided by the digital signature.

D. If the hash sum is different from what is expected, it implies that the message has been altered. This is an integrity test.

A5-245 The computer security incident response team of an organization disseminates detailed descriptions of recent threats. An IS auditor's **GREATEST** concern should be that the users may:

A. use this information to launch attacks.
B. forward the security alert.
C. implement individual solutions.
D. fail to understand the threat.

A is the correct answer.

Justification:

A. An organization's computer security incident response team (CSIRT) should disseminate recent threats, security guidelines and security updates to the users to assist them in understanding the security risk of errors and omissions. However, this introduces the risk that the users may use this information to launch attacks, directly or indirectly. An IS auditor should ensure that the CSIRT is actively involved with users to assist them in mitigation of risk arising from security failures and to prevent additional security incidents resulting from the same threat.
B. Forwarding the security alert is not harmful to the organization.
C. Implementing individual solutions is unlikely and inefficient, but not a serious risk.
D. Users failing to understand the threat would not be a serious concern.

A5-246 Which of the following would be an indicator of the effectiveness of a computer security incident response team?

A. Financial impact per security incident
B. Number of security vulnerabilities that were patched
C. Percentage of business applications that are being protected
D. Number of successful penetration tests

A is the correct answer.

Justification:

A. The most important indicator is the financial impact per security incident. It may not be possible to prevent incidents entirely, but the team should be able to limit the cost of incidents through a combination of effective prevention, detection and response.
B. Patching of security vulnerabilities is important but not a direct responsibility of the computer security incident response team (CSIRT).
C. The CSIRT is not responsible for the protection of systems. That is the responsibility of the security team.
D. The number of penetration tests measures the effectiveness of the security team and the patch management process, but not the effectiveness of the CSIRT.

A5-247 A benefit of quality of service is that the:

A. entire network's availability and performance will be significantly improved.
B. telecom carrier will provide the company with accurate service-level compliance reports.
C. participating applications will have bandwidth guaranteed.
D. communications link will be supported by security controls to perform secure online transactions.

C is the correct answer.

Justification:
A. Quality of service (QoS) will not guarantee that the communication itself will be improved. While the speed of data exchange for specific applications could be faster, availability will not be improved.
B. The QoS tools that many carriers are using do not provide reports of service levels; however, there are other tools that will generate service-level reports.
C. The main function of QoS is to optimize network performance by assigning priority to business applications and end users through the allocation of dedicated parts of the bandwidth to specific traffic.
D. Even when QoS is integrated with firewalls, virtual private networks (VPNs), encryption tools and others, the tool itself is not intended to provide security controls.

A5-248 Which of the following procedures would **MOST** effectively detect the loading of illegal software packages onto a network?

A. The use of diskless workstations
B. Periodic checking of hard drives
C. The use of current antivirus software
D. Policies that result in instant dismissal if violated

B is the correct answer.

Justification:
A. Diskless workstations act as a preventive control and are not totally effective in preventing users from accessing illegal software over the network.
B. The periodic checking of hard drives would be the most effective method of identifying illegal software packages loaded onto the network.
C. Antivirus software will not necessarily identify illegal software, unless the software contains a virus.
D. Policies are a preventive control to lay out the rules about loading the software, but will not detect the actual occurrence.

A5-249 An online stock trading firm is in the process of implementing a system to provide secure email exchange with its customers. What is the **BEST** option to ensure confidentiality, integrity and nonrepudiation?

A. Symmetric key encryption
B. Digital signatures
C. Message digest algorithms
D. Digital certificates

D is the correct answer.

Justification:

A. Symmetric key encryption uses a single pass phrase to encrypt and decrypt the message. While this type of encryption is strong, it suffers from the inherent problem of needing to share the pass phrase in a secure manner and does not address integrity and nonrepudiation.
B. Digital signatures provide message integrity and nonrepudiation; however, confidentiality is not provided.
C. Message digest algorithms are a way to design hashing functions to verify the integrity of the message/data. Message digest algorithms do not provide confidentiality or nonrepudiation.
D. A digital certificate contains the public key and identifying information about the owner of the public key. The associated private key pair is kept secret with the owner. These certificates are generally verified by a trusted authority, with the purpose of associating a person's identity with the public key. Email confidentiality and integrity are obtained by following the public key-private key encryption. With the digital certificate verified by the trusted third party, nonrepudiation of the sender is obtained.

A5-250 An IS auditor reviewing the authentication controls of an organization should be **MOST** concerned if:

A. user accounts are not locked out after five failed attempts.
B. passwords can be reused by employees within a defined time frame.
C. system administrators use shared login credentials.
D. password expiration is not automated.

C is the correct answer.

Justification:

A. If user accounts are not locked after multiple failed attempts, a brute force attack could be used to gain access to the system. While this is a risk, a typical user would have limited system access compared to an administrator.
B. The reuse of passwords is a risk. However, the use of shared login credentials by administrators is a more severe risk.
C. The use of shared login credentials makes accountability impossible. This is especially a risk with privileged accounts.
D. If password expiration is not automated, it is most likely that employees will not change their passwords regularly. However, this is not as serious as passwords being shared, and the use of shared login credentials by administrators is a more severe risk.

A5-251 The IS auditor is reviewing the implementation of a storage area network (SAN). The SAN administrator indicates that logging and monitoring is active, hard zoning is used to isolate data from different business units and all unused SAN ports are disabled. The administrator implemented the system, performed and documented security testing during implementation, and is the only user with administrative rights to the system. What should the IS auditor's initial determination be?

A. There is no significant potential risk.
B. Soft zoning presents a potential risk.
C. Disabling of unused ports presents a potential risk.
D. The SAN administrator presents a potential risk.

D is the correct answer.

Justification:

A. While the storage area network (SAN) may have been implemented with good controls, there is risk created by the combination of roles held by the SAN administrator.
B. Hard zoning is more secure than soft zoning.
C. Unused ports should generally be disabled to increase security.
D. The potential risk in this scenario is posed by the SAN administrator. One concern is having a "single point of failure." Because only one administrator has the knowledge and access required to administer the system, the organization is susceptible to risk. For example, if the SAN administrator decided to quit unexpectedly, or was otherwise unavailable, the company may not be able to adequately administer the SAN. In addition, having a single administrator for a large, complex system such as a SAN also presents a segregation of duties risk. The organization currently relies entirely on the SAN administrator to implement, maintain, and validate all security controls; this means that the SAN administrator could modify or remove those controls without detection.

A5-252 Which of the following exposures associated with the spooling of sensitive reports for offline printing should an IS auditor consider to be the **MOST** serious?

A. Sensitive data might be read by operators.
B. Data might be amended without authorization.
C. Unauthorized report copies might be printed.
D. Output might be lost in the event of system failure.

C is the correct answer.

Justification:

A. Operators often have high-level access as a necessity to perform their job duties. To the extent that this is a risk, it exists for any form of non-local printing and is not specifically tied to spooled reports.
B. Data on spool files are no easier to amend without authority than any other file.
C. Spooling for offline printing may enable additional copies to be printed unless adequate safeguards exist as compensating controls.
D. Loss of data at the spooler level would only require reprinting.

A5-253 Web and email filtering tools are valuable to an organization **PRIMARILY** because they:

A. protect the organization from viruses and nonbusiness materials.
B. maximize employee performance.
C. safeguard the organization's image.
D. assist the organization in preventing legal issues.

A is the correct answer.

Justification:

A. The main reason for investing in web and email filtering tools is that they significantly reduce risk related to viruses, spam, mail chains, recreational surfing and recreational email.
B. Maximizing employee performance could be true in some circumstances (i.e., it would need to be implemented along with an awareness program so that employee performance can be significantly improved). However, the primary benefit is protecting the organization from viruses and nonbusiness activity.
C. Safeguarding the organization's image is a secondary benefit.
D. Preventing legal issues is important, but not the primary reason for filtering.

A5-254 Which of the following types of firewalls provide the **GREATEST** degree and granularity of control?

A. Screening router
B. Packet filter
C. Application gateway
D. Circuit gateway

C is the correct answer.

Justification:

A. Screening routers and packet filters work at the protocol, service and/or port level. This means that they analyze packets from layers 3 and 4 and not from higher levels.
B. A packet filter works at too low of a level of the communication stack to provide granular control.
C. The application gateway is similar to a circuit gateway, but it has specific proxies for each service. To handle web services, it has a Hypertext Transmission Protocol (HTTP) proxy that acts as an intermediary between externals and internals but is specifically for HTTP. This means that it not only checks the packet Internet Protocol (IP) addresses (Open Systems Interconnection [OSI] Layer 3) and the ports it is directed to (in this case port 80, or layer 4), it also checks every HTTP command (OSI Layers 5 and 7). Therefore, it works in a more detailed (granularity) way than the other choices.
D. A circuit gateway is based on a proxy or program that acts as an intermediary between external and internal accesses. This means that, during an external access, instead of opening a single connection to the internal server, two connections are established—one from the external server to the proxy (which conforms the circuit-gateway) and one from the proxy to the internal server. OSI Layers 3 and 4 (IP and Transmission Control Protocol) and some general features from higher protocols are used to perform these tasks.

A5-255 After installing a network, an organization implemented a vulnerability assessment tool = to identify possible weaknesses. Which type of reporting poses the **MOST** serious risk associated with such tools?

A. Differential
B. False-positive
C. False-negative
D. Less-detail

C is the correct answer.

Justification:

A. Differential reporting function provided by this tool compares scan results over a period of time.
B. False-positive reporting is one in which the system falsely reports a vulnerability. Controls may be in place, but are evaluated as weak, which should prompt a rechecking of the controls.
C. False-negative reporting on weaknesses means the control weaknesses in the network are not identified and, therefore, may not be addressed, leaving the network vulnerable to attack.
D. Less-detail reporting would require additional tools or analysis to determine the existence and severity of vulnerabilities.

A5-256 Which of the following is the **MOST** reliably effective method for dealing with the spread of a network worm that exploits vulnerability in a protocol?

A. Install the latest vendor security patches immediately.
B. Block the protocol traffic in the perimeter firewall.
C. Block the protocol traffic between internal network segments.
D. Stop the services that the protocol uses.

D is the correct answer.

Justification:

A. Installing the latest patches will improve the situation only if a patch has been released that addresses the particular vulnerability in the protocol. Also, patches should not be installed prior to testing, because patching systems can create new vulnerabilities or impact performance.
B. Blocking the protocol on the perimeter does not stop the worm from spreading if it is introduced via portable media.
C. Blocking the protocol helps to slow the spread, but also prohibits any software that uses it from working between segments.
D. Stopping the services is the most effective way to prevent a worm from spreading, because it directly addresses the means of propagation at the lowest practical level.

A5-257 An IS auditor is reviewing an organization's controls related to email encryption. The company's policy states that all sent email must be encrypted to protect the confidentiality of the message because the organization shares nonpublic information through email. In a public-key infrastructure implementation properly configured to provide confidentiality, email is:

A. encrypted with the sender's private key and decrypted with the sender's public key.
B. encrypted with the recipient's private key and decrypted with the sender's private key.
C. encrypted with the sender's private key and decrypted with the recipient's private key.
D. encrypted with the recipient's public key and decrypted with the recipient's private key.

D is the correct answer.

Justification:

A. Encrypting a message with the sender's private key and decrypting it with the sender's public key ensures that the message came from the sender; however, it does not guarantee message confidentiality. With public key infrastructure, a message encrypted with a private key must be decrypted with the responding public key, and vice versa.
B. The sender would not have access to the receiver's private key.
C. A message encrypted with the sender's private key could not be decrypted using the recipient's private key.
D. Encrypting a message with the recipient's public key and decrypting it with the recipient's private key ensures message confidentiality, because only the intended recipient has the correct private key to decrypt the message.

A5-258 Which of the following types of firewalls would **BEST** protect a network from an Internet attack?

A. Screened subnet firewall
B. Application filtering gateway
C. Packet filtering router
D. Circuit-level gateway

A is the correct answer.

Justification:

A. A screened subnet firewall would provide the best protection. The screening router can be a commercial router or a node with routing capabilities and the ability to allow or avoid traffic between nets or nodes based on addresses, ports, protocols, interfaces, etc. The subnet would isolate Internet-based traffic from the rest of the corporate network.
B. Application-level gateways are mediators between two entities that want to communicate, also known as proxy gateways. The application level (proxy) works at the application level, not just at a packet level. This would be the best solution to protect an application but not a network.
C. A packet filtering router examines the header of every packet or data traveling between the Internet and the corporate network. This is a low-level control.
D. A circuit level gateway, such as a Socket Secure server, will protect users by acting as a proxy but is not the best defense for a network.

A5-259 Neural networks are effective in detecting fraud because they can:

A. discover new trends because they are inherently linear.
B solve problems where large and general sets of training data are not obtainable.
C address problems that require consideration of a large number of input variables.
D. make assumptions about the shape of any curve relating variables to the output.

C is the correct answer.

Justification:
A. Neural networks are inherently nonlinear.
B. Neural networks will not work well at solving problems for which sufficiently large and general sets of training data are not obtainable.
C. Neural networks can be used to attack problems that require consideration of numerous input variables. They are capable of capturing relationships and patterns often missed by other statistical methods, but they will not discover new trends.
D. Neural networks make no assumption about the shape of any curve relating variables to the output.

A5-260 Which of the following **BEST** encrypts data on mobile devices?

A. Elliptical curve cryptography
B. Data encryption standard
C. Advanced encryption standard
D. The Blowfish algorithm

A is the correct answer.

Justification:
A. Elliptical curve cryptography (ECC) requires limited bandwidth resources and is suitable for encrypting mobile devices.
B. Data encryption standard uses less processing power when compared with advanced encryption standard (AES), but ECC is more suitable for encrypting data on mobile devices.
C. AES is a symmetric algorithm and has the problem of key management and distribution. ECC is an asymmetric algorithm and is better suited for a mobile environment.
D. The use of the Blowfish algorithm consumes too much processing power.

A5-261 Confidentiality of transmitted data can best be delivered by encrypting the:

A. Message digest with the sender's private key.
B. Session key with the sender's public key.
C. Messages with the receiver's private key.
D. Session key with the receiver's public key.

D is the correct answer.

Justification:
A. This will ensure authentication and nonrepudiation.
B. This will make the message accessible to only the sender.
C. A message encrypted with a receiver's private key could be decrypted by anyone using the receiver's public key.
D. This will ensure that the session key can only be obtained using the receiver's private key, retained by the receiver.

A5-262 The risk of dumpster diving is **BEST** mitigated by:

A. Implementing security awareness training.
B. Placing shred bins in copy rooms.
C. Developing a media disposal policy.
D. Placing shredders in individual offices.

A is the correct answer.

Justification:

A. Dumpster diving is used to steal documents or computer media that were not properly discarded. Users should be educated to know the risk of carelessly discarding sensitive documents and other items.
B. The shred bins may not be properly used if users are not aware of proper security techniques.
C. A media disposal policy is a good idea; however, if users are not aware of the policy it may not be effective.
D. The shredders may not be properly used if users are not aware of proper security techniques.

A5-263 An organization provides information to its supply chain partners and customers through an extranet infrastructure. Which of the following should be the **GREATEST** concern to an IS auditor reviewing the firewall security architecture?

A. A Secure Sockets Layer has been implemented for user authentication and remote administration of the firewall.
B. Firewall policies are updated on the basis of changing requirements.
C Inbound traffic is blocked unless the traffic type and connections have been specifically permitted.
D. The firewall is placed on top of the commercial operating system with all default installation options.

D is the correct answer.

Justification:

A. Using Secure Sockets Layer for firewall administration is important because changes in user and supply chain partners' roles and profiles will be dynamic.
B. It is appropriate to maintain the firewall policies as needed.
C. It is prudent to block all inbound traffic to an extranet unless permitted.
D. The greatest concern when implementing firewalls on top of commercial operating systems is the potential presence of vulnerabilities that could undermine the security posture of the firewall platform itself. In most circumstances, when commercial firewalls are breached, that breach is facilitated by vulnerabilities in the underlying operating system. Keeping all installation options available on the system further increases the risk of vulnerabilities and exploits.

A5-264 An organization is proposing to establish a wireless local area network (WLAN). Management asks the IS auditor to recommend security controls for the WLAN. Which of the following would be the **MOST** appropriate recommendation?

A. Physically secure wireless access points to prevent tampering.
B. Use service set identifiers that clearly identify the organization.
C. Encrypt traffic using the Wired Equivalent Privacy mechanism.
D. Implement the Simple Network Management Protocol to allow active monitoring.

A is the correct answer.

Justification:

A. Physically securing access points such as wireless routers, as well as preventing theft, addresses the risk of malicious parties tampering with device settings. If access points can be physically reached, it is often a simple matter to restore weak default passwords and encryption keys, or to totally remove authentication and encryption from the network.

B. Service set identifiers should not be used to identify the organization because hackers can associate the wireless local area network with a known organization, and this increases both their motivation to attack and, potentially, the information available to do so.

C. The original Wired Equivalent Privacy security mechanism has been demonstrated to have a number of exploitable weaknesses. The more recently developed Wi-Fi Protected Access and Wi-Fi Protected Access 2 standards represent considerably more secure means of authentication and encryption.

D. Installing Simple Network Management Protocol on wireless access points can actually open up security vulnerabilities. If SNMP is required at all, then SNMP v3, which has stronger authentication mechanisms than earlier versions, should be deployed.

A5-265 Which of the following situations would increase the likelihood of fraud?

A. Application programmers are implementing changes to production programs.
B. Administrators are implementing vendor patches to vendor-supplied software without following change control procedures.
C. Operations support staff members are implementing changes to batch schedules.
D. Database administrators are implementing changes to data structures.

A is the correct answer.

Justification:

A. Production programs are used for processing an enterprise's data. It is imperative that controls on changes to production programs are stringent. Lack of control in this area could result in application programs being modified to manipulate the data.

B. The lack of change control is a serious risk—but if the changes are only vendor-supplied patches to vendor software then the risk is minimal.

C. The implementation of changes to batch schedules by operations support staff will affect the scheduling of the batches only; it does not impact the live data unless jobs are run in the wrong sequence.

D. Database administrators are required to implement changes to data structures. This is required for reorganization of the database to allow for additions, modifications or deletions of fields or tables in the database.

A5-266 A consulting firm has created a File Transfer Protocol (FTP) site for the purpose of receiving financial data and has communicated the site's address, user ID and password to the financial services company in separate email messages. The company is to transmit its data to the FTP site after manually encrypting the data. The IS auditor's **GREATEST** concern with this process is that:

A. The users may not remember to manually encrypt the data before transmission.
B. The site credentials were sent to the financial services company via email.
C. Personnel at the consulting firm may obtain access to sensitive data.
D. The use of a shared user id to the ftp site does not allow for user accountability.

A is the correct answer.

Justification:
A. If the data is not encrypted, an unauthorized external party may download sensitive company data.
B. Even though the possibility exists that the logon information was captured from the emails, data should be encrypted, so the theft of the data would not allow the attacker to read it.
C. Some of the employees at the consulting firm will have access to the sensitive data and the consulting firm must have procedures in place to protect the data.
D. Tracing accountability is of minimal concern compared to the compromise of sensitive data.

A5-267 Java applets and Active X controls are distributed programs that execute in the background of a client web browser. This practice is considered reasonable when:

A. A firewall exists.
B. A secure web connection is used.
C. The source of the executable file is certain.
D. The host web site is part of the organization.

C is the correct answer.

Justification:
A. There should always be a firewall on an Internet connection; however, whether to allow active models is a decision made depending on the source of the module.
B. A secure web connection provides confidentiality. Neither a secure web connection nor a firewall can identify an executable file as friendly.
C. Acceptance of these mechanisms should be based on established trust. The control is provided by only knowing the source and then allowing the acceptance of the applets. Hostile applets can be received from anywhere.
D. Hosting the web site as part of the organization is impractical. The client will accept the program if the parameters are established to do so.

A5-268 Which of the following controls will **MOST** effectively detect the presence of bursts of errors in network transmissions?

A. Parity check
B. Echo check
C. Block sum check
D. Cyclic redundancy check

D is the correct answer.

Justification:

A. Parity check (known as vertical redundancy check) also involves adding a bit (known as the parity bit) to each character during transmission. In this case, where there is a presence of bursts of errors (i.e., impulsing noise during high transmission rates), it has a reliability of approximately 50 percent. In higher transmission rates, this limitation is significant.
B. Echo checks detect line errors by retransmitting data to the sending device for comparison with the original transmission.
C. A block sum check is a form of parity checking and has a low level of reliability.
D. The cyclic redundancy check (CRC) can check for a block of transmitted data. The workstations generate the CRC and transmit it with the data. The receiving workstation computes a CRC and compares it to the transmitted CRC. If both of them are equal, then the block is assumed error free. In this case (such as in parity error or echo check), multiple errors can be detected. In general, CRC can detect all single-bit and double-bit errors.

A5-269 Which of the following types of transmission media provide the **BEST** security against unauthorized access?

A. Copper wire
B. Shielded twisted pair
C. Fiber-optic cables
D. Coaxial cables

C is the correct answer.

Justification:

A. Twisted pair, coaxial and copper wire traffic can be monitored with inexpensive equipment.
B. Twisted pair cabling is a form of copper wire, and while shielding affords some degree of protection from interference, it does not improve security against unauthorized access.
C. Fiber-optic cables have proven to be more secure and more difficult to tap than the other media.
D. Coaxial cable can be monitored with relative ease.

A5-270 Which of the following is the **BEST** audit procedure to determine if a firewall is configured in compliance with an organization's security policy?

A. Review the parameter settings.
B. Interview the firewall administrator.
C. Review the actual procedures.
D. Review the device's log file for recent attacks.

A is the correct answer.

Justification:

A. A review of the parameter settings will provide a good basis for comparison of the actual configuration to the security policy and will provide audit evidence documentation.
B. An interview with the firewall administrator will not ensure that the firewall is configured correctly.
C. Reviewing the actual procedures is good but will not ensure that the firewall rules are correct and compliant with policy.
D. Recent attacks may indicate problems with the firewall but will not ensure that it is correctly configured.

A5-271 An IS auditor is reviewing the network infrastructure of a call center and determines that the internal telephone system is based on Voice-over Internet Protocol technology. Which of the following is the **GREATEST** concern?

A. Voice communication uses the same equipment that is used for data communication.
B. Ethernet switches are not protected by uninterrupted power supply units.
C. Voice communication is not encrypted on the local network.
D. The team that supports the data network also is responsible for the telephone system.

B is the correct answer.

Justification:

A. Voice-over Internet Protocol (VoIP) telephone systems use the local area network (LAN) infrastructure of a company for communication, which can save on wiring cost and simplify both the installation and support of the telephone system. This use of shared infrastructure is a benefit of VoIP and therefore is not a concern.

B. VoIP telephone systems use the LAN infrastructure of a company for communication, typically using Ethernet connectivity to connect individual phones to the system. Most companies have a backup power supply for the main servers and systems, but typically do not have uninterrupted power supply units for the LAN switches. In the case of even a brief power outage, not having backup power on all network devices makes it impossible to send or receive phone calls, which is a concern, particularly in a call center.

C. VoIP devices do not normally encrypt the voice traffic on the local network, so this is not a concern. Typically, a VoIP phone system connects to a telephone company voice circuit, which would not normally be encrypted. If the system uses the Internet for connectivity, then encryption is required.

D. VoIP telephone systems use the LAN infrastructure of a company for communication, so the personnel who support and maintain that infrastructure are now responsible for both the data and voice network by default. Therefore, this would not be a concern.

A5-272 Which of the following would **BEST** ensure continuity of a wide area network across the organization?

A. Built-in alternative routing
B. Complete full system backup daily
C. A repair contract with a service provider
D. A duplicate machine alongside each server

A is the correct answer.

Justification:

A. Alternative routing would ensure that the network would continue if a communication device fails or if a link is severed because message rerouting could be automatic.

B. System backup will not afford protection for a networking failure.

C. The repair contract will almost always result in some lost time and is not as effective as permanent alternative routing.

D. Standby servers will not provide continuity if a link is severed.

A5-273 An organization is planning to deploy an outsourced cloud-based application that is used to track job applicant data for the human resources department. Which of the following should be the **GREATEST** concern to an IS auditor?

A. The service level agreement (SLA) ensures strict limits for uptime and performance.
B. The cloud provider will not agree to an unlimited right-to-audit as part of the SLA.
C. The SLA is not explicit regarding the disaster recovery plan capabilities of the cloud provider.
D. The cloud provider's physical data centers are in multiple cities and countries.

D is the correct answer.

Justification:
A. Although this application may have strict requirements for availability, it is assumed that the service level agreement (SLA) would contain these same elements; therefore, this is not a concern.
B. The right-to-audit clause is good to have, but there are limits on how a cloud service provider may interpret this requirement. The task of reviewing and assessing all the controls in place at a multinational cloud provider would likely be a costly and time-consuming exercise; therefore, such a requirement may be of limited value.
C. Because the SLA would normally specify uptime requirements, the means used to achieve those goals (which would include the specific disaster recovery plan capabilities of the provider) are typically not reviewed in-depth by the customer, nor are they typically specified in a SLA.
D. Having data in multiple countries is the greatest concern because human resources (HR) applicant data could contain personally identifiable information. There may be legal compliance issues if these data are stored in a country with different laws regarding data privacy. While the organization would be bound by the privacy laws where it is based, it may not have legal recourse if a data breach happens in a jurisdiction where the same laws do not apply.

A5-274 An organization is reviewing its contract with a cloud computing provider. For which of the following reasons would the organization want to remove a lock-in clause from the cloud service contract?

A. Availability
B. Portability
C. Agility
D. Scalability

B is the correct answer.

Justification:
A. Removing the customer lock-in clause will not secure availability of the systems resources stored in a cloud computing environment.
B. When drawing up a contract with a cloud service provider, the ideal practice is to remove the customer lock-in clause. It may be important for the client to secure portability of their system assets (i.e., the right to transfer from one vendor to another).
C. Agility refers to efficiency of solutions enabling organizations to respond to business needs faster. This is a desirable quality of cloud computing.
D. Scalability is the strength of cloud computing through the ability to adjust service levels according to changing business circumstances. Therefore, this is not the best option.

A5-275 Which of the following is an object-oriented technology characteristic that permits an enhanced degree of security over data?

A. Inheritance
B. Dynamic warehousing
C. Encapsulation
D. Polymorphism

C is the correct answer.

Justification:

A. In object-oriented systems an object is called by another module and inherits its data from the calling module. This does not affect security.
B. Dynamic warehousing is not related to the security of object-oriented technology.
C. Encapsulation is a property of objects, and it prevents accessing either properties or methods that have not been previously defined as public. This means that any implementation of the behavior of an object is not accessible. An object defines a communication interface with the exterior and only that which belongs to that interface can be accessed.
D. Polymorphism is the principle of creating different objects that will behave differently depending on the input. This is not a security feature.

A5-276 A review of wide area network (WAN) usage discovers that traffic on one communication line between sites, synchronously linking the master and standby database, peaks at 96 percent of the line capacity. An IS auditor should conclude that:

A. analysis is required to determine if a pattern emerges that results in a service loss for a short period of time.
B. WAN capacity is adequate for the maximum traffic demands because saturation has not been reached.
C. the line should immediately be replaced by one with a larger capacity to provide approximately 85 percent saturation.
D. users should be instructed to reduce their traffic demands or distribute them across all service hours to flatten bandwidth consumption.

A is the correct answer.

Justification:

A. The peak at 96 percent could be the result of a one-off incident (e.g., a user downloading a large amount of data); therefore, analysis to establish whether this is a regular pattern and what causes this behavior should be carried out before expenditure on a larger line capacity is recommended.
B. A peak traffic load of 96 percent is approaching a critical level, and the auditor should not assume that capacity is adequate at this time or for the foreseeable future. Further investigation is required.
C. If the peak is established to be a regular occurrence without any other opportunities for mitigation (usage of bandwidth reservation protocol or other types of prioritizing network traffic), the line should be replaced because there is the risk of loss of service as the traffic approaches 100 percent. At this point, further research is required.
D. If the peak traffic load is a rare one-off occurrence or if traffic can be reengineered to transfer at other time frames, then user education may be an option. Further investigation will be required.

A5-277 Which of the following **BEST** limits the impact of server failures in a distributed environment?

A. Redundant pathways
B. Clustering
C. Dial backup lines
D. Standby power

B is the correct answer.

Justification:

A. Redundant pathways will minimize the impact of channel communications failures but will not address the problem of server failure.
B. Clustering allows two or more servers to work as a unit so that when one of them fails, the other takes over.
C. Dial backup lines will minimize the impact of channel communications failures but not a server failure.
D. Standby power provides an alternative power source in the event of an energy failure but does not address the problem of a server failure.

A5-278 The **MAIN** reason for requiring that all computer clocks across an organization are synchronized is to:

A. Prevent omission or duplication of transactions.
B. Ensure smooth data transition from client machines to servers.
C. Ensure that email messages have accurate time stamps.
D. Support the incident investigation process.

D is the correct answer.

Justification:

A. The possibility of omission or duplication of transactions will not happen due to lack of clock synchronization.
B. Data transfer has nothing to do with the time stamp.
C. Although the time stamp on an email may not be accurate, this is not a significant issue.
D. During an investigation of incidents, audit logs are used as evidence, and the time stamp information in them is useful. If the clocks are not synchronized, investigations will be more difficult, because a time line of events occurring on different systems might not be easily established.

A5-279 When reviewing the configuration of network devices, an IS auditor should **FIRST** identify:

A. The good practices for the type of network devices deployed
B. Whether components of the network are missing
C. The importance of the network devices in the topology
D. Whether subcomponents of the network are being used appropriately

C is the correct answer.

Justification:

A. After understanding the devices in the network, a good practice for using the device should be reviewed to ensure that there are no anomalies within the configuration.
B. Identification of which component is missing can only be known after reviewing and understanding the topology and a good practice for deployment of the device in the network.
C. The first step is to understand the importance and role of the network device within the organization's network topology.
D. Identification of which subcomponent is being used inappropriately can only be known after reviewing and understanding the topology and a good practice for deployment of the device in the network.

A5-280 Which of the following will **BEST** maintain the integrity of a firewall log?

A. Granting access to log information only to administrators
B. Capturing log events in the operating system layer
C. Writing dual logs onto separate storage media
D. Sending log information to a dedicated third-party log server

D is the correct answer.

Justification:

A. To enforce segregation of duties, administrators should not have access to log files. This primarily contributes to the assurance of confidentiality rather than integrity.
B. There are many ways to capture log information—through the application layer, network layer, operating systems layer, etc. However, there is no log integrity advantage in capturing events in the operating systems layer.
C. If it is a highly mission-critical information system, it may be nice to run the system with a dual log mode. Having logs in two different storage devices will primarily contribute to the assurance of the availability of log information, rather than maintaining its integrity.
D. Establishing a dedicated third-party log server and logging events in it is the best procedure for maintaining the integrity of a firewall log. When access control to the log server is adequately maintained, the risk of unauthorized log modification is mitigated, therefore improving the integrity of log information.

A5-281 An IS auditor reviewing a cloud computing environment that is managed by a third party should be **MOST** concerned when:

A. The organization is not permitted to assess the controls in the participating vendor's site.
B. The service level agreement does not address the responsibility of the vendor in the case of a security breach.
C. Laws and regulations are different in the countries of the organization and the vendor.
D. The organization is using an older version of a browser and is vulnerable to certain types of security risk.

B is the correct answer.

Justification:

A. The IS auditor has no role to play if the contract between the parties does not provide for assessment of controls in the other vendor's site.
B. Administration of cloud computing occurs over the Internet and involves more than one participating entity. It is the responsibility of each of the partners in the cloud computing environment to take care of security issues in their own environments. When there is a security breach, the party responsible for the breach should be identified and made accountable. This is not possible if the service level agreement (SLA) does not address the responsibilities of the partners during a security breach.
C. The IS auditor should ensure that the contract addresses the differing laws and regulations in the countries of the organization and the vendor, but having different laws and regulations is not a problem.
D. The IS auditor can make suggestions to the audited entity to use appropriate patches or switch over to safer browsers, and then the IS auditor can follow up on the action taken.

A5-282 Which one of the following can be used to provide automated assurance that proper data files are being used during processing?

A. File header record
B. Version usage
C. Parity checking
D. File security controls

A is the correct answer.

Justification:

A. A file header record provides assurance that proper data files are being used, and it allows for automatic checking.

B. Although version usage provides assurance that the correct file and version are being used, it does not allow for automatic checking.

C. Parity checking is a data integrity validation method typically used by a data transfer program. Although parity checking may help to ensure that data and program files are transferred successfully, it does not help to ensure that the proper data or program files are being used.

D. File security controls cannot be used to provide assurance that proper data files are being used and cannot allow for automatic checking.

A5-283 A cyclic redundancy check is commonly used to determine the:

A. Accuracy of data input.
B. Integrity of a downloaded program.
C. Adequacy of encryption.
D. Validity of data transfer.

D is the correct answer.

Justification:

A. Accuracy of data input can be enforced by data validation controls, such as picklists, cross checks, reasonableness checks, control totals and allowed character checks.

B. A checksum or digital signature is commonly used to validate the integrity of a downloaded program or other transferred data.

C. Encryption adequacy is driven by the sensitivity of the data to be protected and algorithms that determine how long it will take to break a specific encryption method.

D. The accuracy of blocks of data transfers, such as data transfer from hard disks, is validated by a cyclic redundancy check.

A5-284 An IS auditor is performing a review of a network. Users report that the network is slow and web pages periodically time out. The IS auditor confirms the users' feedback and reports the findings to the network manager. The most appropriate action for the network management team should be to **FIRST**:

A. Use a protocol analyzer to perform network analysis and review error logs of local area network equipment.
B. Take steps to increase the bandwidth of the connection to the Internet.
C. Create a baseline using a protocol analyzer and implement quality of service to ensure that critical business applications work as intended.
D. Implement virtual local area networks to segment the network and ensure performance.

A is the correct answer.

Justification:

A. In this case, the first step is to identify the problem through review and analysis of network traffic. Using a protocol analyzer and reviewing the log files of the related switches or routers will determine whether there is a configuration issue or hardware malfunction.
B. Although increasing Internet bandwidth may be required, this may not be needed if the performance issue is due to a different problem or error condition.
C. Although creating a baseline and implementing quality of service will ensure that critical applications have the appropriate bandwidth, in this case, the performance issue may be related to misconfiguration or equipment malfunction.
D. Although implementing virtual local area networks may be good practice for ensuring adequate performance, in this case, the issue may be related to misconfigurations or equipment malfunction.

A5-285 In a small organization, an employee performs computer operations and, when the situation demands, program modifications. Which of the following should the IS auditor recommend?

A. Automated logging of changes to development libraries
B. Additional staff to provide separation of duties
C. Procedures that verify that only approved program changes are implemented
D. Access controls to prevent the operator from making program modifications

C is the correct answer.

Justification:

A. Logging of changes to production libraries is good practice, but because the administrator can alter the logs, this is not a sufficient control.
B. Although adherence to separation of duties and recruitment of additional staff are preferred , this practice is not always possible in small organizations.
C. An IS auditor must consider recommending a better process. An IS auditor should recommend a formal change control process that manages and can detect changes to production source and object code, such as code comparisons, so the changes can be reviewed on a regular basis by a third party. This is a compensating control process.
D. Requiring a third party to do the changes may not be practical in a small organization where another person with adequate expertise may not be available.

Page intentionally left blank

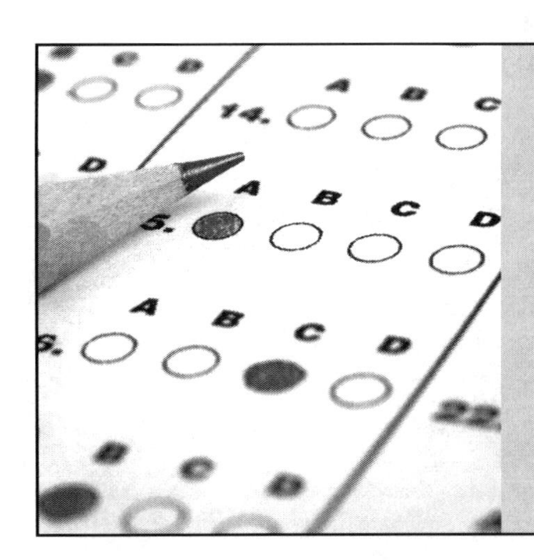

POSTTEST

If you wish to take a posttest to determine strengths and weaknesses, the Sample Exam begins on page 469 and the posttest answer sheet begins on page 495. You can score your posttest with the Sample Exam Answer and Reference Key on page 491.

Page intentionally left blank

CISA Review Questions, Answers & Explanations Manual 12th Edition

Sample Exam

1. The **PRIMARY** purpose of implementing Redundant Array of Inexpensive Disks (RAID) level 1 in a file server is to:

    A. achieve performance improvement.
    B. provide user authentication.
    C. ensure availability of data.
    D. ensure the confidentiality of data.

2. A financial institution that processes millions of transactions each day has a central communications processor (switch) for connecting to automated teller machines. Which of the following would be the **BEST** contingency plan for the communications processor?

    A. Reciprocal agreement with another organization
    B. Alternate processor in the same location
    C. Alternate processor at another network node
    D. Duplex communication links

3. During a compliance audit of a small bank, the IS auditor notes that the IT and accounting functions are being performed by the same user of the financial system. Which of the following reviews that are conducted by the user's supervisor represents the **BEST** compensating control?

    A. Audit trails that show the date and time of the transaction
    B. A daily report with the total numbers and dollar amounts of each transaction
    C. User account administration
    D. Computer log files that show individual transactions

4. Which of the following should an IS auditor use to detect duplicate invoice records within an invoice master file?

    A. Attribute sampling
    B. Computer-assisted audit techniques
    C. Compliance testing
    D. Integrated test facility

5. Over the long term, which of the following has the greatest potential to improve the security incident response process?

    A. A walk-through review of incident response procedures
    B. Simulation exercises performed by incident response team
    C. Ongoing security training for users
    D. Documenting responses to an incident

6. As part of the business continuity planning process, which of the following should be identified **FIRST** in the business impact analysis?

    A. Risk such as single point-of-failure and infrastructure risk
    B. Threats to critical business processes
    C. Critical business processes for ascertaining the priority for recovery
    D. Resources required for resumption of business

7. A review of wide area network (WAN) usage discovers that traffic on one communication line between sites, synchronously linking the master and standby database, peaks at 96 percent of the line capacity. An IS auditor should conclude that:

A. analysis is required to determine if a pattern emerges that results in a service loss for a short period of time.
B. WAN capacity is adequate for the maximum traffic demands because saturation has not been reached.
C. the line should immediately be replaced by one with a larger capacity to provide approximately 85 percent saturation.
D. users should be instructed to reduce their traffic demands or distribute them across all service hours to flatten bandwidth consumption.

8. An IS auditor discovers that developers have operator access to the command line of a production environment operating system. Which of the following controls would **BEST** mitigate the risk of undetected and unauthorized program changes to the production environment?

A. Commands typed on the command line are logged.
B. Hash keys are calculated periodically for programs and matched against hash keys calculated for the most recent authorized versions of the programs.
C. Access to the operating system command line is granted through an access restriction tool with preapproved rights.
D. Software development tools and compilers have been removed from the production environment.

9. An IS auditor performing a review of application controls would evaluate the:

A. efficiency of the application in meeting the business processes.
B. impact of any exposures discovered.
C. business processes served by the application.
D. application's optimization.

10. The **PRIMARY** purpose for meeting with auditees prior to formally closing a review is to:

A. Confirm that the auditors did not overlook any important issues.
B. Gain agreement on the findings.
C. Receive feedback on the adequacy of the audit procedures.
D. Test the structure of the final presentation.

11. Authorizing access to application data is the responsibility of the:

A. data custodian.
B. application administrator.
C. data owner.
D. security administrator.

12. An IS auditor finds that not all employees are aware of the enterprise's information security policy. The IS auditor should conclude that:

A. This lack of knowledge may lead to unintentional disclosure of sensitive information.
B. Information security is not critical to all functions.
C. Is audit should provide security training to the employees.
D. The audit finding will cause management to provide continuous training to staff.

13. During fieldwork, an IS auditor experienced a system crash caused by a security patch installation. To provide reasonable assurance that this event will not recur, the IS auditor should ensure that:

A. only systems administrators perform the patch process.
B. the client's change management process is adequate.
C. patches are validated using parallel testing in production.
D. an approval process of the patch, including a risk assessment, is developed.

14. While performing an audit of an accounting application's internal data integrity controls, an IS auditor identifies a major control deficiency in the change management software supporting the accounting application. The **MOST** appropriate action for the IS auditor to take is to:

A. Continue to test the accounting application controls and inform the IT manager about the control deficiency and recommend possible solutions.
B. Complete the audit and not report the control deficiency because it is not part of the audit scope.
C. Continue to test the accounting application controls and include the deficiency in the final report.
D. Cease all audit activity until the control deficiency is resolved.

15. Which of the following manages the digital certificate life cycle to ensure adequate security and controls exist in digital signature applications related to ecommerce?

A. Registration authority
B. Certificate authority
C. Certification revocation list
D. Certification practice statement

16. A financial services enterprise has a small IT department, and individuals perform more than one role. Which of the following practices represents the **GREATEST** risk?

A. The developers promote code into the production environment.
B. The business analyst writes the requirements and performs functional testing.
C. The IT manager also performs systems administration.
D. The database administrator also performs data backups.

17. The **MOST** likely effect of the lack of senior management commitment to IT strategic planning is:

A. Lack of investment in technology
B. Lack of a methodology for systems development
C. Technology not aligning with organization objectives
D. Absence of control over technology contracts

18. When reviewing the configuration of network devices, an IS auditor should **FIRST** identify:

A. The good practices for the type of network devices deployed
B. Whether components of the network are missing
C. The importance of the network devices in the topology
D. Whether subcomponents of the network are being used appropriately

19. In a review of the human resources policies and procedures within an organization, an IS auditor is **MOST** concerned with the absence of a:

A. requirement for periodic job rotations.
B. process for formalized exit interviews.
C. termination checklist.
D. requirement for new employees to sign a nondisclosure agreement.

20. Digital signatures require the:

A. signer to have a public key and the receiver to have a private key.
B. signer to have a private key and the receiver to have a public key.
C. signer and receiver to have a public key.
D. signer and receiver to have a private key.

21. Determining the service delivery objective should be based **PRIMARILY** on:

A. the minimum acceptable operational capability.
B. the cost-effectiveness of the restoration process.
C. meeting the recovery time objectives.
D. the allowable interruption window.

22. When developing a risk management program, what is the **FIRST** activity to be performed?

A. Threat assessment
B. Classification of data
C. Inventory of assets
D. Criticality analysis

23. Which of the following would an IS auditor consider to be **MOST** helpful when evaluating the effectiveness and adequacy of a preventive computer maintenance program?

A. A system downtime log
B. Vendors' reliability figures
C. Regularly scheduled maintenance log
D. A written preventive maintenance schedule

24. Which of the following is the **MOST** important criterion when selecting a location for an offsite storage facility for IS backup files? The offsite facility must be:

A. physically separated from the data center and not subject to the same risk.
B. given the same level of protection as that of the computer data center.
C. outsourced to a reliable third party.
D. equipped with surveillance capabilities.

25. Which of the following is an appropriate test method to apply to a business continuity plan?

A. Pilot
B. Paper
C. Unit
D. System

26. Which of the following **BEST** limits the impact of server failures in a distributed environment?

A. Redundant pathways
B. Clustering
C. Dial backup lines
D. Standby power

27. Information for detecting unauthorized input from a user workstation would be **BEST** provided by the:

A. console log printout.
B. transaction journal.
C. automated suspense file listing.
D. user error report.

28. Which of the following provides the **MOST** relevant information for proactively strengthening security settings?

A. Bastion host
B. Intrusion detection system
C. Honeypot
D. Intrusion prevention system

29. An IS auditor finds that a database administrator (DBA) has read and write access to production data. The IS auditor should:

A. accept the DBA access as a common practice.
B. assess the controls relevant to the DBA function.
C. recommend the immediate revocation of the DBA access to production data.
D. review user access authorizations approved by the DBA.

30. Which of the following inputs adds the **MOST** value to the strategic IT initiative decision-making process?

A. The maturity of the project management process
B. The regulatory environment
C. Past audit findings
D. The IT project portfolio analysis

31. During the planning stage of an IS audit, the **PRIMARY** goal of an IS auditor is to:

A. Address audit objectives.
B. Collect sufficient evidence.
C. Specify appropriate tests.
D. Minimize audit resources.

32. In a public key infrastructure, a registration authority:

A. verifies information supplied by the subject requesting a certificate.
B. issues the certificate after the required attributes are verified and the keys are generated.
C. digitally signs a message to achieve nonrepudiation of the signed message.
D. registers signed messages to protect them from future repudiation.

33. The editing/validation of data entered at a remote site is performed **MOST** effectively at the:

A. central processing site after running the application system.
B. central processing site during the running of the application system.
C. remote processing site after transmission of the data to the central processing site.
D. remote processing site prior to transmission of the data to the central processing site.

34. An IS auditor is determining the appropriate sample size for testing the existence of program change approvals. Previous audits did not indicate any exceptions, and management has confirmed that no exceptions have been reported for the review period. In this context, the IS auditor can adopt a:

A. lower confidence coefficient, resulting in a smaller sample size.
B. higher confidence coefficient, resulting in a smaller sample size.
C. higher confidence coefficient, resulting in a larger sample size.
D. lower confidence coefficient, resulting in a larger sample size.

35. When selecting audit procedures, an IS auditor should use professional judgment to ensure that:

A. Sufficient evidence will be collected.
B. Significant deficiencies will be corrected within a reasonable period.
C. All material weaknesses will be identified.
D. Audit costs will be kept at a minimum level.

36. Which of the following results in a denial-of-service attack?

A. Brute force attack
B. Ping of death
C. Leapfrog attack
D. Negative acknowledgment attack

37. The **MAIN** criterion for determining the severity level of a service disruption incident is:

A. cost of recovery.
B. negative public opinion.
C. geographic location.
D. downtime.

38. An IS auditor is performing a review of an organization's governance model. Which of the following should be of **MOST** concern to the auditor?

A. The information security policy is not periodically reviewed by senior management.
B. A policy ensuring systems are patched in a timely manner does not exist.
C. The audit committee did not review the organization's mission statement.
D. An organizational policy related to information asset protection does not exist.

39. An IS auditor discovers that devices connected to the network are not included in a network diagram that had been used to develop the scope of the audit. The chief information officer explains that the diagram is being updated and awaiting final approval. The IS auditor should **FIRST**:

A. expand the scope of the IS audit to include the devices that are not on the network diagram.
B. evaluate the impact of the undocumented devices on the audit scope.
C. note a control deficiency because the network diagram has not been approved.
D. plan follow-up audits of the undocumented devices.

40. An IS auditor analyzing the audit log of a database management system (DBMS) finds that some transactions were partially executed as a result of an error and have not been rolled back. Which of the following transaction processing features has been violated?

A. Consistency
B. Isolation
C. Durability
D. Atomicity

41. Which of the following will **MOST** successfully identify overlapping key controls in business application systems?

A. Reviewing system functionalities that are attached to complex business processes
B. Submitting test transactions through an integrated test facility
C. Replacing manual monitoring with an automated auditing solution
D. Testing controls to validate that they are effective

42. Which of the following criteria are **MOST** needed to ensure that log information is admissible in court? Ensure that data have been:

A. independently time stamped.
B. recorded by multiple logging systems.
C. encrypted by the most secure algorithm.
D. verified to ensure log integrity.

43. The waterfall life cycle model of software development is **MOST** appropriately used when:

A. requirements are well understood and are expected to remain stable, as is the business environment in which the system will operate.
B. requirements are well understood and the project is subject to time pressures.
C. the project intends to apply an object-oriented design and programming approach.
D. the project will involve the use of new technology.

44. Which of the following would **BEST** support 24/7 availability?

A. Daily backup
B. Offsite storage
C. Mirroring
D. Periodic testing

45. Responsibility and reporting lines cannot always be established when auditing automated systems because:

A. diversified control makes ownership irrelevant.
B. staff traditionally changes jobs with greater frequency.
C. ownership is difficult to establish where resources are shared.
D. duties change frequently in the rapid development of technology.

46. An IS auditor observes that an enterprise has outsourced software development to a third party that is a startup company. To ensure that the enterprise's investment in software is protected, which of the following should be recommended by the IS auditor?

A. Due diligence should be performed on the software vendor.
B. A quarterly audit of the vendor facilities should be performed.
C. There should be a source code escrow agreement in place.
D. A high penalty clause should be included in the contract.

47. When performing a review of a business process reengineering (BPR) effort, which of the following is of **PRIMARY** concern?

A. Controls are eliminated as part of the streamlining BPR effort.
B. Resources are not adequate to support the BPR process.
C. The audit department does not have a consulting role in the BPR effort.
D. The BPR effort includes employees with limited knowledge of the process area.

48. Which of the following is the **MOST** efficient and sufficiently reliable way to test the design effectiveness of a change control process?

A. Test a sample population of change requests
B. Test a sample of authorized changes
C. Interview personnel in charge of the change control process
D. Perform an end-to-end walk-through of the process

49. The **BEST** method of confirming the accuracy of a system tax calculation is by:

A. review and analysis of the source code of the calculation programs.
B. recreating program logic using generalized audit software to calculate monthly totals.
C. preparing simulated transactions for processing and comparing the results to predetermined results.
D. automatic flowcharting and analysis of the source code of the calculation programs.

50. Which of the following audit techniques **BEST** helps an IS auditor in determining whether there have been unauthorized program changes since the last authorized program update?

A. Test data run
B. Code review
C. Automated code comparison
D. Review of code migration procedures

51. Involvement of senior management is **MOST** important in the development of:

A. Strategic plans.
B. IT policies.
C. IT procedures.
D. Standards and guidelines.

52. To prevent Internet Protocol (IP) spoofing attacks, a firewall should be configured to drop a packet for which the sender of a packet:

A. specifies the route that a packet should take through the network (the source routing field is enabled).
B. puts multiple destination hosts (the destination field has a broadcast address).
C. indicates that the computer should immediately stop using the TCP connection (a reset flag is turned on).
D. allows use of dynamic routing instead of static routing (Open Shortest Path First protocol is enabled).

53. An organization sells books and music online at its secure web site. Transactions are transferred to the accounting and delivery systems every hour to be processed. Which of the following controls **BEST** ensures that sales processed on the secure web site are transferred to both the delivery and accounting systems?

A. Transaction totals are recorded on a daily basis in the sales systems. Daily sales system totals are aggregated and totaled.
B. Transactions are automatically numerically sequenced. Sequences are checked and gaps in continuity are accounted for.
C. Processing systems check for duplicated transaction numbers. If a transaction number is duplicated (already present), it is rejected.
D. System time is synchronized hourly using a centralized time server. All transactions have a date/time stamp.

54. An IS auditor is evaluating a newly developed IT policy for an organization. Which of the following factors does the IS auditor consider **MOST** important to facilitate compliance with the policy upon its implementation?

A. Existing IT mechanisms enabling compliance
B. Alignment of the policy to the business strategy
C. Current and future technology initiatives
D. Regulatory compliance objectives defined in the policy

55. During a security audit of IT processes, an IS auditor finds that documented security procedures do not exist. The IS auditor should:

A. Create the procedures document based on the practices.
B. Issue an opinion of the current state and end the audit.
C. Conduct compliance testing on available data.
D. Identify and evaluate existing practices.

56. An IS auditor is reviewing access controls for a manufacturing organization. During the review, the IS auditor discovers that data owners have the ability to change access controls for a low-risk application. The **BEST** course of action for the IS auditor is to:

A. recommend that mandatory access control be implemented.
B. report this as a finding to upper management
C. report this to the data owners to determine whether it is an exception.
D. not report this issue because discretionary access controls are in place.

57. Which of the following factors is **MOST** critical when evaluating the effectiveness of an IT governance implementation?

A. Ensure that assurance objectives are defined.
B. Determine stakeholder requirements and involvement.
C. Identify relevant risk and related opportunities.
D. Determine relevant enablers and their applicability.

58. There is a concern that the risk of unauthorized access may increase after implementing a single sign-on process. To prevent unauthorized access, the **MOST** important action is to:

A. monitor failed authentication attempts.
B. review log files regularly.
C. deactivate unused accounts promptly.
D. mandate a strong password policy.

59. Which of the following is the **MAIN** reason an organization should have an incident response plan? The plan helps to:

A. ensure prompt communication of adverse events to relevant management.
B. contain costs related to maintaining disaster recovery plan capabilities.
C. ensure that customers are promptly notified of issues such as security breaches.
D. minimize the duration and impact of system outages and security incidents.

60. Due to a reorganization, a business application system will be extended to other departments. Which of the following should be of the **GREATEST** concern for an IS auditor?

A. Process owners have not been identified.
B. The billing cost allocation method has not been determined.
C. Multiple application owners exist.
D. A training program does not exist.

61. The purpose of code signing is to provide assurance that:

A. the software has not been subsequently modified.
B. the application can safely interface with another signed application.
C. the signer of the application is trusted.
D. the private key of the signer has not been compromised.

62. As an outcome of information security governance, strategic alignment provides:

A. Security requirements driven by enterprise requirements.
B. Baseline security following good practices.
C. Institutionalized and commoditized solutions.
D. An understanding of risk exposure.

63. Email message authenticity and confidentiality is **BEST** achieved by signing the message using the:

A. sender's private key and encrypting the message using the receiver's public key.
B. sender's public key and encrypting the message using the receiver's private key.
C. receiver's private key and encrypting the message using the sender's public key.
D. receiver's public key and encrypting the message using the sender's private key.

64. Which of the following is the **PRIMARY** objective of an IT performance measurement process?

A. Minimize errors
B. Gather performance data
C. Establish performance baselines
D. Optimize performance

65. A financial enterprise has had difficulties establishing clear responsibilities between its IT strategy committee and its IT steering committee. Which of the following responsibilities would **MOST** likely be assigned to its IT steering committee?

A. Approving IT project plans and budgets
B. Aligning IT to business objectives
C. Advising on IT compliance risk
D. Promoting IT governance practices

66. An organization is planning to replace its wired networks with wireless networks. Which of the following would **BEST** secure the wireless network from unauthorized access?

A. Implement Wired Equivalent Privacy.
B. Permit access to only authorized media access control addresses.
C. Disable open broadcast of service set identifiers.
D. Implement Wi-Fi Protected Access 2.

67. Why does an audit manager review the staff's audit papers, even when the IS auditors have many years of experience?

A. Internal quality requirements
B. The audit guidelines
C. The audit methodology
D. Professional standards

68. Which of the following sampling methods would be the **MOST** effective to determine whether purchase orders issued to vendors have been authorized as per the authorization matrix?

A. Variable sampling
B. Stratified mean per unit
C. Attribute sampling
D. Unstratified mean per unit

69. An IS auditor is reviewing a manufacturing company and finds that mainframe users at a remote site connect to the mainframe at headquarters over the Internet via Telnet. Which of the following offers the **STRONGEST** security?

A. Use of a point-to-point leased line
B. Use of a firewall rule to allow only the Internet Protocol address of the remote site
C. Use of two-factor authentication
D. Use of a nonstandard port for Telnet

70. A business application system accesses a corporate database using a single ID and password embedded in a program. Which of the following would provide efficient access control over the organization's data?

A. Introduce a secondary authentication method such as card swipe.
B. Apply role-based permissions within the application system.
C. Have users input the ID and password for each database transaction.
D. Set an expiration period for the database password embedded in the program.

71. An Internet-based attack using password sniffing can:

A. enable one party to act as if they are another party.
B. cause modification to the contents of certain transactions.
C. be used to gain access to systems containing proprietary information.
D. result in major problems with billing systems and transaction processing agreements.

72. Which of the following should an IS auditor review to gain an understanding of the effectiveness of controls over the management of multiple projects?

A. Project database
B. Policy documents
C. Project portfolio database
D. Program organization

73. An IS auditor is reviewing a software-based firewall configuration. Which of the following represents the **GREATEST** vulnerability?

A. An implicit deny rule as the last rule in the rule base.
B. Installation on an operating system configured with default settings.
C. Rules permitting or denying access to systems or networks.
D. Configuration as a virtual private network endpoint.

74. Which of the following techniques would **BEST** help an IS auditor gain reasonable assurance that a project can meet its target date?

A. Estimation of the actual end date based on the completion percentages and estimated time to complete, taken from status reports
B. Confirmation of the target date based on interviews with experienced managers and staff involved in the completion of the project deliverables
C. Extrapolation of the overall end date based on completed work packages and current resources
D. Calculation of the expected end date based on current resources and remaining available project budget

75. An IS auditor wants to analyze audit trails on critical servers to discover potential anomalies in user or system behavior. Which of the following is the **MOST** suitable for performing that task?

A. Computer-aided software engineering tools
B. Embedded data collection tools
C. Trend/variance detection tools
D. Heuristic scanning tools

76. An organization has implemented an online customer help desk application using a software as a service (SaaS) operating model. An IS auditor is asked to recommend the best control to monitor the service level agreement (SLA) with the SaaS vendor as it relates to availability. What is the **BEST** recommendation that the IS auditor can provide?

A. Ask the SaaS vendor to provide a weekly report on application uptime.
B. Implement an online polling tool to monitor the application and record outages.
C. Log all application outages reported by users and aggregate the outage time weekly.
D. Contract an independent third party to provide weekly reports on application uptime.

77. Which of the following sampling methods is **MOST** useful when testing for compliance?

A. Attribute sampling
B. Variable sampling
C. Stratified mean-per-unit sampling
D. Difference estimation sampling

78. A substantive test to verify that tape library inventory records are accurate is:

A. Determining whether bar code readers are installed.
B. Determining whether the movement of tapes is authorized.
C. Conducting a physical count of the tape inventory.
D. Checking whether receipts and issues of tapes are accurately recorded.

79. When auditing the proposed acquisition of a new computer system, an IS auditor should **FIRST** ensure that:

A. A clear business case has been approved by management.
B. Corporate security standards will be met.
C. Users will be involved in the implementation plan.
D. The new system will meet all required user functionality.

80. An enterprise's risk appetite is **BEST** established by:

A. The chief legal officer
B. Security management
C. The audit committee
D. The steering committee

81. Electromagnetic emissions from a terminal represent a risk because they:

A. could damage or erase nearby storage media.
B. can disrupt processor functions.
C. could have adverse health effects on personnel.
D. can be detected and displayed.

82. The **GREATEST** benefit of having well-defined data classification policies and procedures is:

A. a more accurate inventory of information assets.
B. a decreased cost of controls.
C. a reduced risk of inappropriate system access.
D. an improved regulatory compliance.

83. Which of the following types of risk is **MOST** likely encountered in a software as a service environment?

A. Noncompliance with software license agreements
B. Performance issues due to Internet delivery method
C. Higher cost due to software licensing requirements
D. Higher cost due to the need to update to compatible hardware

84. In the process of evaluating program change controls, an IS auditor uses source code comparison software to:

A. Examine source program changes without information from IS personnel.
B. Detect a source program change made between acquiring a copy of the source and the comparison run.
C. Confirm that the control copy is the current version of the production program.
D. Ensure that all changes made in the current source copy are tested.

85. An organization is considering connecting a critical PC-based system to the Internet. Which of the following would provide the **BEST** protection against hacking?

A. An application-level gateway
B. A remote access server
C. A proxy server
D. Port scanning

86. During an audit of a small company that provides medical transcription services, an IS auditor observes several issues related to the backup and restore process. Which of the following should be the auditor's **GREATEST** concern?

A. Restoration testing for backup media is not performed; however, all data restore requests have been successful.
B. The policy for data backup and retention has not been reviewed by the business owner for the past three years.
C. The company stores transcription backup tapes offsite using a third-party service provider, which inventories backup tapes annually.
D. Failed backup alerts for the marketing department data files are not followed up on or resolved by the IT administrator.

87. An IS auditor has been assigned to review an organization's information security policy. Which of the following issues represents the **HIGHEST** potential risk?

A. The policy has not been updated in more than one year.
B. The policy includes no revision history.
C. The policy is approved by the security administrator.
D. The company does not have an information security policy committee.

88. During an audit of a telecommunications system, an IS auditor finds that the risk of intercepting data transmitted to and from remote sites is very high. The **MOST** effective control for reducing this exposure is:

A. encryption.
B. callback modems.
C. message authentication.
D. dedicated leased lines.

89. A Transmission Control Protocol/Internet Protocol (TCP/IP)-based environment is exposed to the Internet. Which of the following **BEST** ensures that complete encryption and authentication protocols exist for protecting information while transmitted?

A. Work is completed in tunnel mode with IP security.
B. A digital signature with RSA has been implemented.
C. Digital certificates with RSA are being used.
D. Work is being completed in TCP services.

90. Which of the following is the **BEST** reason to implement a policy that places conditions on secondary employment for IT employees?

A. To prevent the misuse of corporate resources
B. To prevent conflicts of interest
C. To prevent employee performance issues
D. To prevent theft of IT assets

91. An IS auditor examining the security configuration of an operating system should review the:

A. transaction logs.
B. authorization tables.
C. parameter settings.
D. routing tables.

92. Which of the following would contribute **MOST** to an effective business continuity plan?

A. The document is circulated to all interested parties.
B. Planning involves all user departments.
C. The plan is approved by senior management.
D. An audit is performed by an external IS auditor.

93. A new application has been purchased from a vendor and is about to be implemented. Which of the following choices is a key consideration when implementing the application?

A. Preventing the compromise of the source code during the implementation process
B. Ensuring that vendor default accounts and passwords have been disabled
C. Removing the old copies of the program from escrow to avoid confusion
D. Verifying that the vendor is meeting support and maintenance agreements

94. An IS auditor reviewing the application change management process for a large multinational company should be **MOST** concerned when:

A. test systems run different configurations than do production systems.
B. change management records are paper based.
C. the configuration management database is not maintained.
D. the test environment is installed on the production server.

95. An IS auditor was asked to review a contract for a vendor being considered to provide data center services. Which is the **BEST** way to determine whether the terms of the contract are adhered to after the contract is signed?

A. Require the vendor to provide monthly status reports.
B. Have periodic meetings with the client IT manager.
C. Conduct periodic audit reviews of the vendor.
D. Require that performance parameters be stated within the contract.

96. An IS auditor reviewing a network log discovers that an employee ran elevated commands on their PC by invoking the task scheduler to launch restricted applications. This is an example what type of attack?

A. A race condition
B. A privilege escalation
C. A buffer overflow
D. An impersonation

97. Which of the following is an advantage of elliptic curve encryption over RSA encryption?

A. Computation speed
B. Ability to support digital signatures
C. Simpler key distribution
D. Message integrity controls

98. While reviewing the IT infrastructure, an IS auditor notices that storage resources are continuously being added. The IS auditor should:

A. recommend the use of disk mirroring.
B. review the adequacy of offsite storage.
C. review the capacity management process.
D. recommend the use of a compression algorithm.

99. During a data center audit, an IS auditor observes that some parameters in the tape management system are set to bypass or ignore tape header records. Which of the following is the **MOST** effective compensating control for this weakness?

A. Staging and job setup
B. Supervisory review of logs
C. Regular backup of tapes
D. Offsite storage of tapes

100. Which of the following has the **MOST** significant impact on the success of an application systems implementation?

A. The prototyping application development methodology
B. Compliance with applicable external requirements
C. The overall organizational environment
D. The software reengineering technique

101. The **MOST** effective audit practice to determine whether the operational effectiveness of controls is properly applied to transaction processing is:

A. control design testing.
B. substantive testing.
C. inspection of relevant documentation.
D. perform tests on risk prevention.

102. When testing program change requests for a remote system, an IS auditor finds that the number of changes available for sampling does not provide a reasonable level of assurance. What is the **MOST** appropriate action for the IS auditor to take?

A. Develop an alternate testing procedure.
B. Report the finding to management.
C. Perform a walkthrough of the change management process.
D. Create additional sample data to test additional changes.

103. Management observed that the initial phase of a multiphase implementation was behind schedule and over budget. Prior to commencing with the next phase, an IS auditor's **PRIMARY** suggestion for a postimplementation focus should be to:

A. assess whether the planned cost benefits are being measured, analyzed and reported.
B. review control balances and verify that the system is processing data accurately.
C. review the impact of program changes made during the first phase on the remainder of the project.
D. determine whether the system's objectives were achieved.

104. An IS auditor reviewing access controls for a client-server environment should **FIRST**:

A. evaluate the encryption technique.
B. identify the network access points.
C. review the identity management system.
D. review the application level access controls.

105. Which of the following provides the **BEST** evidence of an organization's disaster recovery capability readiness?

A. A disaster recovery plan (DRP)
B. Customer references for the alternate site provider
C. Processes for maintaining the DRP
D. Results of tests and exercises

106. Which of the following would be the **BEST** overall control for an Internet business looking for confidentiality, reliability and integrity of data?

A. Secure Sockets Layer
B. Intrusion detection system
C. Public key infrastructure
D. Virtual private network

107. The feature of a digital signature that ensures the sender cannot later deny generating and sending the message is called:

A. data integrity.
B. authentication.
C. nonrepudiation.
D. replay protection.

108. Which of the following would normally be the **MOST** reliable evidence for an IS auditor?

A. A confirmation letter received from a third party verifying an account balance
B. Assurance from line management that an application is working as designed
C. Trend data obtained from Internet sources
D. Ratio analysis developed by the IS auditor from reports supplied by line management

109. The **MOST** important point of consideration for an IS auditor while reviewing an enterprise's project portfolio is that it:

A. Does not exceed the existing IT budget.
B. Is aligned with the investment strategy.
C. Has been approved by the IT steering committee.
D. Is aligned with the business plan.

110. When preparing a business case to support the need of an electronic data warehouse solution, which of the following choices is the **MOST** important to assist management in the decision-making process?

A. Discuss a single solution.
B. Consider security controls.
C. Demonstrate feasibility.
D. Consult the audit department.

111. A company has contracted with an external consulting firm to implement a commercial financial system to replace its existing system developed in-house. In reviewing the proposed development approach, which of the following would be of **GREATEST** concern?

A. Acceptance testing is to be managed by users.
B. A quality plan is not part of the contracted deliverables.
C. Not all business functions will be available on initial implementation.
D. Prototyping is being used to confirm that the system meets business requirements.

112. The **MAIN** reason for requiring that all computer clocks across an organization are synchronized is to:

A. Prevent omission or duplication of transactions.
B. Ensure smooth data transition from client machines to servers.
C. Ensure that email messages have accurate time stamps.
D. Support the incident investigation process.

113. A batch transaction job failed in production; however, the same job returned no issues during user acceptance testing (UAT). Analysis of the production batch job indicates that it was altered after UAT. Which of the following ways would be the **BEST** to mitigate this risk in the future?

A. Improve regression test cases.
B. Activate audit trails for a limited period after release.
C. Conduct an application user access review.
D. Ensure that developers do not have access to code after testing.

114. Which of the following is the initial step in creating a firewall policy?

A. A cost-benefit analysis of methods for securing the applications
B. Identification of network applications to be externally accessed
C. Identification of vulnerabilities associated with network applications to be externally accessed
D. Creation of an application traffic matrix showing protection methods

115. Which of the following controls would be **MOST** effective in ensuring that production source code and object code are synchronized?

A. Release-to-release source and object comparison reports
B. Library control software restricting changes to source code
C. Restricted access to source code and object code
D. Date and time-stamp reviews of source and object code

116. Confidentiality of the data transmitted in a wireless local area network is **BEST** protected if the session is:

A. restricted to predefined media access control addresses.
B. encrypted using static keys.
C. encrypted using dynamic keys.
D. initiated from devices that have encrypted storage.

117. Which of the following does a lack of adequate security controls represent?

A. Threat
B. Asset
C. Impact
D. Vulnerability

118. Which of the following is the **MOST** secure and economical method for connecting a private network over the Internet in a small- to medium-sized organization?

A. Virtual private network
B. Dedicated line
C. Leased line
D. Integrated services digital network

119. The decisions and actions of an IS auditor are **MOST** likely to affect which of the following types of risk?

A. Inherent
B. Detection
C. Control
D. Business

120. When using public key encryption to secure data being transmitted across a network:

A. both the key used to encrypt and decrypt the data are public.
B. the key used to encrypt is private, but the key used to decrypt the data is public.
C. the key used to encrypt is public, but the key used to decrypt the data is private.
D. both the key used to encrypt and decrypt the data are private.

121. Which of the following is the **MOST** critical step when planning an IS audit?

A. Review findings from prior audits
B. Executive management's approval of the audit plan
C. Review information security policies and procedures
D. Perform a risk assessment

122. The technique used to ensure security in virtual private networks is called:

A. data encapsulation.
B. data wrapping.
C. data transformation.
D. data hashing.

123. While reviewing the IT governance processes of an organization, an IS auditor discovers the firm has recently implemented an IT balanced scorecard (BSC). The implementation is complete; however, the IS auditor notices that performance indicators are not objectively measurable. What is the **PRIMARY** risk presented by this situation?

A. Key performance indicators are not reported to management and management cannot determine the effectiveness of the BSC.
B. IT projects could suffer from cost overruns.
C. Misleading indications of IT performance may be presented to management.
D. IT service level agreements may not be accurate.

124. During an assessment of software development practices, an IS auditor finds that open source software components were used in an application designed for a client. What is the **GREATEST** concern the auditor would have about the use of open source software?

A. The client did not pay for the open source software components.
B. The organization and client must comply with open source software license terms.
C. Open source software has security vulnerabilities.
D. Open source software is unreliable for commercial use.

125. Functionality is a characteristic associated with evaluating the quality of software products throughout their life cycle, and is **BEST** described as the set of attributes that bear on the:

A. existence of a set of functions and their specified properties.
B. ability of the software to be transferred from one environment to another.
C. capability of software to maintain its level of performance under stated conditions.
D. relationship between the performance of the software and the amount of resources used.

126. A database administrator (DBA) who needs to make emergency changes to a database after normal working hours should log in:

A. with their named account to make the changes.
B. with the shared DBA account to make the changes.
C. to the server administrative account to make the changes.
D. to the user's account to make the changes.

127. Which of the following is an advantage of prototyping?

A. The finished system normally has strong internal controls.
B. Prototype systems can provide significant time and cost savings.
C. Change control is often less complicated with prototype systems.
D. Prototyping ensures that functions or extras are not added to the intended system.

128. Applying a retention date on a file will ensure that:

A. data cannot be read until the date is set.
B. data will not be deleted before that date.
C. backup copies are not retained after that date.
D. datasets having the same name are differentiated.

129. When implementing an application software package, which of the following presents the **GREATEST** risk?

A. Uncontrolled multiple software versions
B. Source programs that are not synchronized with object code
C. Incorrectly set parameters
D. Programming errors

130. In wireless communication, which of the following controls allows the receiving device to verify that the received communications have not been altered in transit?

A. Device authentication and data origin authentication
B. Wireless intrusion detection and intrusion prevention systems
C. The use of cryptographic hashes
D. Packet headers and trailers

131. Which of the following is an object-oriented technology characteristic that permits an enhanced degree of security over data?

A. Inheritance
B. Dynamic warehousing
C. Encapsulation
D. Polymorphism

132. An IS auditor selects a server for a penetration test that will be carried out by a technical specialist. Which of the following is **MOST** important?

A. The tools used to conduct the test
B. Certifications held by the IS auditor
C. Permission from the data owner of the server
D. An intrusion detection system is enabled

133. An IS auditor is reviewing an organization to ensure that evidence related to a data breach case is preserved. Which of the following choices would be of **MOST** concern to the IS auditor?

A. End users are not aware of incident reporting procedures.
B. Log servers are not on a separate network.
C. Backups are not performed consistently.
D. There is no chain of custody policy.

134. Which of the following is **MOST** critical when creating data for testing the logic in a new or modified application system?

A. A sufficient quantity of data for each test case
B. Data representing conditions that are expected in actual processing
C. Completing the test on schedule
D. A random sample of actual data

135. The **GREATEST** risk from an improperly implemented intrusion prevention system is:

A. too many alerts for system administrators to verify.
B. decreased network performance due to additional traffic.
C. blocking of critical systems or services due to false triggers.
D. reliance on specialized expertise within the IT organization.

136. An IS auditor is reviewing access to an application to determine whether recently added accounts were appropriately authorized. This is an example of:

A. variable sampling.
B. substantive testing.
C. compliance testing.
D. stop-or-go sampling.

137. An IS auditor is testing employee access to a large financial system, and the IS auditor selected a sample from the current employee list provided by the auditee. Which of the following evidence is the **MOST** reliable to support the testing?

A. A spreadsheet provided by the system administrator
B. Human resources access documents signed by employees' managers
C. A list of accounts with access levels generated by the system
D. Observations performed onsite in the presence of a system administrator

138. The extent to which data will be collected during an IS audit should be determined based on the:

A. Availability of critical and required information.
B. Auditor's familiarity with the circumstances.
C. Auditee's ability to find relevant evidence.
D. Purpose and scope of the audit being done.

139. Which of the following is responsible for the approval of an information security policy?

A. IT department
B. Security committee
C. Security administrator
D. Board of directors

140. Which technique will **BEST** test for the existence of dual control when auditing the wire transfer systems of a bank?

A. Analysis of transaction logs
B. Reperformance
C. Observation
D. Interviewing personnel

141. Which of the following is the **MOST** effective method for disposing of magnetic media that contains confidential information?

A. Degaussing
B. Defragmenting
C. Erasing
D. Destroying

142. An organization's IS audit charter should specify the:

A. plans for IS audit engagements.
B. objectives and scope of IS audit engagements.
C. detailed training plan for the IS audit staff.
D. role of the IS audit function.

143. A certificate authority (CA) can delegate the processes of:

A. revocation and suspension of a subscriber's certificate.
B. generation and distribution of the CA public key.
C. establishing a link between the requesting entity and its public key.
D. issuing and distributing subscriber certificates.

144. Which of the following should be included in an organization's information security policy?

A. A list of key IT resources to be secured
B. The basis for access control authorization
C. Identity of sensitive security assets
D. Relevant software security features

145. An IS auditor finds that user acceptance testing of a new system is being repeatedly interrupted by defect fixes from the developers. Which of the following would be the **BEST** recommendation for an IS auditor to make?

A. Consider the feasibility of a separate user acceptance environment
B. Schedule user testing to occur at a given time each day
C. Implement a source code version control tool
D. Only retest high-priority defects

146. In planning an IS audit, the **MOST** critical step is the identification of the:

A. areas of significant risk
B. skill sets of the audit staff
C. test steps in the audit
D. time allotted for the audit

147. An IS auditor finds that database administrators (DBAs) have access to the log location on the database server and the ability to purge logs from the system. What is the **BEST** audit recommendation to ensure that DBA activity is effectively monitored?

A. Change permissions to prevent DBAs from purging logs.
B. Forward database logs to a centralized log server to which the DBAs do not have access.
C. Require that critical changes to the database are formally approved.
D. Back up database logs to tape.

148. Which of the following is a function of an IT steering committee?

A. Monitoring vendor-controlled change control and testing
B. Ensuring a separation of duties within the information's processing environment
C. Approving and monitoring the status of IT plans and budgets
D. Liaising between the IT department and end users

149. During a risk analysis, an IS auditor identifies threats and potential impacts. Next, the IS auditor should:

A. Ensure the risk assessment is aligned to management's risk assessment process.
B. Identify information assets and the underlying systems.
C. Disclose the threats and impacts to management.
D. Identify and evaluate the existing controls.

150. Which of the following situations could impair the independence of an IS auditor? The IS auditor:

A. implemented specific functionality during the development of an application.
B. designed an embedded audit module for auditing an application.
C. participated as a member of an application project team and did not have operational responsibilities.
D. provided consulting advice concerning application good practices.

CISA® Review Questions, Answers & Explanations Manual 12th Edition

SAMPLE EXAM ANSWER AND REFERENCE KEY

Exam Question #	Key	Ref. #	Exam Question #	Key	Ref. #	Exam Question #	Key	Ref. #
1	C	A4-156	51	A	A2-26	101	B	A1-23
2	C	A4-66	52	A	A5-174	102	A	A1-18
3	D	A1-85	53	B	A3-119	103	C	A3-49
4	B	A1-73	54	A	A2-22	104	B	A5-173
5	B	A5-149	55	D	A1-29	105	D	A4-67
6	C	A4-232	56	D	A5-46	106	A	A5-72
7	A	A5-276	57	B	A2-5	107	C	A5-89
8	B	A4-118	58	D	A5-176	108	A	A1-31
9	B	A1-44	59	D	A5-104	109	D	A2-146
10	B	A1-63	60	A	A3-16	110	C	A3-65
11	C	A4-252	61	A	A4-92	111	B	A3-64
12	A	A2-50	62	A	A2-75	112	D	A5-278
13	B	A4-81	63	A	A5-105	113	D	A4-82
14	C	A1-91	64	D	A2-74	114	B	A2-54
15	B	A5-86	65	A	A2-150	115	D	A4-47
16	A	A2-149	66	D	A5-226	116	C	A5-147
17	C	A2-23	67	D	A1-120	117	D	A2-73
18	C	A5-279	68	C	A1-42	118	A	A5-107
19	C	A2-4	69	A	A5-175	119	B	A1-11
20	B	A5-88	70	B	A5-205	120	C	A5-96
21	A	A4-185	71	C	A5-99	121	D	A1-12
22	C	A1-74	72	C	A3-90	122	A	A5-97
23	A	A4-28	73	B	A5-227	123	C	A2-52
24	A	A4-157	74	C	A3-120	124	B	A4-49
25	B	A4-231	75	C	A1-90	125	A	A3-66
26	B	A5-277	76	B	A4-29	126	A	A4-48
27	B	A3-29	77	A	A1-17	127	B	A3-51
28	C	A5-148	78	C	A1-57	128	B	A4-30
29	B	A4-253	79	A	A3-17	129	C	A3-50
30	D	A2-72	80	D	A2-148	130	C	A5-225
31	A	A1-55	81	D	A5-47	131	C	A5-275
32	A	A5-146	82	B	A5-207	132	C	A5-206
33	D	A3-31	83	B	A3-18	133	D	A5-45
34	A	A1-119	84	A	A1-62	134	B	A3-89
35	A	A1-56	85	A	A5-106	135	C	A5-228
36	B	A5-70	86	C	A4-184	136	C	A1-10
37	D	A4-120	87	C	A2-7	137	C	A1-84
38	A	A2-25	88	A	A5-98	138	D	A1-24
39	B	A1-83	89	A	A5-87	139	D	A2-51
40	D	A4-93	90	B	A2-6	140	C	A1-121
41	C	A1-92	91	C	A4-107	141	D	A4-254
42	D	A5-208	92	B	A4-233	142	D	A1-72
43	A	A3-88	93	B	A4-119	143	C	A5-69
44	C	A4-155	94	C	A4-186	144	B	A2-53
45	C	A4-94	95	C	A2-71	145	A	A3-121
46	C	A2-147	96	B	A5-44	146	A	A1-22
47	A	A2-8	97	A	A5-71	147	B	A4-68
48	D	A4-80	98	C	A4-109	148	C	A2-24
49	C	A1-43	99	A	A4-108	149	D	A1-30
50	C	A1-64	100	C	A3-30	150	A	A1-19

Reference example: A4-22= See domain 4, question 22 for explanation of the answer.

Page intentionally left blank

CISA® Review Questions, Answers & Explanations Manual 12th Edition

SAMPLE EXAM ANSWER SHEET (PRETEST)

(side 1)

Please use this answer sheet to take the sample exam as a pretest to determine strengths and weaknesses. The answer key/reference grid is on page 491.

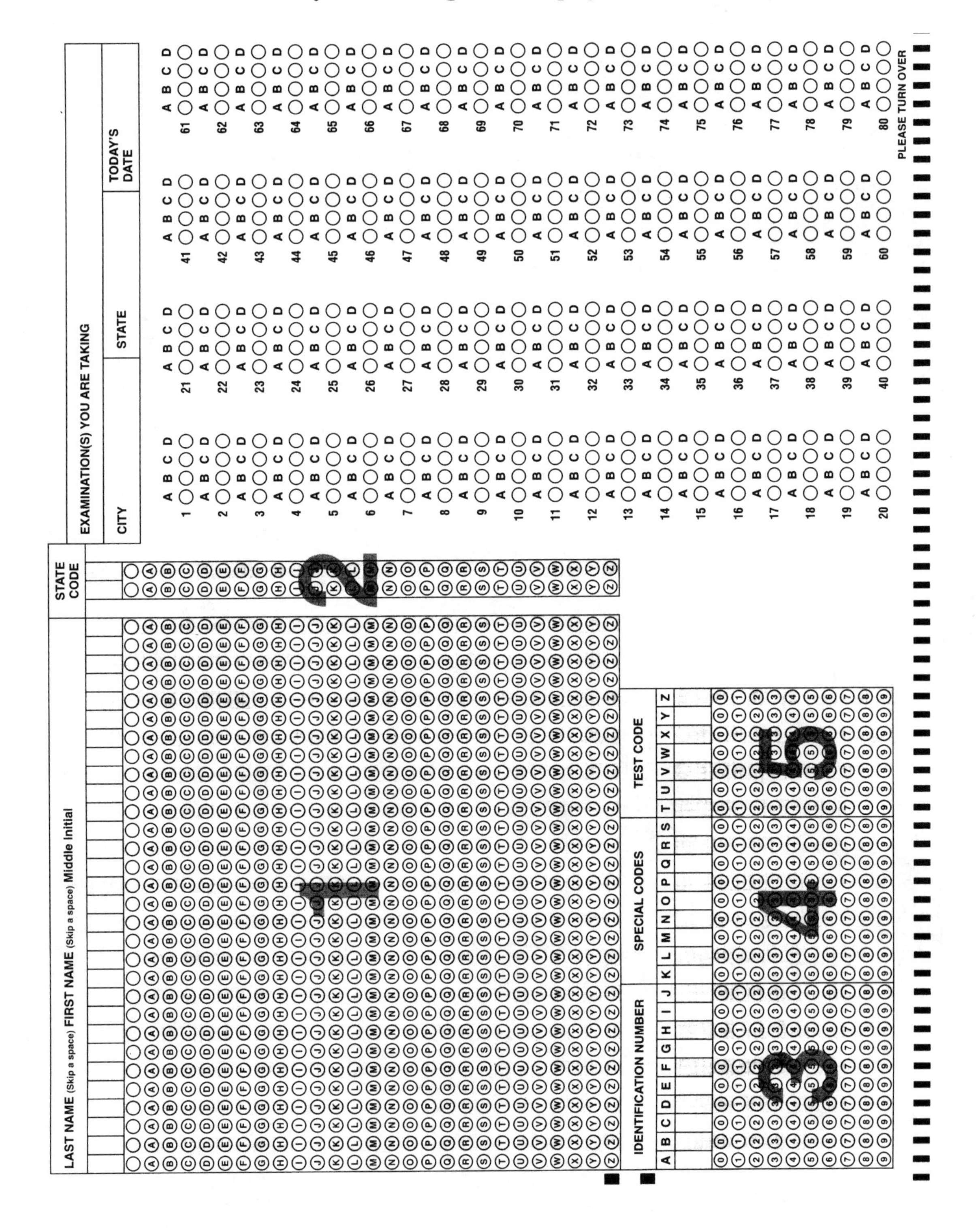

(side 2)

Please use this answer sheet to take the sample exam as a pretest to determine strengths and weaknesses. The answer key/reference grid is on page 491.

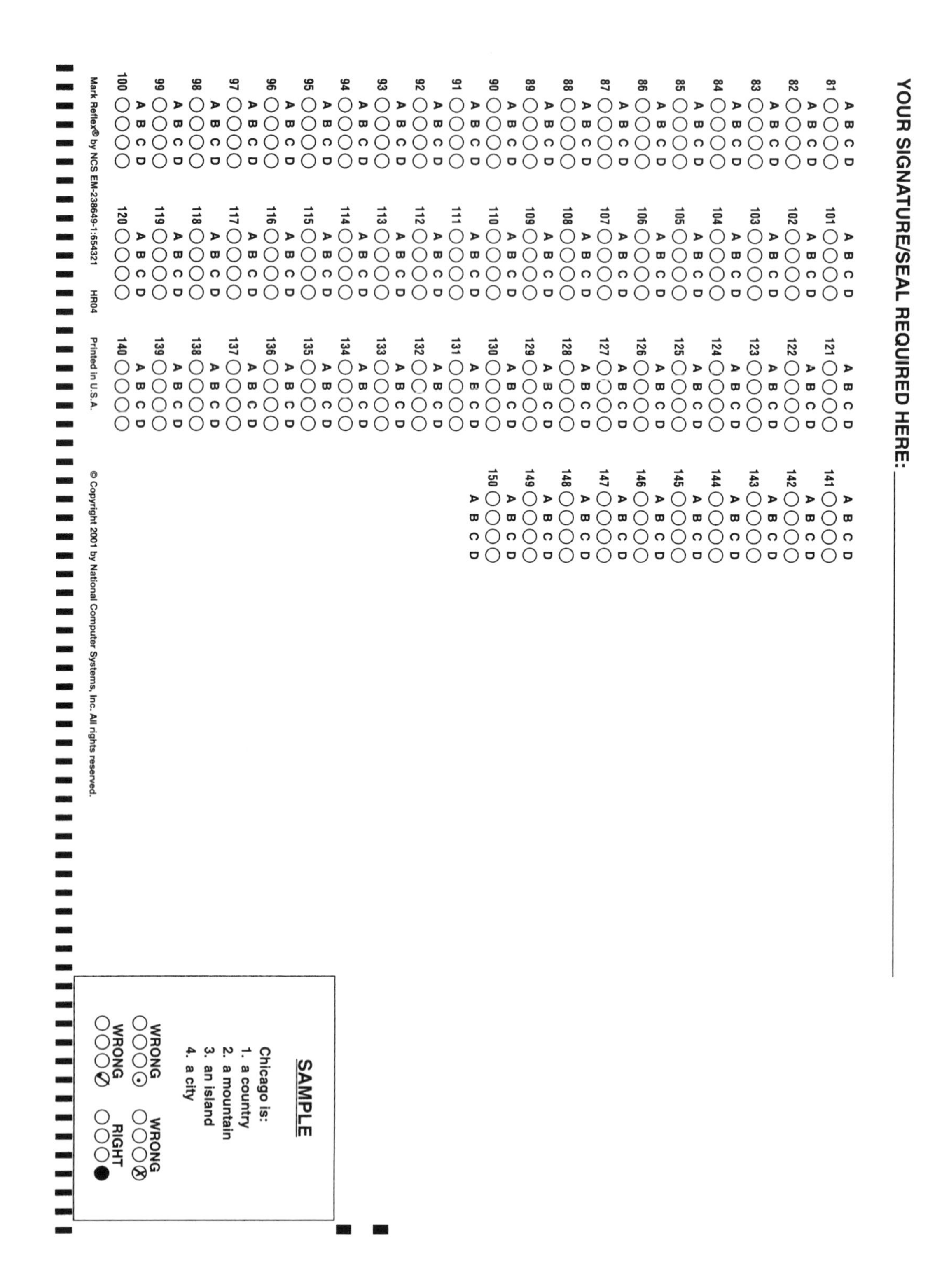

CISA® Review Questions, Answers & Explanations Manual 12th Edition

SAMPLE EXAM ANSWER SHEET (POSTTEST)

(side 1)

Please use this answer sheet to take the sample exam as a posttest to determine strengths and weaknesses. The answer key/reference grid is on page 491.

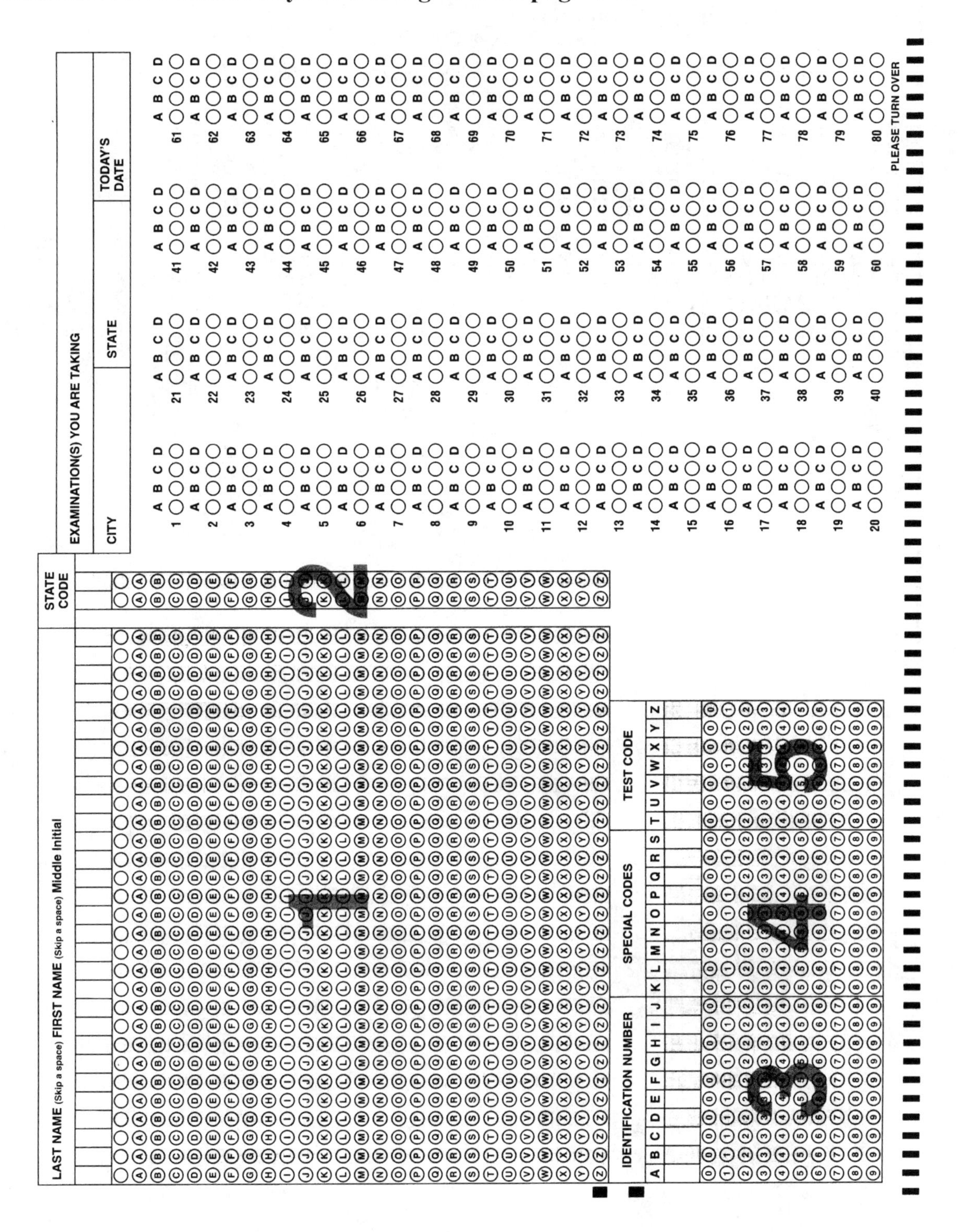

(side 2)

Please use this answer sheet to take the sample exam as a posttest to determine strengths and weaknesses. The answer key/reference grid is on page 491.

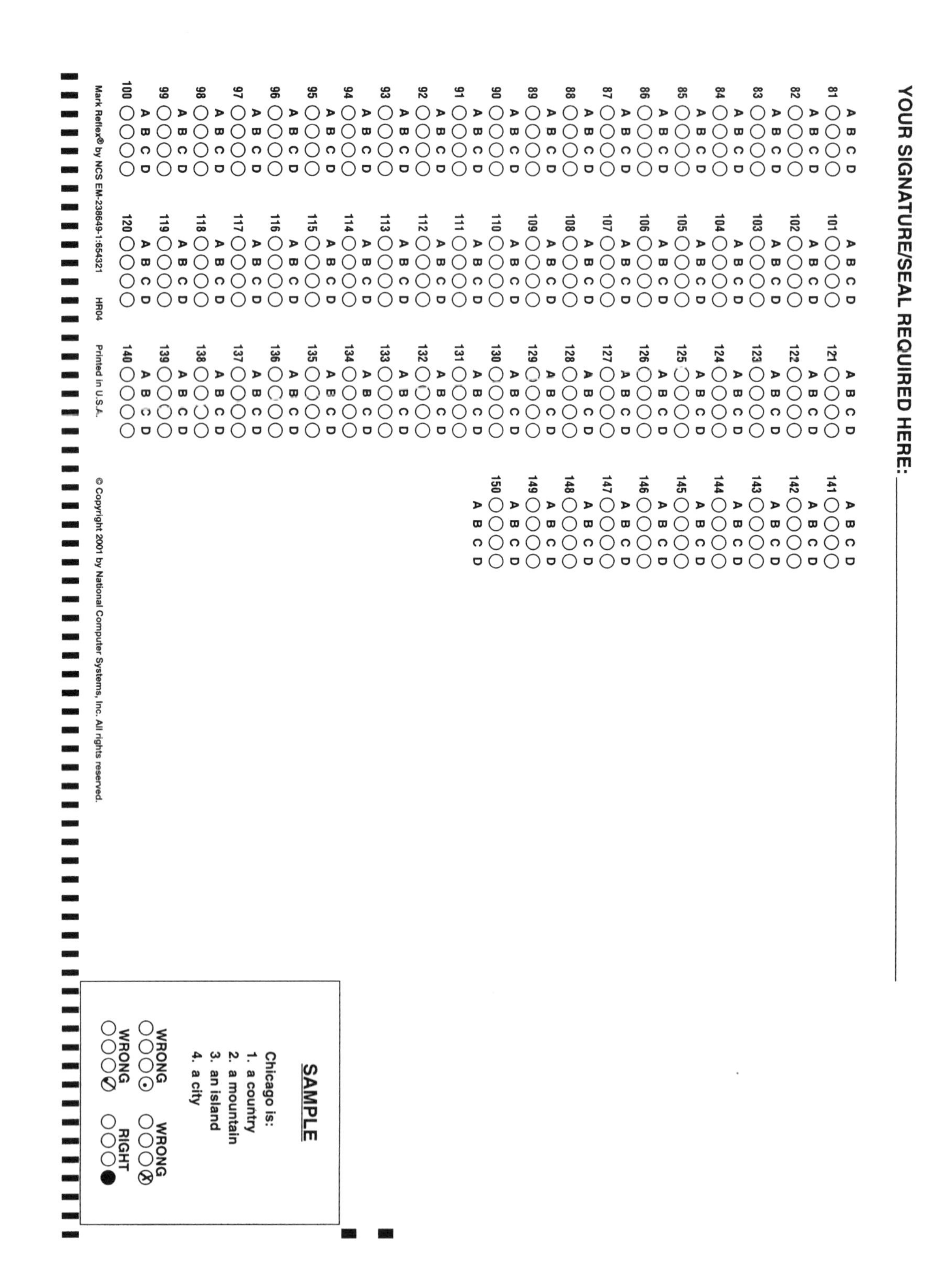

EVALUATION

ISACA continuously monitors the swift and profound professional, technological and environmental advances affecting the IS audit and control professions. Recognizing these rapid advances, ISACA updates CISA review manuals annually.

To assist ISACA in keeping abreast of these advances, please take a moment to evaluate the *CISA® Review Questions, Answers & Explanations Manual 12th Edition*. Your feedback is valuable to fully serve the profession and future CISA exam registrants.

To submit your feedback, please visit support.isaca.org.